Statistical Practice in Business and Industry

STATISTICS IN PRACTICE

Founding Editor

Statistics in Practice is an important international series of texts which provide detailed coverage of statistical concepts, methods and worked case studies in specific fields of investigation and study.

With sound motivation and many worked practical examples, the books show in down-to-earth terms how to select and use an appropriate range of statistical techniques in a particular practical field within each title's special topic area.

The books provide statistical support for professionals and research workers across a range of employment fields and research environments. Subject areas covered include medicine and pharmaceutics; industry, finance and commerce; public services; the earth and environmental sciences, and so on.

The books also provide support to students studying statistical courses applied to the above areas. The demand for graduates to be equipped for the work environment has led to such courses becoming increasingly prevalent at universities and colleges.

It is our aim to present judiciously chosen and well-written workbooks to meet everyday practical needs. Feedback of views from readers will be most valuable to monitor the success of this aim.

A complete list of titles in this series appears at the end of the volume.

Statistical Practice in Business and Industry

Edited by

Shirley Coleman

Industrial Statistics Research Unit, Newcastle University, UK

Tony Greenfield

Greenfield Research, UK

Dave Stewardson

Industrial Statistics Research Unit, Newcastle University, UK

Douglas C. Montgomery

Arizona State University, USA

John Wiley & Sons, Ltd

Telephone (+44) 1243 779777

Email (for orders and customer service enquiries): cs-books@wiley.co.uk
Visit our Home Page on www.wileyeurope.com or www.wiley.com

Other Wiley Editorial Offices

John Wiley & Sons Inc., 111 River Street, Hoboken, NJ 07030, USA

Jossey-Bass, 989 Market Street, San Francisco, CA 94103-1741, USA

Wiley-VCH Verlag GmbH, Boschstr. 12, D-69469 Weinheim, Germany

John Wiley & Sons Australia Ltd, 42 McDougall Street, Milton, Queensland 4064, Australia

John Wiley & Sons (Asia) Pte Ltd, 2 Clementi Loop #02-01, Jin Xing Distripark, Singapore 129809

John Wiley & Sons Canada Ltd, 6045 Freemont Blvd, Mississauga, Ontario, L5R 4J3, Canada

Wiley also publishes its books in a variety of electronic formats. Some content that appears in print may not be available in electronic books.

Library of Congress Cataloging-in-Publication Data

Statistical practice in business and industry / edited by Shirley Coleman ... [*et al.*]
p. cm.
Includes bibliographical references and index.
ISBN 978-0-470-01497-4 (cloth)
1. Commercial statistics. I. Coleman, Shirley.
HF1016.S7 2008
658.4′033 – dc22

2007045558

British Library Cataloguing in Publication Data

A catalogue record for this book is available from the British Library

ISBN 978-0-470-01497-4 (HB)

Typeset in 10/12pt Times by Laserwords Private Limited, Chennai, India
Printed and bound in Great Britain by TJ International, Padstow, Cornwall
This book is printed on acid-free paper responsibly manufactured from sustainable forestry in which at least two trees are planted for each one used for paper production.

Contents

*Roman numerals are used to identify sub-chapters, of chapters 3 and 12, to indicate different authors within the chapter.

Contributors

GIULIO BARBATO Dipartimento di Sistemi di Produzione ed Economia dell'Azienda, Politecnico di Torino, Italy

CONNIE BORROR Mathematical Sciences and Applied Computing, Arizona State University, Phoenix, AZ 85069, USA

ROLAND CAULCUTT Caulcutt Associates, 1 Calders Garden, Sherbourne, Dorset, DT9 3GA, UK

PAOLA CERCHIELLO Department of Statistics and Applied Economics 'Libero Lenti', University of Pavia, Via Strada Nuova 65, 27100 Pavia, Italy

SHIRLEY Y. COLEMAN Technical Director, Industrial Statistics Research Unit, Newcastle University, Newcastle upon Tyne NE1 7RU, UK

ANNE DE FRENNE Math-X sprl, Brussels, Belgium

JEROEN DE MAST Institute for Business and Industrial Statistics of the University of Amsterdam (IBIS UvA), Plantage Muidergracht 24, 1018 TV Amsterdam, Netherlands

RONALD J.M.M. DOES Managing Director, Institute for Business and Industrial Statistics of the University of Amsterdam (IBIS UvA), Plantage Muidergracht 24, 1018 TV Amsterdam, Netherlands

ØYSTEIN EVANDT DNV, P.O. Box 300, 1322 Hovik, Norway, and Industrial Statistics Research Unit, Newcastle University, Newcastle upon Tyne NE1 7RU, UK

SILVIA FIGINI Department of Statistics and Applied Economics 'Libero Lenti', University of Pavia, Via Strada Nuova 65, 27100 Pavia, Italy

TONY FOUWEATHER Industrial Statistics Research Unit, Newcastle University, Newcastle upon Tyne NE1 7RU, UK

PAOLO GIUDICI Department of Statistics and Applied Economics 'Libero Lenti', University of Pavia, Via Strada Nuova 65, 27100 Pavia, Italy

TONY GREENFIELD Greenfield Research, Little Hucklow, Buxton SK17 8RT, UK

JULIO HOLGADO Computing Services, Universidad Rey Juan Carlos, Spain

RON S. KENETT KPA Ltd., P.O. Box 2525, Raanana 43100, Israel, and University of Torino, Torino, Italy

MURAT KULAHCI Institute for Informatics and Mathematical Modelling, Technical University of Denmark, Lyngby, Denmark

RAFFAELLO LEVI Dipartimento di Sistemi di Produzione ed Economia dell'Azienda, Politecnico di Torino, Italy

JOHN LOGSDON, Quantex Research Ltd, Manchester, UK

CHRISTOPHER MCCOLLIN Nottingham Trent University, Burton Street, Nottingham NG1 4BU, UK

DOUGLAS C. MONTGOMERY Regents' Professor of Industrial Engineering & Statistics, ASU Foundation Professor of Engineering, Department of Industrial Engineering Arizona State University Tempe, AZ 85287-5906, USA

RAÚL MORENO Department of Statistics and Operations Research, Universidad Rey Juan Carlos, Madrid, Spain

JORGE MURUZÁBAL Formerly of Department of Statistics and Operations Research, Universidad Rey Juan Carlos, Spain († 5 August 2006)

IRENA OGRAJENŠEK University of Ljubljana, Faculty of Economics, Statistical Institute (SIEF), Kardeljeva ploščad 17, 1000 Ljubljana, Slovenia

JESUS PALOMO Department of Statistics and Operations Research, Universidad Rey Juan Carlos, Spain, and SAMSI-Duke University, Research Triangle Park, NC, USA

M. F. RAMALHOTO Instituto Superior Técnico, Technical University of Lisbon, Portugal

MARCO P. SEABRA DOS REIS Chemical Engineering Department, University of Coimbra, Polo II, Rua Sílvio Lima, 3030-790 Coimbra, Portugal

DAVID RÍOS INSUA Department of Statistics and Operations Research, Universidad Rey Juan Carlos, Spain

TIMOTHY J. ROBINSON Associate Professor, Department of Statistics, University of Wyoming, Laramie, WY 82070, USA

FABRIZIO RUGGERI CNR-IMATI, Via Bassini 15, I-20133 Milano, Italy

PEDRO M. SARAIVA Chemical Engineering Department, University of Coimbra, Polo II, Rua Sílvio Lima, 3030-790 Coimbra, Portugal

D. J. STEWARDSON Industrial Statistics Research Unit, Newcastle University, Newcastle upon Tyne NE1 7RU, UK

XAVIER TORT-MARTORELL Technical University of Catalonia (UPC), Barcelona, Spain

ALBERT TRIP Institute for Business and Industrial Statistics of the University of Amsterdam (IBIS UvA), Plantage Muidergracht 24, 1018 TV Amsterdam, Netherlands

GRAZIA VICARIO Dipartimento di Matematica, Politecnico di Torino, Italy

DONALD J. WHEELER SPC Press, Suite C, 5908 Toole Drive, Knoxville, TN 37919, USA

ANDRÁS ZEMPLÉNI Department of Probability Theory and Statistics, Eötvös Loránd University, Budapest, Hungary

Preface

This book was originally conceived as a handbook arising from the European Commission funded Fifth Framework project entitled European Network for Promotion of Business and Industrial Statistics (pro-ENBIS).[1] The book was a natural consequence of the vibrant and productive three-year working partnership between many of the foremost players in the field of practical statistics. Contributions for the book have been volunteered from each of the Pro-ENBIS work package leaders and from most of the other project partners and members. Many of the contributions are collaborative efforts, not only ensuring that the subjects covered are dealt with from more than one perspective, but also carrying on the initial drive of the project in promoting collaborations to further the subject as a whole. ENBIS, the European Network of Business and Industrial Statistics, is part of a world-wide movement to promote statistical practice, and as a reflection of this viewpoint the book also includes contributions from people who were not directly involved in Pro-ENBIS but who are acknowledged experts and supporters of the ENBIS mission of promoting the application and understanding of statistical methods. The resulting book provides a thorough overview of the state of the art in business and industrial statistics.

The subject areas are dealt with in a logical manner, starting with a brief overview of the history of industrial statistics, before tackling a variety of subjects which have become vital to many areas of business and industry throughout the world. The first part of the book deals with statistical consultancy including process improvement and Six Sigma initiatives, which seek to implement statistical methods and tools in businesses and industry. Chapters on management statistics and service quality are followed by data mining. The important area of process modelling is dealt with in two chapters followed by some fundamental areas of statistics, including designing experiments, multivariate analysis and simulation.

The book covers methodology and techniques currently in use and those about to be launched, and includes references to miscellaneous sources of information, including websites. No book on statistical practice would be complete without giving due time and attention to communication. The last chapter is dedicated to this most important part of statistical practice.

Shirley Coleman

[1]Project funded by the European Community under the 'Competitive and Sustainable Growth' Programme (1998–2002), contract no. G6RT-CT-2001-05059.

1

Introduction: ENBIS, Pro-ENBIS and this book

Shirley Coleman, Tony Fouweather and Dave Stewardson

The practical application of statistics has fascinated statisticians for many years. Business and industrial processes are rich with information that can be used to understand, improve and change the way things are done. This book arose from the work carried out by members of the European Network for Business and Industrial Statistics (ENBIS) which was strengthened by a three-year thematic network called Pro-ENBIS funded by the European Commission under the Fifth Framework Programme (FP5). Contributions have been made by statistical practitioners from Europe and from other parts of the world. The aim of the book is to promote the wider understanding and application of contemporary and emerging statistical methods. The book is aimed at people working in business and industry worldwide. It can be used for the professional development of statistical practitioners (a statistical practitioner is any person using statistical methods, whether formally trained or not) and to foster best practice for the benefit of business and industry.

This introductory chapter gives a brief description of ENBIS followed by an overview of the Pro-ENBIS project.

In August 1999 a group of about 20 statistical practitioners met in Linköping, Sweden, at the end of the First International Symposium on Industrial Statistics to initiate the creation of the European Society for Industrial and Applied Statistics (ESIAS). As a first step, a list of interested members was drawn up and a dialogue started. The internet was identified as a feasible platform cheaply and efficiently to co-ordinate the activities of any body formed out of these meetings. A network of people from industry and academia from all European nations interested in

Statistical Practice in Business and Industry Edited by S.Y. Coleman, T. Greenfield, D.J. Stewardson and D.C. Montgomery

applying, promoting and facilitating the use of statistics in business and industry was created to address two main observations:

- As with many disciplines, applied statisticians and statistical practitioners often work in professional environments where they are rather isolated from interactions and stimulation from like-minded professionals.

- Statistics is vital for the economic and technical development and improved competitiveness of European industry.

By February 2000 an executive committee was formed which held a founding meeting at EURANDOM in Eindhoven, the Netherlands, on 26–27 February. The name ENBIS was adopted as the defining name for the society. It was decided to have a founding conference on 11 December 2000 in Amsterdam, and during that meeting ENBIS was formally launched. The conference was accompanied by a three-day workshop on design of experiments lead by Søren Bisgaard, a renowned European statistician.

The mission of ENBIS is to:

- foster and facilitate the application and understanding of statistical methods to the benefit of European business and industry;

- provide a forum for the dynamic exchange of ideas and facilitate networking among statistical practitioners;

- nurture interactions and professional development of statistical practitioners regionally and internationally.

ENBIS has adopted the following points as its vision:

- to promote the widespread use of sound science driven, applied statistical methods in European business and industry;

- that membership consists primarily of statistical practitioners from business and industry;

- to emphasise multidisciplinary problem-solving involving statistics;

- to facilitate the rapid transfer of statistical methods and related technologies to and from business and industry;

- to link academic teaching and research in statistics with industrial and business practice;

- to facilitate and sponsor continuing professional development;

- to keep its membership up to date in the field of statistics and related technologies;

- to seek collaborative agreements with related organisations.

ENBIS is a web-based society, and its activities can be found at http://www.enbis.org.

ENBIS has arranged annual and occasional conferences at various locations around Europe which have allowed the showcasing of a broad spectrum of applications and generated discussion about the use of statistics in a wide range of business and industrial areas. Ideas for new projects often arise from these meetings, and ENBIS provides an ideal forum for gathering project partners and cementing working relationships. Conferences organised by ENBIS members have been held in Oslo (2001), Rimini (2002), Barcelona (2003), Copenhagen (2004), Newcastle (2005), Wroclaw (2006) and Dortmund (2007). In addition to the annual conferences, ENBIS runs spring meetings dedicated to particular topics. These include design of experiments in Cagliari (2004), data mining in Gengenbach (2006) and computer experiments versus physical experiments in Turin (2007).

The Pro-ENBIS thematic network sought to build on the success of ENBIS and to develop partnerships within Europe to support selected projects at the forefront of industrial and business statistics, with the specific mission 'to promote the widespread use of sound science driven, applied statistical methods in European business and industry'.

Pro-ENBIS was contracted for three years until 31 December 2004 by the European Commission with a budget of €800 000. The project was co-ordinated by the University of Newcastle upon Tyne (UK) and had contractors and members from across Europe. There were a total of 37 partners from 18 countries, 10 of which were principal contractors. There was also an invited expert, Søren Bisgaard, who helped with strategic thinking and dissemination.

The thematic network was funded so that it could achieve specific outcomes, including promoting industrial statistics through workshops and industrial visits and through publishing both academic papers and articles in the popular press. These activities relate to the network's aim to provide a forum for the dissemination of business and industrial statistical methodology directly from statisticians and practitioners to European business and industry. They were very successful in generating considerable interest in ENBIS and in statistical practice generally.

The deliverables were grouped around statistical themes, with eight work packages:

- WP1 Design of experiments
- WP2 Data mining/warehousing
- WP3 General statistical modelling, process modelling and control
- WP4 Reliability, safety and quality improvement
- WP5 Discovering European resources and expertise
- WP6 Drafting and initiating further activity

- WP7 Management statistics

- WP8 Network management

Work packages 1, 2, 3, 4 and 7 each focused on a particular area of statistics. As a thematic network, the outcomes for these work packages included organizing workshops, events and network meetings on relevant topics as well as publishing papers, articles and notes. The project deliverables agreed with the European Commission were all met or exceeded.

Many papers co-authored by partners were published in the academic press and some of these are listed below. The partners were enthusiastic to publish relevant articles in the popular press. These were typically written in languages other than English and appeared in local newspapers, press releases and other outlets. The articles published in the popular press have been important in drawing attention to activity in statistical practice across Europe as well as publicizing the Pro-ENBIS project and raising awareness of the ENBIS organisation and its activities. A lot of thoughtful responses and comments relating to these articles from people in business and industry have been received by the authors. Discussions continue within the framework of ENBIS.

Work packages 5, 6 and 8 were concerned with networking, information gathering and project management.

The chapters in the book cover all of the areas addressed in the work packages.

Work package 1: Design of experiments

- *Mission*. Dedicated to furthering the use of (DoE) in an industrial and commercial context. The emphasis is on developing concepts, to be of use daily in quality and productivity improvement.

Work package motivation

International competition is getting tougher, product development cycles shorter, manufacturing processes more complex, and customers' expectations of quality higher. Dealing with these problems and generating new knowledge often require extensive experimentation. Research and development scientists and engineers are therefore under pressure to be more effective in conducting experiments for quality, productivity and yield optimisation. In particular, it is recognised that quality cannot economically be inspected into products and processes, but must be designed upstream in the design and development phases. Design of experiments is a powerful and economically efficient strategy employing modern statistical principles for solving design and manufacturing problems, for the discovery of important factors influencing product and process quality and for experimental optimisation. Experimental design is also an important tool for developing new products that are robust (insensitive) to environment and internal component variation. Carefully planned statistical studies can remove hindrances to high quality and productivity at every stage from concept development to production, saving time and money. The emphasis is on discussing and developing concepts, to be of use daily in quality and productivity improvement.

A key contribution of WP1 was a series of very successful workshops. There was good industrial participation and they also provided a forum to discuss connected issues. The area of application of DoE covered was satisfyingly wide and newer areas provided much interest: for example, applications to the service sector, market research and sensory evaluation.

Work package outcomes

WP1 achieved more than the minimum numbers of visits, workshops, papers and articles during the project. Four industrial visits were required but six were actually carried out. The requirement of two journal papers to be submitted was surpassed with five actually submitted for this work package. Seven workshops were carried out, five more than specified in the contract. Four articles were published in the popular press, two more than required by the contract.

Sample papers produced by Pro-ENBIS network

D. Romano and A. Giovagnoli (2004) Optimal experiments for software production in the presence of a learning effect: a problem suggested by software production. *Statistical Methods & Applications*, **13**(2), 227–239.

A. Bucchianico, T. Figarella, G. Hulsken, M.H. Jansen and H.P. Wynn (2004) A Multi-scale approach to functional signature analysis for product end of life management. *Quality Reliability Engineering International*, **20**(5), 457–467.

D. Romano, M. Varetto and G. Vicario (2004) Multiresponse robust design: a general framework based on combined array. *Journal of Quality Technology*, **36**(1), 27–37.

Work package 2: Data mining/warehousing

- *Mission*. Dedicated to exposing researchers and industrialists in the field of statistics to data mining problems and tools, so that the natural skills of statisticians, such as the ability to model real observational data, can be applied to data mining as well.

Work package motivation

Data mining is not only the statistical analysis of large databases, using established statistical techniques, but also a new challenging field, which involves:

- sampling appropriately the available massive data;

- learning the data generating mechanism underlying the data at hand;
- being able to deliver results of statistical data analysis in ways that are efficient and communicable to practitioners;
- working at the interplay between statistical modelling and computationally intensive methods.

Data mining tools can be classified into three broad areas: association rules, classification problems, and predictive problems. In the first area would fit methodologies such as descriptive multivariate measures of association, log-linear models and graphical models. In the second area regression methods, classification trees, neural networks and cluster analysis seem the most obvious choices. Finally, probabilistic expert systems (Bayesian networks), regression methods and neural networks seem to fit in the third class. Concerning applications, the network decided to confine itself to applications that relate to business and industry, leaving alone others related to areas such as epidemiology or genetics.

Work package outcomes

WP2 achieved more than the minimum numbers of workshops and published papers during the project. Four industrial visits were carried out and two articles were published in the popular press for this work package as required by the contract. Three journal papers were submitted, one more than required, and six workshops were run, four more than required by the contract.

Sample papers produced by Pro-ENBIS network

R.J.M.M. Does, E.R. van den Heuvel, J. de Mast and S. Bisgaard (2003) Comparing non-manufacturing with traditional applications of Six Sigma. *Quality Engineering*, **15**(1), 177–182.

R. Kenett, M. Ramalhoto and J. Shade (2003) A proposal for managing and engineering knowledge of stochastics in the quality movement. In T. Bedford and P.H.A.J.M. van Gelder (eds), *Safety and Reliability*. A.A. Balkema, Lisse, Netherlands.

P. Johansson, B. Bergman, S. Barone and A. Chakhunashvili (2006) Variation mode and effect analysis: a practical tool for quality improvement. *Quality and Reliability Engineering International*, **22**(8), 865–876.

Workshops and papers were beneficial because of the information dissemination to professional users and/or improving the body of scientific knowledge through constructive comments and scientific discourse in the peer review of the papers. The general benefits of the published articles are to stimulate interest in the project and to advertise its use as a resource to industry.

Work package 3: General statistical modelling

- *Mission.* To improve the use of process monitoring and control methods for both industry and commerce and to help companies introduce advanced techniques via workshops and visits.

Work package motivation

Methods such as control charts, cusums, range charts, variance charts, means charts, exponentially weighted moving average (EWMA) charts and the newer multivariate charts are well known. There is a need to develop and further the application of these and associated methods in European industry – in particular, the use of multivariate methods using principal components and factor analysis. The advanced use of statistical modelling, using generalised linear models and similar applications, needs to made better known.

Work package outcomes

WP3 achieved more than the minimum numbers of visits, workshops, papers and articles during the project. For example, a manufacturing process from raw materials to finished products can be modelled by simulation. Statistical process control (SPC) and DoE can be used for inspection and improvements. This work package prepared a template, using these tools, to apply to a wide variety of manufacturing contexts. It was used for short (half-hour) demonstrations for works visits up to workshops of several days.

Overall eight industrial visits were completed, twice as many as the four required. Six journal papers were submitted, four more than required, and 12 workshops were run, 10 more than the contract required. Four articles were published in the popular press for this work package, two more than required by the contract.

Sample papers produced by Pro-ENBIS network

A. Zempléni, M. Véber, B. Duarte and P. Saraiva (2004) Control charts: a cost-optimisation approach for processes with random shifts. *Journal of Applied Stochastic Models in Business & Industry*, **20**(3), 185–200.

L. Marco, X. Tort-Martorell, J.A. Cuadrado and L. Pozueta (2004) Optimisation of a car brake prototype as a consequence of successful DOE training. *Quality and Reliability Engineering International*, **20**(5), 469–480.

D.J. Stewardson, M. Ramalhoto, L. Da Silva and L. Drewett (2003) Establishing steel rail reliability by combining fatigue tests, factorial experiments and data transformations. In T. Bedford and P.H.A.J.M. van Gelder (eds), *Safety and Reliability*. A.A. Balkema, Lisse, Netherlands.

Much of the first 18 months of the contract period was devoted to developing material, including instructional software for workshops and works visits which was used more fully in the second 18 months of the contract period. The following programmes have been developed:

1. Simulation of production processes (aluminium wheels and oil filters) for teaching DoE and SPC. The concepts behind the simulation apply to most processes and production. Aluminium wheels were chosen as they are part of the automotive industry, which is well recognised as being in the forefront of industrial statistics applications. Oil filters were chosen, as the co-ordinators have had long-term connections with a company manufacturing filters and could advise on details to make the simulation realistic. This package is currently called Process Training. Details can be found on http://www.greenfieldresearch.co.uk.

2. Simulation of clinical trials, for teaching the design of trials protocols and the analysis of simulated data. This package is called MetaGen. It will enable the user to develop and compare protocols for clinical trials to discover the best protocol for any situation. Details can be found at http://www.greenfieldresearch.co.uk.

Work package 4: Reliability, safety and quality improvement

- *Mission.* To disseminate help on improving the reliability of processes, products and systems and help facilitate the philosophy of statistical thinking as applied to the world of work, via workshops and visits.

Work package motivation

There is a need to further the use of reliability modelling – in particular to help companies understand better the whole-lifetime issues involved in product design and how to measure this. Warranty data analysis, lifetimes, failure rates of products and systems all need scrutiny. This work package focused on helping people to understand how to measure these and more complex issues in an operational sense.

The work package investigated ideas relating to the concept of stochastics for quality movement (SQM). SQM brings together several recent research results to develop a structural approach to analysis of reliability data, and introduces some other aspects of reliability analysis, for example data manipulation prior to analysis and analysis of multivariate and covariate structures.

Work package outcomes

WP 4 achieved more than the minimum numbers of visits, workshops, papers and articles during the project. Six industrial visits were completed, two more than the four required. Five journal papers were submitted, three more than required, and six workshops were run, four more than the contract required. Three articles were published in the popular press, one more than required for this work package.

Sample papers produced by Pro-ENBIS network

D. Romano and G. Vicario (2003) Assessing part conformance by coordinate measuring machines. In T. Bedford and P.H.A.J.M. van Gelder (eds), *Safety and Reliability*. A.A. Balkema, Lisse, Netherlands.

O. Evandt, S.Y. Coleman, M.F. Ramalhoto and C. van Lottum (2004) A little-known robust estimator of the correlation coefficient and its use in a robust graphical test for bivariate normality with applications in the aluminium industry. *Quality and Reliability Engineering International*, **20**(5), 433–456.

C. McCollin, C. Buena and M. Ramalhoto (2003) Exploratory data analysis approaches to reliability: some new directions. In T. Bedford and P.H.A.J.M. van Gelder (eds), *Safety and Reliability*. A.A. Balkema, Lisse, Netherlands.

Work package 5: Discovering european resources and expertise

- *Mission.* To find the statistical resources already available in the European Community, preparing lists of journals, providing a knowledge base of all resources available for the mentoring of European company staff, developing a European database of relevant teaching resources, and organizing a pan-European audit of statistical expertise.

Work package motivation

One of the main purposes of this work package was to investigate what European statistical resources are available. Each country has its own unique provision and very few people are familiar with those of other countries. The various lists produced helped collaboration between members and also led to the creation of the ENBIS Academic Publication Panel, which advises members on how and where to publish their work, and the ENBIS Academy, which co-ordinates the various short courses and workshops run by members at ENBIS conferences.

Work package outcomes

The knowledge base of resources includes four categories:

- A collection of organisations. These are specialist consultancies, institutes, companies, that concentrate on industrial statistics. The database lists the organisations by country. which makes searching for expertise in a particular country very easy.

- University departments. These statistics departments may or may not concentrate on industrial statistics applications. The database lists the organisations by country. The list also includes entries for non-European countries, such as Australia, Brazil, Canada, China, India, Mexico, New Zealand, South Africa, South Korea, Taiwan, Thailand and United States.
- The national statistics bureaux. These are generally European government statistics agencies that may have useful information.
- Statistical societies. These are a useful contact point, and are listed by country.

The list of European statistical expertise was created by the use of an on-line questionnaire facility provided by a Pro-ENBIS partner. The survey builder can be found at http://www.kpa.co.il/english/surveys/index.php. The website also includes instructions for use and how to launch the survey when constructed.

Work package 6: Drafting and initiating further activity

- *Mission*. Dedicated to drafting and initiating outcomes such as: a series of conferences around Europe on various applications of statistical methodology to measurement in industry and commerce; a new journal that concentrates on application-based case-study material; a European business-to-business mentoring scheme for helping small and medium-sized enterprises; a European measurement system audit programme; and a European training network for helping young graduates to gain advanced statistical skills.

Work package motivation

There was a lot of enthusiasm from partners and members of Pro-ENBIS for working together, and ideas were widely contributed for each of the deliverables. Typically a draft was prepared by the co-ordinator and circulated via the web and/or discussed at a project meeting. The final copy was then completed and presented on the Pro-ENBIS website. A series of meetings and discussions were conducted with relevant standards agencies, in particular in relation to measurement and measurement systems. Other work involved discussions and preparations for introducing a British Standard on Six Sigma.

Work package outcomes

WP6 achieved the deliverables that it set out to complete during the project. A draft for holding a series of conferences around Europe was produced during the project. Members of Pro-ENBIS have extensive experience of holding conferences. The draft begins with a summary of the conferences held, the numbers attending and the Pro-ENBIS members directly involved in the organisation. It continues with plans for future conferences. Guidelines for costing conferences and a roadmap for organizing them are given, followed by a summary of important points.

The issue of starting a journal produced a very interesting discussion with vigorous debate at a project meeting and thereafter in email exchanges. Broadly there were two camps. There were some people who considered a journal vital to the continued success of ENBIS, even though it would mean that ENBIS members had to pay to join what is at the moment a free-to-join society. Others took the view that there are already too many journals, that it would be difficult to obtain enough material consistently and that ENBIS is better as a rather unusual free-to-join society. In the end an excellent compromise was reached and ENBIS produced a four-page magazine within *Scientific Computing World* (SCW). SCW is posted every 2 months to all ENBIS members who request it. The editors have plenty of material for this and there been no need to raise a subscription.

A list of partners for the European business-to-business mentoring scheme was drawn up in consultation with ENBIS members, as was a list of partners for a European measurement system audit programme and a list of potential partners for a European training network.

Work package 7: Management statistics

- *Mission.* Dedicated to furthering the use of statistics in strategic planning, operational control and marketing in an industrial and commercial context. The emphasis will be on developing concepts for ease of application within the management capabilities of European industry.

Work package motivation

The central question for research in this work package is: What are the contributions of statistics to management processes in organisations? This question is approached from different viewpoints: the quality management viewpoint, the economics viewpoint, an empirical viewpoint, a historical viewpoint, and the statistical programmes viewpoint. The research strategy is to study the central question from these more limited perspectives.

This work package aimed to improve, and in many cases introduce for the first time, the use of process investigation, monitoring and control methods for the better strategic management and operational control of business. Many of the techniques used to monitor processes in industry are directly transferable to the improvement of business processes. Methods such as control charts, cusums, range charts, variance charts, means charts, EWMA charts and the newer multivariate charts are applicable. In addition, a combination of these with other quantitative methods, such as risk analysis, and management tools can greatly improve business performance. The use of key performance indicators, if properly founded, can enhance the outcomes of most commercial enterprises.

Work package outcomes

The network has developed and furthered the application of management statistics and associated methods in European commerce during the project – in particular, the use of risk analysis within strategic decision-making, advanced methods within operational control and statistical methods to help marketing. The use of conjoint analysis and market analysis via multivariate methods has been promoted throughout the project. The advanced use of statistical modelling, using generalised linear models and similar applications, has been disseminated widely during the project.

Sample papers produced by Pro-ENBIS network

S. Bisgaard and P. Thyregod (2003) Quality quandaries: a method for identifying which tolerances cause malfunction in assembled products. *Quality Engineering*, **15**(4), 687–692.

I. Ograjenšek and P. Thyregod (2004) Qualitative vs. quantitative methods. *Quality Progress*, **37**(1), 82–85.

S. Bisgaard, R.J.M.M Does and D.J. Stewardson (2002) European Statistics Network grows rapidly: Aims to increase understanding, idea exchange, networking and professional development. *Quality Progress*, **35**(12), 100–101.

Work package 8: Network management

This work package was responsible for providing and maintaining the network website, contacting standardization bodies in Europe and reporting and preparing published material. All partners contributed. All periodic reports, including the final project report, were completed on time and the Pro-ENBIS project was successfully concluded on time to the satisfaction of the funding European Commission.

In conclusion

The value of the network was in helping to release the true effectiveness of the many facets of business and industrial statistics for the benefit of Europe. It also helped to introduce these quality methods into non-manufacturing sectors of the economy. Some particular benefits from more effective dissemination of the methods include:

- greater involvement by industry generally;
- better scope for cross-sector learning;
- more intra-state collaborations;

- better-quality products;
- less material and energy wastage;
- technology transfer from wealthier member states to the less well off;
- improved business processes;
- better job creation.

Members are involved directly with the European Committee for Standardization (CEN), the International Organisation for Standardisation (ISO), the Royal Statistical Society (RSS) and most of the major European statistical and quality organisations.

Achievements and the future

- *ENBIS Magazine*, within *Scientific Computing World*, posted free to ENBIS members every 2 months;
- George Box medal for exceptional contributions to industrial statistics established and awarded annually at ENBIS conference;
- prizes for best presentation, young statistician and supporting manager awarded annually at ENBIS conference;
- establish local networks in most European countries;
- have ENBIS members in the top 10 companies in each European country;
- continue Pro-ENBIS type workshop and research activities;
- PROMOTE ENBIS membership widely.

Presidents of ENBIS

2000 Henry Wynn
2001 Dave Stewardson
2002 Tony Greenfield
2003 Poul Thyregod
2004 Shirley Coleman
2005 Fabrizio Ruggeri
2006 Ron Kenett
2007 Andrea Ahlemeyer-Stubbe

2

A history of industrial statistics and quality and efficiency improvement

Jeroen de Mast

The twentieth century witnessed enormous increases in product quality, while in the same period product prices dropped dramatically. These important improvements in quality and efficiency in industry were the result of innovations in management and engineering. But these developments were supported by industrial statistics, a new discipline that both contributed and owed its development to the emergence of modern industry. This chapter gives an overview of the emergence of industrial statistics, placed in the context of the rise of modern industry, and in particular in the context in which it has made its most important contributions: quality and efficiency improvement. In addition, the scientific nature of the discipline will be discussed.

2.1 The old paradigm: craftsmanship and guilds

For many centuries, 'industry' was a small-scale and very local affair. Products were manufactured by craftsmen who worked in small workshops and sold to customers who as a rule lived in the immediate vicinity. Boys learned a trade from a master while serving as his apprentice. In many European countries, tradesmen were organised in guilds, which would to a large extent determine what was manufactured and how. Efficiency was hardly an issue. Quality assurance was based on the craftsman's reputation, on guild rules, and on acceptance inspection by customers.

Statistical Practice in Business and Industry Edited by S.Y. Coleman, T. Greenfield, D.J. Stewardson and D.C. Montgomery

2.2 The innovation of mass fabrication

The last decades of the eighteenth century saw the rise of a new industrial system: that of the factory. Originating in England, the Industrial Revolution saw the emergence of firms of up to a hundred employees, especially in the textile and pottery industries. Masters and apprentices became foremen and workers. Mechanization offered huge possibilities for new products, but above all for large volumes and high productivity, and innovations in transportation and communication allowed firms to trade far beyond their immediate region. By the last decades of the nineteenth century, the factory system had spread to countries such as Germany and the United States, with steel and electricity related industries as the key industries. Giant companies, such as General Electric and Siemens, emerged. Products became more and more complex, and it became vitally important that parts were 'interchangeable': it was no longer feasible to touch up and adjust parts to make them fit. Instead, parts should be close enough to specification to fit without adjustment.

Around 1900, the United States broke with the centuries old European manufacturing tradition by separating planning and management from labour work. The complete manufacturing process was broken down in single tasks, each of which was executed by different workers. Two innovations were to establish the standard manufacturing paradigm in the West up until the 1980s. In 1911, Frederick Taylor published *The Principles of Scientific Management*. Taylor set the scene for a systematic endeavour to improve productivity (output per man-hour). Productivity improvement was achieved by carefully measuring and observing how each task was done, and, based on the results designing and experimenting with more efficient ways to do it. Taylor's basic tenet was: never assume that the best way to do something is the way it has always been done.

The second innovation that revolutionised the manufacturing paradigm was Henry Ford's system of mass fabrication. In 1913 he introduced the first assembly line, transgressing the limitations of scale and low productivity of batch and single piece production. In the years to come, Ford managed to decimate the cost of an automobile by exploiting economies of scale and extreme standardization.

The economical focal points of mass fabrication are volume and productivity. Companies improved efficiency by mechanization, standardization, productivity improvement (Taylor), and exploitation of economies of scale. Quality issues centred around interchangeability of parts: a product was good if its characteristics were within agreed tolerances. Because of the complexity of processes and products, specialised quality functions arose, such as quality inspectors, supervisors and engineers. Quality assurance focused on inspection of outgoing products: it was considered more economical to single out defective products 'at the gate' than to invest in process improvement to prevent defects.

2.3 Early contributions from industrial statistics

Parallel to and interwoven with the development of mass fabrication, the disciplines of industrial statistics and quality engineering emerged. William Gosset, who worked for Guinness, is often considered the first industrial statistician. Guinness, at that time the world's largest brewery, strove to make the brewing business more scientific. They hired some of the brightest young men they could find, mostly trained in chemistry, and gave them top management positions. Their task was to discover the determinative properties of barley and hops, and the important factors in the brewing process that determined yield and quality. Until that time, choices of raw materials and process settings were based on experience – for what it was worth – and perception, but Gosset and his colleagues were given a research laboratory. The young researchers found the existing knowledge on the subject inadequate, and therefore decided to conduct experiments to figure out how to grow barley, which varieties to use, how to treat hops, and many more issues.

Soon, Gosset and his colleagues came across the problem that they lacked an adequate methodology for data analysis, especially when sample sizes were fairly small. Besides the factors that they studied, there were many more influences, such as farm-to-farm differences, and it was hard to decide which of the observed effects were due to the factors studied, and which were random noise. Gosset took up the challenge to work out a methodology to draw reliable inferences in the presence of uncertainty and random noise. In 1908, he published his famous paper 'The Probable Error of the Mean'. In this paper, which Gosset was forced to publish under the pseudonym 'Student', he laid down the basics of the familiar t-test. The work would start a revolution in the then infant discipline of statistics, and it inspired men such as Sir Ronald Fisher, Egon Pearson and Jerzy Neyman to develop much of the now standard machinery for analysing the results of comparative experiments (hypothesis testing, the analysis of variance, regression analysis). In the 1920s and 1930s, the statistical theory of experimentation and data analysis would be developed further, mainly in the field of agricultural research. See Box (1987) for a description of the life and work of Gosset.

But it was not only in process improvement and experimentation that statistical methods were needed. In controlling production processes, too, problems abounded that cried out for methods for data analysis. If production processes ran off target – due to technical malfunctions, inferior raw materials, human mistakes, or whatever reason – it was up to the operators to intervene: detect what was wrong and make a suitable adjustment to the process. In order to know whether the process was still in control, operators made periodic measurements. It appeared to be problematic for operators to decide, on the basis of these measurements, whether to intervene in the process or not: the human eye tends to mistake random fluctuations for systematic trends or shifts. As a consequence, operators typically adjusted the process much too often, thus making a process less consistent instead of better.

In the 1920s Walter Shewhart – who worked for the Western Electric company (now Lucent Technologies), a manufacturer of telephony hardware for Bell Telephone – introduced the notions of assignable cause of variation and chance cause. Assignable cause refers to patterns in process measurements that are to be interpreted as a signal that something in the process has changed. All other patterns in data are to be interpreted as random noise, which operators should ignore. Shewhart developed an easy-to-use tool that helps operators discern between the two: the well-known control chart. The foundations that Shewhart laid for statistical process control – or SPC as it is perhaps even better known – still represent standard practice in quality control.

The importance of Shewhart – and his followers, such as William Edwards Deming – goes beyond the control chart. Shewhart took up the scientific study of quality control quite seriously. His focus shifted from inspection of products to control of production processes. This is an important innovation: not only is process control closer to the source of problems – thus enabling a quicker and more pointed response – but also process control is a necessary condition for performance levels that can never be reached by inspection alone.

Another important element of Shewhart's process control philosophy is that measurements and the use of simple statistical tools for data analysis enable a team of operators to learn from their process. This innovative idea prepares for the later shift of focus from singling out defective products to actually improving the performance of the production process. The standard reference for Shewhart's work is his landmark book, *Economic Control of Quality of Manufactured Product* (1931).

Despite Shewhart's emphasis on process control, quality inspection continued to be the dominant way to ensure quality (in fact, Shewhart's ideas would largely be ignored in the West until the 1980s, when the Japanese had demonstrated their value). Because 100 % inspection became very costly, time-consuming, and often was simply impossible (tests would destroy products), quality inspection based on samples became a necessity. Accepting or rejecting batches of products based on inspection of a sample creates risks: the sample could give too optimistic or too pessimistic an impression of the batch. Probability theory was called upon to find a balance between sample size, the consumer's risk, and the producer's risk. In the 1930s, Harold Dodge, who worked – like Shewhart and Deming – for the Bell company, developed standard schemes for sampling inspection and published tables that related sample sizes to risks. Furthermore, Dodge was one of the initiators of the movement that came to be known as statistical quality control, of which the present American Society for Quality was one of the results.

2.4 The innovation of lean manufacturing

After the Second World War, Western companies continued to work within the mass fabrication paradigm: although they focused on productivity and volume, not on quality and efficiency of operations, with millions of consumers eager to buy they got away with it. There were some problems, though, with the prevailing paradigm.

Firstly, operators were completely deprived of a sense of responsibility or pride in their craftsmanship. Consequently, they often assumed a lax attitude, not intervening when processes derailed, let alone making an effort to improve the process ('if a product is bad enough, the inspector will sort it out'). The quality departments, on the other hand, were too isolated from the shop floor to really understand the daily practice and daily problems.

Secondly, economic conditions had changed since the days of Ford, and as a consequence some choices that made sense in Ford's day were no longer economically optimal. Due to innovations in transportation and law the prevailing ideas about supply chain management and optimal degrees of vertical integration had become outdated. Operators were better educated than in Ford's day and could cope with more responsibility, and because of the increased reliability of transportation and suppliers, huge inventory levels were no longer needed.

Post-war Japan was forced to invent a new paradigm. For companies like Toyota it was impossible to compete with the huge US automotive companies on factors such as productivity and volume. Toyota did not have the capital to maintain large inventory levels, nor the market to produce volumes even close to those of General Motors. Instead, Toyota decided to make the best out of a bad situation and created a production system based on low inventory levels, speed and flexibility. In order for the system to work, Toyota needed manufacturing processes that ran like clockwork: optimised changeovers to enable low inventory levels, aggressive defect reduction to eliminate inefficiencies and enable short cycle times, and partnerships with suppliers. Having made all processes more reliable, Toyota did not need excessive buffers of inventory: suppliers delivered the exact number of components needed just in time.

The prevailing Western organisational structures – which were basically derived from Ford's command-and-control structure – did not suit Toyota's needs. Running one's operation like clockwork implied delegating authority to the operators to intervene when a problem arose. Furthermore, problem-solving was put in the hands of shop-floor workers in the form of quality circles. Instead of seeing management as the source of process improvement, Toyota mobilised shop-floor workers to participate in continuous improvement.

And not only Toyota innovated a completely new manufacturing paradigm: the Japanese government hired Western experts like William Edwards Deming and Joseph Juran to teach them the theory of statistics and quality control, which had been largely ignored in the West. Deming promoted sound problem-solving techniques and statistical methods, such as his plan–do–check–act (PDCA) cycle of continuous improvement; see his book *Out of the Crisis* (1986). Juran introduced his principles of quality management to help integrate quality activities into all layers of an organisation (Juran, 1989). The Japanese implemented Juran's tenet that top management should play an active role in quality activities. Furthermore, Juran introduced the quality trilogy of quality planning, quality improvement and quality control as three complementary (but integrated) aspects of quality management. Among the important new innovations were Juran's project-by-project approach to quality improvement, and his ideas on operator controllability. A flood of new practices arose in Japanese companies, which became known in the West only decades later: kaizen, lean manufacturing, just in time, quality circles, and many more.

2.5 The innovation of quality as a strategic weapon

What Toyota and other Japanese companies had invented was a manufacturing paradigm that was superior to the century-old mass fabrication paradigm of the West. The Japanese had broken with dogmas like the following:

- Managers and staff experts should do the thinking so that workers can concentrate on doing.
- A certain rate of defects is unavoidable.
- Communication and management should proceed through a hierarchical chain of command.
- Inventories should be used to buffer different processes from each other and from suppliers.

One of the consequences of the new paradigm was that the Japanese surpassed Western companies on several dimensions simultaneously, while these dimensions were traditionally seen as trade-offs. Quality versus cost versus responsiveness versus flexibility: the Japanese manufacturing virtuosity made it possible to have them all rather than choose. In the late 1970s the Japanese assaulted the world markets with their clockwork manufacturing machines, and appeared to have huge competitive advantages: they sold similar products to those of Western companies, but at lower cost, with fewer defects, more reliability and better durability. The Japanese had learned that better quality actually reduces costs. Quality gurus such as Philip Crosby brought this message to the US; his 1979 book, *Quality Is Free*, had a big impact, introducing the concept of zero defects. Other authors, such as Armand Feigenbaum, introduced the concept of cost of poor quality, and its constituents prevention costs, appraisal costs and failure costs.

The notion of quality that was generally held in the first half of the twentieth century was that of conformance quality (the extent to which products are free from defects). A wider perspective on the role of quality in business was originally proposed by Peter Drucker in his landmark book, *The Practice of Management* (1954), written after his consultancy work for General Motors in the 1940s. Drucker identified the prevailing paradigm at the time as a *manufacturing mindset:* starting from what you make, you price it based on what it costs you, and you find a customer to buy it. If you are in the automotive business, the way to be a better company is to produce more cars at lower cost. Drucker invented a *marketing mindset*: you do not start with what you produce; you start with asking what your customer wants and how much he would be willing to pay for it. This determines both what you produce and how much you can spend making it. In the new mindset, a company does not produce cars, but strives to provide value for its customers. And what value means is determined by the customer and his perception, not by the company and its management and developers.

Since the 1980s, the notions that better conformance quality reduces costs, and the importance of listening to the voice of the customer, have been generally accepted. The Japanese *company wide quality control* spread to the West as *total quality management* (TQM), which developed into an influential movement, based on principles similar to the Japanese lean manufacturing paradigm: a company-wide commitment to quality, with the motivation that quality is related to customer satisfaction and production cost. The programme prescribed that all layers of the organisation should be trained in the quality philosophy and in statistical and non-statistical problem-solving methods. The actual quality improvement was achieved by projects, run by autonomous teams (sometimes called *process action teams*). TQM followers claimed high competitive advantages for companies that adopted and implemented TQM. But it is not all that clear how much benefit TQM has in fact brought to its followers. It is clear, though, that TQM – in the way it was typically implemented – has some shortcomings:

- Quality was pursued for the sake of quality, but this pursuit was typically poorly integrated in a business strategy, and not linked to economically relevant results such as profits. The notion was that improved quality would 'automatically' result in a better competitive position or superior profitability.

- Related to the previous point is the observation that TQM was highly activity-based (as opposed to results-based): if enough employees were trained in problem-solving techniques, and enough teams were running improvement projects, results would follow automatically.

- Tools, concepts and methods were poorly integrated in a coherent methodology.

2.6 The 1980s and 1990s: Catch-up in the west

The Japanese quality revolution was a major contributor to recession in Western industries. In 1980, NBC broadcasted a documentary with the title 'If Japan can ... why can't we?'. The first reactions to the Japanese assault were confused and often beside the point. For several years in a row, the success of the Japanese competition was attributed to a superior cost structure due to cheap labour, low quality and imitation. When finally the West realised that they were facing a completely different and clearly superior manufacturing paradigm, the first reactions were unfocused, rash and confused: within a few years, the Western business world was swamped by quality gurus (Crosby, Feigenbaum, Ishikawa, Taguchi, Shainin, Shingo), and each month there was a new magic trick such as quality circles, JIT, kanban, pull-systems, kaizen. Juran and Deming, both in their old age, saw a renewed interest in their work.

Some of these tricks appeared to be nothing more than fads; many more had their valuable points, but were not so generic as claimed and failed to endure. Most of the valuable ideas have been integrated into more generic theories, which is probably their right place. Perhaps the most comprehensive of these generic theories is Six Sigma.

2.7 Post-war contributions from industrial statistics

Gosset's work on statistical methodology for experimentation inspired an important development, especially with an eye to applications in agricultural research – the theory of the design and analysis of experiments. It was George Box who made important modifications to the theory to make it applicable in industry. Experimentation in agriculture, where each test can last several months, focuses on large, one-shot experiments. Industry, on the other hand, has a need for sequences of smaller-scale experiments, where each test elaborates on the results of the previous tests. In their seminal paper of 1951, Box and Wilson introduced the theory of response surface methodology, which rapidly became popular in the chemical and other industries for process optimisation and product development. The dissemination of computers made statistical modelling approaches – such as regression analysis – easy to apply, and their use in industry proliferated.

Post-war Japan saw statistics as the secret weapon that had helped the Allies win the war, and they were eager to learn from experts such as Deming. In line with their approach of continuous improvement led by teams of shop-floor workers, the Japanese invested heavily in training their workforce in simple statistical and problem-solving tools. Kaoru Ishikawa developed seven basic tools of quality: the histogram, Pareto chart, cause and effect diagram, run chart, scatter diagram, flow chart and control chart (see Ishikawa, 1982). The ideas of Shewhart and Deming on statistical process control were commonly applied in Japan, including control charts and capability indices.

From the 1950s onwards, the Japanese engineer Genichi Taguchi developed his methods for quality engineering. Unfamiliar with the Western literature on the subject, Taguchi developed a distinctive philosophy and methodology. Taguchi gave a new definition of quality, which considers any deviation from target as a quality loss (as opposed to the traditional Western view, which would regard deviations within tolerances as zero quality loss). Thus, variation came to play a dominant role. Taguchi introduced the notion that a large amount of variation in product characteristics is attributable to sources of variation that he called noise factors. Noise factors could be factors in the manufacturing process that resulted in variation in product properties, but also deterioration due to wear and variability in circumstances and modes of usage of the product. Taguchi's most important innovation was the concept of robust design: products and processes should be designed such that they are as insensitive as possible to the effects of noise variables. This sensitivity should be studied experimentally as early as the design phase, and robustness should be an important requirement in product design.

Along with his philosophy on quality engineering, Taguchi developed a distinctive methodology for product and process design, focusing on experimentation, parameter design and tolerance design. In general, this methodology is less effective than the alternative methods that Western industrial statistics had developed, and sometimes unnecessarily complex. Taguchi's philosophy, on the other hand, is generally considered an important innovation, and although similar ideas can be found in the Western literature, Taguchi at least deserves the credit for getting these ideas widely accepted in industry. Especially in the 1980s, the Taguchi methods were popular in Western industry, although their usage would mostly be limited to engineering and development departments.

With the West's renewed interest in quality, industrial statistics flourished and the arsenal of techniques for quality engineering expanded rapidly. Standard methods for studying the precision of measurement systems were developed, as well as methods for analysing non-normal data, statistical models and methods for reliability engineering, multivariate methods, capability analysis, variance component estimation, and graphical methods. The older theories about design and analysis of experiments were still being extended and adapted to new applications. Research into methodologies for process monitoring, control and adjustment thrived. The development of powerful computers and user-friendly statistical software brought statistical methods within reach of a large public.

2.8 Standards, certification and awards

Following the example of the British Standards Institution (established in 1901), many countries founded similar institutions to help standardise metrics, products, definitions and quality systems. After the Second World War, the US government demanded a sound statistical basis for quality control from its vendors, resulting in the military standard MIL-STD-105A, which determined contractually required sample sizes and maximum tolerable defect rates. In the 1960s, the US military imposed on its vendors a quality standard for military procurement (MIL-Q-9858A), detailing what suppliers had to do to achieve conformance. This standard was technically oriented, and in 1968 NATO adopted the Allied Quality Assurance Procedures, which also provided guidelines for organisational issues in quality assurance.

Because it was widely recognised that it is inefficient if customers have to assess each supplier's quality system individually, the idea of standards for quality assurance systems spread beyond the military. In 1979 the British Standards Institution introduced the BS 5750 series, with the aim of providing a standard contractual document to demonstrate that production was controlled. Based on this standard, the International Organisation for Standardisation (ISO) completed the first version of the ISO 9000 series by 1987. The standards and requirements in the ISO 9000 series offer guidelines to companies for establishing a documented quality assurance system. Moreover, the standards and requirements are used to evaluate a company's quality system. As a consequence, companies gradually reduced the number of supplier assessments they did themselves, and instead accepted certification of suppliers by accredited certification institutions. The latest version of the ISO 9000 series is the ISO 9000:2000 series. This offers definitions of terminology (ISO 9000), standards for certification focusing on quality assurance (ISO 9001), and guidelines for establishing a quality system for quality improvement (ISO 9004). Based on the ISO 9000 series, the automotive industry introduced its own influential quality standards (QS-9000) in 1994. Compliance with QS-9000 is a prerequisite for vendors willing to supply to the US automotive industry. The certification hype has settled down somewhat in recent years, and in retrospect its importance may be put in perspective. It is important to realise what the ISO 9000 series is *not*. ISO 9000 is about the quality of the quality assurance system, not about the quality of products. Furthermore, ISO 9000 describes the least requirements to protect the customer's interests. But these least demands are completely different from approaches that seek to improve quality. Companies that want to compete on quality or efficiency need approaches that are far more ambitious.

Besides standards and certification, the 1980s saw the emergence of quality awards in the West. In response to the continuing loss of market share of US companies, in 1987 the US government established the Malcolm Baldrige National Quality Award to promote best practice sharing and as a benchmark for customer-focused quality systems. The first recipient of the Baldrige Award was Motorola (1988), confirming the success of its Six Sigma programme.

2.9 Six Sigma as the latest innovation

The last decades of the twentieth century witnessed economic developments in many industries which have come to be labelled 'hypercompetition'. Companies in industries such as consumer electronics, automotive and food competed with each other on higher and higher levels of efficiency and quality. The main winner of this race has been the consumer; for the participating companies, the result has been profit margins coming under increasing pressure. Falling behind in the race for quality and efficiency means going out of business.

Against this background, Motorola suffered big losses in the late 1970s as a consequence of fierce foreign competition. As one of its executives diagnosed: 'The real problem at Motorola is that our quality stinks.' Rough estimates of the costs of poor quality ranged from 5 % to 20 % of Motorola's turnover. A few years later, Bill Smith, one of Motorola's senior engineers, demonstrated the relationship between customer complaints and inconsistencies in production processes. Motorola's chief executive officer, Bob Galvin, who had been encouraging quality improvement initiatives since the early 1980s, was enthusiastic and in 1987 he launched the Six Sigma programme, with the intention of having all processes at six sigma level by 1992. The statistical foundations of the sigma metric for conformance quality and the six sigma objective are described in Harry and Stewart (1988). By 1997, 65 000 employees (out of a total of 142 000), spread over 5000 teams, were involved in Six Sigma projects.

In later years, the methodology took shape. The DMAIC structure for projects was developed, as well as the organisational structure (consisting of black, green and other belts). After some successes at ABB and AlliedSignal, it was GE's adoption of the programme in 1995 which gave the approach enormous momentum. Hundreds of companies began implementing the programme. Six Sigma became popular on Wall Street, books and other material flooded the market, and consultancy firms offering Six Sigma training mushroomed.

This chapter has shown how innovations and ideas in industrial statistics have contributed to an unprecedented improvement of the quality to price ratio of products in the twentieth century. Six Sigma is simply the next step in this development. Far from being a completely new approach, it absorbs many valuable aspects of previous approaches. To understand the Six Sigma programme, it is important to realise that many dogmas that it takes for granted were in fact twentieth-century innovations and should be seen as responses to the situation at the time they were introduced. Among these innovations are Taylor's systematic pursuit of more efficient ways to perform tasks; most lessons from Shewhart, Deming, Juran and early statistical quality control; and the Japanese lean manufacturing system. But Six Sigma is more than just a comprehensive integration of previous approaches. Several innovations set the programme apart as a step forward:

- Quality is explicitly linked to cost and other business metrics. One of the reasons why quality engineering methods were largely ignored in the West for several decades is that they were employed and promoted by technically trained people, who were unable or did not bother to translate their message into a language that makes sense in the world of managers.

- Six Sigma is more comprehensive than any previous quality or efficiency improvement programme. Its methodology and toolbox contain virtually all relevant innovations of earlier approaches. But the programme goes far beyond methodology and tools, and offers a defined organisational structure, a deployment strategy, guidelines to integrate it with performance management, and so on. Although not many of the elements of the programme are really new, what is truly innovative is their integration into a comprehensive programme.

- Six Sigma companies choose to invest in a well-trained taskforce consisting of Black Belts and Green Belts, dedicated to the systematic pursuit of quality and efficiency improvement. It is true that the Japanese have mobilised their workforce for continuous improvement for decades, and that Western companies also had dedicated staff to develop improvement opportunities, but the scale and resources committed to Six Sigma projects are unparalleled.

2.10 The future

Six Sigma, more than any initiative before, has brought high-level statistical methods to a wider public. Thousands of Black Belts and Green Belts worldwide are trained in the use of design of experiments, non-normal capability analysis, analysis of variance, and so much more (Hoerl, 2001, gives an overview), supported by easy-to-use software. Speculating about the future, one scenario requires mastery of statistical techniques and thinking as part of every professional's skill set. This changes the role of professional statisticians: it will be more and more the engineers, marketers, managers, and other professionals who plan their own investigations and do their own data analyses. Professional statisticians will, on the other hand, find a greater demand for training and support.

Six Sigma as a label may disappear. Its purport will remain important, though, or even gain in momentum. Efficiency was the focus of Western industry up until the 1980s. Many initiatives in the last decades of the twentieth century sought to optimise quality. Economically speaking, the driver for growth and profitability in the twenty-first century will be innovation (cf. Council on Competitiveness, 2004). As Bisgaard and De Mast (2006) argue, Six Sigma has made this step to the twenty-first century, and should be regarded as a managerial and methodological framework for organizing systematic innovation in organisations. Innovations can be breakthrough, but need not be: it is the many smaller incremental innovations that Six Sigma projects typically achieve that accumulate in productivity growth. Improvements resulting from Six Sigma projects may pertain to quality or efficiency, but they need not do so. They can innovate processes and products, but they can also address sales strategies or accounting policies, or improve a business model. In an economy that is determined more and more by dynamics rather than by static advantages, it is company-wide innovative capabilities that determine a company's competitiveness, and industrial statistics, as provider of the tools and techniques for systematic innovation, might just as well find itself at the heart of economic developments.

References

Bisgaard, S. and De Mast, J. (2006) After Six Sigma – What's Next? *Quality Progress*. January 2006.

Box, G.E.P. and Wilson, K.B. (1951) On the experimental attainment of optimum conditions *Journal of the Royal Statistical Society, Series B*, **13**(1), 1–45.

Box, J.F. (1987) Guinness, Gosset, Fisher, and small samples. *Statistical Science*, **2**(1) 45–52.

Council on Competitiveness (2004) *Innovate America: Thriving in a world of challenge and change*. http://www.innovateamerica.org/files/InnovateAmerica_EXEC%20SUM_WITH %20RECS.pdf (accessed September 2007).

Crosby, P.B. (1979) *Quality Is Free: The Art of Making Quality Certain*. McGraw-Hill, New York.

Deming, W.E. (1986) *Out of the Crisis*. Center for Advanced Engineering
sachusetts Institute of Technology, Cambridge, MA.

Drucker, P.F. (1954) *The Practice of Management*. Harper, New York.

Harry, M.J. and Stewart, R. (1988) Six Sigma mechanical design tolerancing. Motorola Publication No. 6s-2-10/88, Scottsdale, AZ.

Hoerl, R. (2001) Six Sigma Black Belts: What do they need to know? *Journal of Quality Technology*, **33**(4), 391–406.

Ishikawa, K. (1982) *Guide to Quality Control*. Asian Productivity Organisation, Tokyo.

Juran, J.M. (1989) *Juran on Leadership for Quality: An Executive Handbook*. Free Press, New York.

Shewhart, W.A. (1931) *Economic Control of Quality of Manufactured Product*. Van Nostrand Reinhold, Princeton, NJ.

3

Statistical consultancy

3.I

A statistician in industry

Ronald J M M Does and Albert Trip

The role of statisticians in industry has for many years been a source of anxiety. In this chapter, we review some papers on this topic, and give comments based on our own experience as statistical process control and Six Sigma experts and as statistics consultants.

Snee (1991) gave his view on the role of the statistician within the context of total quality, a program in which statistics is only one aspect. He distinguished different roles according to whether or not thc statistician, his client, and the organisation are passive or active (see Table 3.I.1). Snee's view is that in order to survive, the statistician should seek an active role, whether his organisation is passive or active. Especially in organisations that actively pursue total quality, statisticians can be of real help only when they influence management as leaders or collaborators, although good technical work is as important as ever. So they need to learn the techniques of organisational development. According to Snee, 'understanding the

Table 3.I.1 Different roles of the statistician.

Organisation		Passive		Active	
Client		Passive	Active	Passive	Active
Statistician	Passive	None	Helper	Teacher	Data blesser
	Active	Crusader	Colleague	Leader	Collaborator

Statistical Practice in Business and Industry Edited by S.Y. Coleman, T. Greenfield, D.J. Stewardson and D.C. Montgomery © 2008 John Wiley & Sons, Ltd behavioural aspects of statistics and using organisational development techniques are the greatest opportunities and challenges facing statisticians today'.

Unfortunately, many statisticians appear not to be equipped with the skills required to be effective in industry. Hoerl *et al.* (1993) give as most important reasons:

- a broad set of values, attitudes and behaviours that do not support industrial needs;
- a reward and recognition system that reinforces these values;
- an inward focus on the profession;
- a lack of emphasis on the fundamental role of statistics and its use in the scientific method;
- a consuming focus on tools and methods rather than the use of statistics in a systematic approach to solving problems and making improvements.

The authors recommend a set of non-statistical skills:

- solid grounding in the fundamentals of modern quality principles;
- basic understanding of the economic and technical issues of the business;
- knowledge of the field of organisational effectiveness;
- consulting and training skills.

They also give advice on the necessary statistical skills to survive, as well as suggestions on literature to acquire those skills. Related to the statistical skills, their general theme is the need to understand how to apply statistical methods to solve real problems.

In a challenging paper, Banks (1993) reinforces this theme: 'Companies must have access to PhD level statisticians internally, who may not be developing new theory, but can comfortably command the old.' Banks refers to the Japanese practice of dealing with statistical methods:

> I would guess that intelligent use of simple tools will achieve about 95 % of the knowledge that could be obtained through more sophisticated techniques, at much smaller cost. Also the simple tools can be applied more quickly to all problems, whereas the complex tools are unlikely to be ubiquitously used.

His recommendations for industrial statisticians are more or less the same as those of Hoerl *et al.* He also compares industrial statistics with academic statistics, and notes that the greatest gulf of all between the two is one of respect. He noticed that some academic statisticians regard most industrial statisticians as those who were not good enough to succeed at a university, while applied statisticians are often openly sceptical of academics' practicality. He fears that these divisions cannot be repaired easily, but hopes that universities design curricula that take better account of the needs of applied statisticians. After all, that is what the majority of their PhD and master's students will be.

In his comment on Banks's paper, Hahn describes the industrial statistician:

> Graduates suited for industry are likely to become impatient with the academic environment and slow pace of the university. High grades is *one* of the things that we, in industry, look for in candidates – but it is only one of many. We need excellent communicators (who can talk in their customer's language, rather than in statistical jargon), good listeners, hard workers, team players and fast learners. We look for people who are enthusiastic, who are willing to work simultaneously on multiple projects, who are self-confident – without being arrogant, who can rapidly diagnose a problem and see the big picture, who are willing to ask fundamental questions and challenge assumptions, who are good at selling projects and themselves, who can cope with frequent management reviews and changes of direction - and who are still cheerful at the end of the day. It is not an environment that is conductive to proving theorems – but it is hectic and exciting!

With so much attention to non-statistical skills, we might easily forget that statistics is still the core competence of a statistician. Gunter (1998) warned strongly against watered-down quality, and quality professionals without a specific set of core skills. But he blames statisticians for sticking to the 'now archaic and simplistic' control charts, while there are so many new and powerful methods: 'We have become a shockingly ingrown community of mathematical specialists with little interest in the practical applications that give real science and engineers their vitality.' Gunter observes that the computer scientist is eager to step into the breach.

Hahn and Hoerl (1998) also observe that the impact of computer science on statistics is large. The first reason for this is the accessibility of statistical software responsible for the 'democratization of statistics'. Then there is the introduction of various new approaches to data analysis (such as neural networks, fuzzy logic and data mining), often independently of the statistical community. And finally, we now have the opportunity to work with management information systems. In this world of 'statistics without statisticians' the unique value of the statistician may, according to Hahn and Hoerl, bring 'an "improvement" mindset – based on proactive avoidance of problems – to the team and the ability to take a holistic approach to problem definition and resolution, as well as our critical quantitative skills'.

Meyer, Trost and Vukovinsky, commenting on Hahn and Hoerl, pose the question whether we are facing a shortage of effective consulting statisticians:

> Our field has not been wildly successful in attracting individuals who innately have the characteristics it takes to be effective. Statisticians, as a group, tend to be introverted. It is not likely that an introvert

> can be trained to be an extrovert. Therefore, the statistical profession should step up efforts to recruit more dynamic individuals to the field. Unfortunately, there is much competition, and these individuals are likely to be attracted to fields that appear to be more exciting or have better pay.

These authors see conflicts with the theoretical rigours of the typical statistics PhD programme at universities. Many students stop at a master's degree, and others get their statistics doctorate in a more user-friendly department (such as industrial engineering).

We are aware that all the above observations are from American authors. We have a strong feeling, however, that the situation is largely the same in Europe. The perceived inward focus of the statistical community is, in our opinion, perfectly illustrated by a discussion in the Netherlands Society for Statistics on the introduction of a certification system for statisticians. Traditional statisticians felt threatened by the 'democratisation of statistics', and wished to limit the harmful influence of dabblers. We are sure that clients are well able to distinguish good from bad statisticians. Moreover, good and bad might well be defined differently by different types of statisticians. Another aspect of the inward focus of the statistical community is the reaction to new methods from outside. The denouncing in the past of the methods of Taguchi is telling. And these days many statisticians appear to dissociate themselves from the success of the Six Sigma programme of Mikel J Harry (see Harry, 1997), even though several world-class companies use it.

A closer co-operation between industry and academia would be useful in Europe as well. The gap between academic and industrial statistics exists here too. For this reason we think that it might be a good idea to recruit industrial statisticians among industrial engineering students, although this is not without problems. The founding of a consulting unit at a university, such as the Institute for Business and Industrial Statistics of the University of Amsterdam (IBIS UvA BV), is another improvement. This is certainly useful for the education of students, for research in applied statistics, for the potential for its co-workers to improve their skills, and for the industry to get expert statistical support and reap the fruits of the latest developments in the field. We deem ourselves fortunate to have the opportunity to work for IBIS UvA, and we are sure that the employer, Philips Semiconductors in Stadskanaal, of the second author during the period 1994 to 2000 has benefited from his experience. A statistician in industry can usefully coach students in research or application projects, and we believe that giving lectures for students is also very important. That the statistician needs good training and communication skills is beyond doubt; he will not succeed in his own company without them.

We totally agree with all authors who stress the need for non-statistical skills. But Gunter has a valid point when he argues that the core skills are still the basis. Otherwise it would be difficult to put the occasional hypes in the right perspective.

For example, within Philips there was a time when 'world class quality' (Bhote, 1991) was hot. The indiscriminate user would have followed the trend, disregarding the many opportunities from traditional experimental design theory. Especially in the rather complex world of semiconductors industry, this would have led to many disappointments.

Regarding statistical software, our experience is that many problems can be solved with simple tools. The statistical possibilities of (simple) programs such as MINITAB or even Excel might often suffice. In this respect we again fully agree with Banks's comments on simple tools. A statistician needs more specialised statistical software, but the ordinary person does not. The non-statistician may not understand much of the output of specialised software. There is a clear trend for enormous amounts of data to be available for analysis. This requires new methods and skills for statisticians, as Gunter, Hahn and Hoerl rightly observe. Our experience is that it is still a difficult matter to combine the relevant data from the separate databases all over the organisation. Knowledge of database management might well be a useful skill for a statistician.

We conclude this section with some remarks about the career of a statistician. We think that for many it may well be a lifetime job; however, the average industrial statistician occasionally wants new challenges. For some the ever-changing world will offer enough of these, but many (and especially those Meyer *et al.* are hinting at) will need more. We can certainly recommend a part-time job at an outside consultancy agency or university. On the other hand, being a part of a large company also offers opportunities for more or less closely related activities within the company. A truly active statistician will notice that several things can be done in all sorts of disciplines such as quality, organisation, and logistics.

References

Banks, D. (1993) Is industrial statistics out of control? (with Comments). *Statistical Science*, **8**, 356–409.

Bhote, K.R. (1991) *World Class Quality*. Amacom, New York.

Gunter, B.H. (1998) Farewell fusillade: an unvarnished opinion on the state of the quality profession. *Quality Progress*, **32**(4), 111–119.

Hahn, G. and Hoerl, R. (1998) Key challenges for statisticians in business and industry (with Discussion). *Technometrics*, **40**, 195–213.

Harry, M.J. (1997) *The Vision of Six Sigma* (5th edition), 8 volumes. Tri Star Publishing, Phoenix, AZ.

Hoerl, R.W., Hooper, J.H., Jacobs, P.J. and Lucas, J.M. (1993) Skills for industrial statisticians to survive and prosper in the emerging quality environment. *American Statistician*, **47**, 280–292.

Snee, R.D. (1991) Can statisticians meet the challenge of total quality? *Quality Progress*, **24**(1), 60–64.

3.II

Black belt types

Roland Caulcutt

3.II.1 Introduction

An essential component of the Six Sigma approach to business improvement is the training of employees for the role of process improvement specialist. These people usually have the words 'Black Belt' in their job title. In this chapter, I focus on Black Belts: what they do and what skills they require if they are to perform this important role. Their role varies from company to company, but it invariably involves extensive use of statistical techniques. Some of the time devoted to statistics training could usefully be spent considering the human dynamics of improvement projects. More specifically, I explore the benefits for Black Belts of understanding personality types.

3.II.2 Black belts

A Black Belt is a high-potential employee who, after extensive training, works full-time as a leader of process improvement projects, for a period of two to three years. This definition will not be precisely true in every organisation. Not all Black Belts devote 100 % of their time to process improvement, and some may move on before they complete the specified period. Nonetheless, this is a useful working definition.

A few Black Belts may later become Master Black Belts, but most are expected to move into management on completion of their tour of duty, and some may become senior managers in due course.

An important feature of Six Sigma is that Black Belts follow a clearly defined procedure in their efforts to improve processes. In many companies this procedure is known as **DMAIC** (**D**efine, **M**easure, **A**nalyse, **I**mprove and **C**ontrol). In the D phase, the project is defined and the performance metrics are clearly stated. Ideally this project definition is produced by the process owner to express without ambiguity what is expected of the Black Belt and his team. In the M phase, the recent performance of the process is assessed. In the A phase, the root causes of poor performance are identified. In the I phase, actions are proposed and one is chosen in the confident expectation that it will lead to an improvement. In the C phase, action is taken to ensure that the improvement achieved in process performance is sustained beyond the end of the project. In practice, the smooth flow through the M, A, I and C phases will depend on the quality of the project definition and the support given to the Black Belt.

The identification of potential Black Belts, the selection and definition of suitable projects, and the reduction of resistance to Black Belt activities are three support processes that can make a huge difference to the success of a Six Sigma program. If these three processes are not well established before Black Belt training starts, then the initiative may falter. Pande *et al.* (2000) strongly emphasise the importance of project selection: 'We once conducted an informal poll of colleagues who had been involved in Six Sigma and other process improvement initiatives, and found an unanticipated consensus: Each person identified project selection as the most critical, and most commonly mishandled, activity in launching Six Sigma.' Other writers agree that project selection can prove to be the Achilles' heel of Six Sigma. Less is written in the professional literature about the difficulties associated with the identification of potential Black Belts and the reduction of resistance to Black Belt activities. Nevertheless, it is clear to many Six Sigma practitioners that the human dynamics of Black Belt projects are a cause for concern. It is widely agreed, for example, that a project is more likely to fail because of poor communication than poor application of statistics. Perhaps some Six Sigma training devotes too much time to statistics and too little time to human dynamics. This comment applies just as much to the training of managers whose support for projects is essential as to the training of Black Belts. My experience in the training of Black Belts leaves me in no doubt that their effectiveness can be increased by the inclusion in the curriculum of several sessions on human dynamics. More particularly, I believe that an understanding of personality types can greatly improve the ability of a Black Belt to communicate and to overcome resistance.

3.II.3 Personality

There are many personality tests, or personality questionnaires, in widespread use. Each attempts to measure the various dimensions of personality, but each is based on a different model. For example, the 16 PF personality test measures 16 personality factors. By answering the 187 questions in the test, you obtain a score on each of the 12 primary factors and, from these, a score on each of four secondary factors. Studying those primary and secondary factors on which you have more extreme scores may help you to understand why your behaviour differs from that of your friends or work colleagues (see Cattell, 1995). An alternative personality test is the Eysenck Personality Inventory (EPI), which gives scores on three personality factors. These resemble the 16 PF secondary factors (see Eysenck, 1981). Both tests are potentially dangerous instruments and only suitably trained testers should use them, as they may reveal aspects of personality that might disturb the person being tested.

There is a third personality test that is much more suitable for discussing the personality of Black Belts. This is the ***Myers – Briggs Type Indicator*** (MBTI), which is much more user-friendly and is accessible, in shortened form, from various websites. A shortened version of the MBTI can be found at http://www.human-metrics.com, and an even shorter version at http://www.personalitypathways.com.

3.II.4 The MBTI

The Myers – Briggs Type Indicator was developed by Katherine Briggs and her daughter, Isabel Briggs Myers, over an extended period that covered much of the twentieth century. The MBTI is now well established as a reliable and non-threatening questionnairc that can shcd light on how people prefer to take in information and how they prefer to make decisions. The gathering of information and then acting upon it are important features of the Black Belt role.

Briefly, if you prefer to take in information through your senses (sight, hearing or touch) then you are a sensing (S) type, while if you prefer to take in information through intuition then you are an intuition (N) type. N types are not so receptive to the finer details of their immediate environment, being more focused on internal theories that may explain the environment. The words and phrases in Table 3.II.1 may help you decide your type.

The second preference you must identify is concerned with how you make judgements or decisions. Do you prefer to step back from a situation and weigh the evidence in a critical, logical and questioning manner, or do you prefer to step into a situation, trusting your emotions to lead you to a judgement that is fair and compatible with personal values? This is the distinction between the thinking (T) preference and the feeling (F) preference. Table 3.II.2 may help you to decide whether your judging preference is T or F.

Table 3.II.1 The perception preference.

Preference for sensing (S)	Preference for intuition (N)
Focus on details	Look for the overall pattern
Live in the present	Focus on the future
Practical	Imaginative
Enjoyment	Anticipation
Conserve	Change

Table 3.II.2 The judging preference.

Preference for thinking (T)	Preference for feeling (F)
Ruled by your head	Ruled by your heart
Objective	Subjective
Focused on justice	Focused on harmony
Principles	Values
Analyse	Empathize

You have now made two choices to identify your perceiving preference and your judging preference. Thus you would now classify yourself as ST, SF, NT or NF. To complete the identification of your four-letter personality type you must make two further choices. First you must choose J or P, depending on whether you prefer to spend time in the judging mode or in the perceiving mode. Then you must choose E or I, depending on whether you are extroverted or introverted. Tables 3.II.3 and 3.II.4 may help you to make these choices. You have now selected the four letters that are a label for your personality type. There are 16 possibilities. Your type is one of the following: ISTJ, ISFJ, INFJ, INTJ, ISTP, ISFP, INFP, INTP, ESTP, ESFP, ENFP, ENTP, ESTJ, ESFJ, ENFJ or ENTJ. For a fuller description of your type you could visit one of the two websites mentioned above or consult one of the texts that discuss the MBTI. These include Pearman and Albritton (1997), Kroeger and Thuesen (1988) and Myers and Myers (1993).

You may not be confident that you have correctly identified your personality type. Perhaps you were unsure of one or more of the choices you made. I stress that Tables 3.II.1–3.II.4 are a poor substitute for a full MBTI questionnaire. I also stress that you should consider carefully how you will react to what you have learned about your personality type. Will you think 'I am a——, therefore I cannot do this or that'? Or will you think 'Because I am a——, I must take extra care to do this because it is a task I will not feel comfortable performing'? I strongly recommend the latter. As Henry Ford often said: 'Whether you think you can, or you think you can't, you are probably right.'

If your four-letter personality type contains an N, this discussion of the MBTI might invigorate you. It opens up new possibilities. The discovery of new ideas, new models or new theories can appeal greatly to an N type. If, on the other hand,

Table 3.II.3 The judging – perceiving preference.

Preference for judging (J)	Preference for perceiving (P)
Organised	Flexible
Control the situation	Experience the situation
Decisive	Curious
Deliberate	Spontaneous
Deadlines	Discoveries

Table 3.II.4 The extroversion – introversion preference.

Extraversion (E)	Introversion (I)
Active	Reflective
Focused outward	Focused inward
Sociable	Reserved
People	Privacy
Many friends	Few friends

your four-letter description contains an S, you may be thinking 'Yes, OK, but so what? How will this help me to work better?' Let us turn to practical issues.

3.II.5 Black belt types

What are the personality types of Black Belts? Perhaps a better question to ask is 'Which of the 16 personality types are likely to be more successful in the Black Belt role?'. Among the hundreds of Black Belts I have trained I have found all 16 types. I am not able to quantify the performance of each trainee during their term as a Black Belt, so am not able to say which of the personality types are more successful. However, I am confident that success as a Black Belt comes to those who recognize their strengths and weaknesses, then devote extra effort to those tasks that put demands on the weaknesses. Success also comes to those who adapt their approach to the preferred working styles of the many personality types they meet. In the Black Belt role, as in many other jobs, it is essential to use both of the perceiving functions (sensing and intuition) and both of the judging functions (thinking and feeling).

Think of yourself as a Black Belt and consider what occurs in each of the four MAIC phases. In the M phase, you must focus on the facts as you assess the recent performance of the process. In the A phase, you must consider possible causes of poor performance; but each possible cause must be tested against the facts and verified or rejected. In the I phase, you must reach a decision about which of several possible actions will be taken. It is important to compare the alternatives objectively, but you must not ignore the feelings and possible reactions of those who will be affected by the change. In the C phase, you must put in place monitoring and control systems to ensure that your hard-won gains are sustained.

Now ask yourself which of the four functions are used in each of the four phases. You will agree that S is an important function in the M phase, but later phases require the use of two or more functions, with all four being required and none being dominant for long. MBTI can usefully explain the often noted tendency to rush to the next phase before finishing the previous. For an N, especially an NJ, the temptation to jump from the M phase to the A phase must be very strong.

In every phase of a project the Black Belt needs to work cooperatively with other team members. So, you have the difficult task of trying to balance the preferences and strengths of team members with the needs of the moment. The MBTI can suggest specific pointers to how you might modify your approach when you need to work with people whose personality types differ from your own. If you are an introvert working with extroverts:

- try to respond more quickly;
- smile more often;
- remember Es think aloud;
- speak up in meetings – Es cannot read your thoughts.

If you are an extrovert working with introverts:

- wait at least five seconds after asking a question;
- distribute agendas and papers well before meetings;
- remember that Is respond more readily in small groups, preferably one-to-one;
- give an I time to think – consider writing to them or sending an email.

If you are a sensing type working with intuitive types:

- give them the big picture before the detail;
- try not to be negative – delay your criticism of new ideas;
- try to share Ns' concern with the possible long-term effect of today's action;
- do not overwhelm Ns with long lists or excess detail.

If you are an intuitive type working with sensing types:

- tell them how big, how many, how cold, how it looks and how it sounds;
- do not see Ss' request for facts as resistance;
- present your new idea as a gradual rather than a radical change;
- be brief and explain things in a step-by-step way.

If you are a feeling type working with thinking types:

- try not to take criticism personally – see it as useful feedback;
- get to the point – do not spend too long on personal issues;
- try to appear logical – list the pros and cons;
- present feelings and reactions as facts to be considered.

If you are a thinking type working with feeling types:

- do not see relationship building as a waste of time;
- say what you agree with before expressing disagreement;
- it is not what you say, but the way you say it;
- think how your suggestions will affect other people.

If you are a judging type working with perceiving types:

- allow enough time for discussion – do not rush to decision;
- do not issue arbitrary or artificial deadlines;

- trust Ps – they will finish the job, even if they are behind schedule;
- be more flexible – permit spontaneity and change.

If you are a perceiving type working with judging types:

- get organised and look organised;
- time is important – be punctual, respect deadlines;
- do not start a second job unless you have a clear plan for finishing the first;
- do not surprise Js – give them advance notice of possible change.

If you want to adapt to the preferred working styles of others, these lists may help useful. But they are no substitute for deeper study of the MBTI, which can be gained by reading the references and practicing behaviours with which you are less familiar. Your growing understanding and your deeper appreciation may help you to:

- gain support and avoid resistance;
- support Black Belt projects;
- be a more effective member of a project team;
- build more effective teams.

3.II.6 Summary

The Black Belt role is demanding. Perhaps this is why successful Six Sigma companies select only their high-potential employees for Black Belt training. This training invariably covers DMAIC, statistics and several project reviews, but I believe it should also include human dynamics, including some personality theory.

The MBTI is a proven, reliable, non-threatening personality questionnaire. By studying the MBTI, you can develop an appreciation of personality types that will help Black Belts to communicate more effectively, gain co-operation and overcome resistance.

References

Cattell, R.B. (1995) Personality structure and the new fifth edition of the 16PF. *Educational & Psychological Measurement*, **55**(6), 926–937.

Eysenk, H.J. (1981) *A Model of Personality*. Springer-Verlag, Berlin.

Kroeger, O. and Thuesen, J.M. (1988) *Type Talk*. Dell Publishing: New York.

Myers, I.B. and Myers P.B. (1993) *Gifts Differing*. Consulting Psychologists Press, Palo Alto, CA.

Pande, P.S., Neuman, P. and Cavanagh, R.R. (2000) *The Six Sigma Way*. McGraw-Hill, New York.

Pearman, R.R. and Albritton, S.C. (1997) *I'm Not Crazy; I'm Just Not You*. Davis-Black Publishing, Palo Alto, CA.

Further reading

Carnell, M. (2003) The project pipeline. *ASQ Six Sigma Forum Magazine*, **2**(3), 28–32.

Caulcutt, R. (2001) Why is Six Sigma so successful? *Journal of Applied Statistics*, **28**(3/4), 301–306.

Eckes, G. (2001) *Making Six Sigma Last*. Wiley, New York.

Harry, M. and Schroeder, R. (2000) *Six Sigma*. Currency, New York.

Kelly, W.M. (2002) Three steps to project selection. *ASQ Six Sigma Forum Magazine*, **2**(1), 29–32.

Lynch, D.P., Bertolino, S. and Cloutier, E. (2003) How to scope DMAIC projects. *Quality Progress*, **36**(1), 37–41.

Wright, D.S., Taylor, A., Davies, D.R., Slukin, W., Lee, S.G.M. and Reason, J.T. (1970) *Introducing Psychology: An Experimental Approach*. Penguin, London.

3.III

Statistical consultancy units at universities

Ronald J M M Does and András Zempléni

3.III.1 Introduction

In this chapter we give a personal view on the role of a statistical consultancy unit, especially one within a university. Both of us work in universities where, besides research, we do consultancy work, and this enables us to explore the issue from an inside point of view. This work is an updated and revised version of Does and Zempléni (2001).

Statistical consultancy can mean anything from resident experts within departments to a dedicated bureau or consultancy unit. Why is there a need for consultancy units? Why cannot the engineers or other local experts solve their own problems? One reason is that problems that occur randomly in time may be solved more readily by someone with an external viewpoint. But our main point is that theory and methods are constantly developing and a university-based group is more likely to be able to keep up with research. This may not be obvious to those with the problems, since the ubiquity of statistical packages creates the illusion that anybody can find the correct answers to problems. Unfortunately, the results from statistical packages are not always correct; the obvious choice is not always the best one to choose from the abundance of available methods, and people may not see the need to check analytical conditions.

So it is necessary to promote the services of the consultancy unit, and this is a tricky issue because it is knowledge rather than a product that has to be sold. Probably one of the best ways to make potential users aware of the existence of the consultancy unit is to put on a special course, which may show the benefits of consulting the unit.

The next question may be: what is the difference between university-based and other consultancy units? The distinction may be blurred because university-based units are quite often transformed into general business enterprises, and then the same personnel and approach are to be found in both types of unit. Rather than become enmeshed in detail concerning these differences, this section confines its investigation to university-based consultancy units.

Minton and Freund (1977) suggested that a consultancy unit located within a statistics or mathematics department probably has the most advantages – for example, recent advances in statistical methodology can be applied if needed and, in return, service courses are revitalised by the consultancy problems. But before explaining our point of view on the role a consultancy unit can play at the host university, we give a brief summary of the available literature on statistical consultancy units.

There is an extensive literature about university-based statistical consultancy units, containing valuable information about such centres at universities in the United States; see Boen (1982) and the references therein, or Khamis and Mann (1994) for a more recent review. The existence of these units is based on the fact that statistics is a special science, as its methodology is used in almost all other sciences and its use is booming in industry (see Bacon, 1999). There are several examples showing that an informal 'service' provided by members of staff to colleagues is not effective. An organisation is needed for performing excellent consultancy.

The literature about European practice is more sparse. Some examples of successful university-based consultancy units are presented in Zempléni (1997). Does (1999) is a recent paper with many examples of research problems initiated by consultancy work.

Within the university, statistical consultancy units can be either commercial, charging a viable rate for their work, or non-commercial, working for no extra payment. We have found more examples of non-commercial (or mostly non-commercial) units than commercial ones; see Carter *et al.* (1986) and Khamis and Mann (1994) for examples. That does not prove that there are only a few commercial units – it could be the case that their consultants are too busy conducting a commercially viable business to disseminate their experiences. A nice example of a self-supporting unit is presented by Boen (1982) in the area of biometrics. Our aim is to add new examples from Europe, where we observe differences in traditions.

An important point is that consulting activities are not usually supported by the traditional university evaluation system, in which research is considered the most important activity (several grants and personal rewards depend on the research activities of the given department), followed by teaching – but in this case quite often the quantitative characteristics are more important than the qualitative ones. Consultancy services – provided to colleagues from other departments or to interested people from the outside world – are considered far less important. This being the case, the establishment of a consultancy unit does not only have the usual organisational difficulties, but there is a danger that the participants themselves will be unsure about their preferences and so the consultancy work might be done less enthusiastically and effectively. In Section 3.III.2 we give a list of differences between commercial and non-commercial units, which should be taken into consideration when deciding whether to establish a unit of either type. Section 3.III.3 summarises the opportunities and the problems when a commercial consultancy unit is already in operation. We summarise both sections in the form of a table. Section 3.III.4 contains our examples, showing how we tried to overcome the observed difficulties when establishing consultancy units in the Netherlands and in Hungary. Section 3.III.5 gives our conclusions.

3.III.2 Comparison of commercial and non-commercial statistical consultancy units

In practice, different traditions at universities have resulted in certain departments or faculties having their own experts in statistics. It is quite clear that such local experts are not able to cover all possible statistical consultancy needs. So there is often a unit (either formal or informal) within the department of mathematics or statistics serving the needs of other departments and/or industry.

There are two main types of such statistical consultancy units: the non-commercial unit, serving mainly the needs of the other faculties of the university; and the commercial unit, which is open to clients both from inside and outside the university on a fee-for-service basis. These pure, extreme cases are rare: often one finds mixtures with emphasis on either the non-commercial or on the commercial aspects.

3.III.2.1 Finance

The most important (and probably the most decisive) question when thinking about starting a consultancy unit is the following: is the university (or any other institution) in a position to offer long-term support to the statistical department (or analogous unit) for providing statistical consulting services to other departments or not? As can be seen from the literature (see Carter *et al.*, 1986), the mission statements of non-commercial units emphasise the positive effect of consultancy on teaching and research. But consultancy is of course a time-consuming activity, so the participating members of staff should get a reduction in their teaching load (financed by the host university). This non-commercial type of unit is less stressful for the members of staff if there is little pressure from the clients for very strict deadlines.

On the other hand, if there is no – or only limited – direct financial support available from the host university, then the unit needs to be self-supporting. There are examples (see below and Zempléni, 1997) where the commercial unit not only became self-supporting, but succeeded even in providing financial support to its faculty and/or university.

The different financial status also has an important effect on the scientific merits of the projects undertaken. If one is responsible for keeping to the annual budget of a commercial unit, then there is strong pressure to obtain income even if the project is not interesting from a scientific point of view. Such pressure is rare in the case of non-commercial units.

The pricing policy is also different for the two types of units: the non-commercial unit is not so heavily dependent on making profits, so there is a possibility to apply flexible charges, depending on the project and the client. On the other hand, commercial units cannot afford to incur losses, so usually a fixed price per consultant per day is charged.

3.III.2.2 Personnel

There are some important points about the skills to be possessed by the members of staff which should also be taken into consideration when thinking about starting a statistical consultancy unit.

Different clienteles require different communication skills: if one only deals with university graduates (which is usually the case for a non-commercial unit), it tends to be easier to achieve respect and to make oneself understandable. However, when communicating with people from the shop floor of a manufacturing company, one has to possess the skills of a good facilitator (see Snee, 1996; see Section 3.II above).

Moreover, the commercial unit has to have a competent head, who is capable not only of understanding the statistical or technical nature of the problems and estimating the difficulty of the projects, but also of managing the whole unit. This requires the skills of an entrepreneur, because he/she has to negotiate project fees and details with other managers. These skills are rarely found in statisticians from a university, and recruitment from outside is not an easy task. The lack of a suitable manager is often the reason for rejecting the idea of establishing a commercial unit.

The organisation, planning and documentation of the work are also different for the two types of units. At a non-commercial unit, in the worst case, the scientific reputation of the author is at risk. To be a consultant at a commercial unit involves taking more responsibility, since in this case legal consequences can follow erroneous advice. Hence the work – and the responsibilities – should always be documented during the whole process of working on a project, (see Kirk, 1991). There is a related question of deadlines. Whilst in the case of non-commercial units the deadlines are rarely strict, this is not so for projects of the commercial consultancy units.

3.III.2.3 Operational structure

As the consultancy units are usually part of the department of mathematics or statistics, the rules of the university apply. Some peculiarities are worth mentioning. For instance, the commercial consultancy unit should have a well-chosen name in order to be distinguishable from the 'old-fashioned' academic department.

Usually, there is no need for major investment when it is decided to start a consultancy unit and, if the consultants already have positions in the department, the financial risk is limited. However, this comfortable situation has its dangers, too: if the head of the unit does not have enough motivation or skills to arrange new contracts, then there is no future for the commercial unit. The non-commercial unit does not need to be so well managed, as the clients usually come on their own initiative.

If the commercial unit is successful, after a few years in operation the unit's growth makes it possible for the university to relax control. Hence, the unit gains more independence. In this case the host department/faculty can create a supervisory board controlling the policy of the unit and the use of the accumulated profit, but not its everyday life.

3.III.2.4 Typical projects

Examples from the practice of the Institute for Business and Industrial Statistics of the University of Amsterdam (IBIS UvA) in the Netherlands show (see below) that the total involvement approach in projects is effective (see Marquardt, 1979). Here the consultant is actively involved in the complete trajectory, from project formulation through understanding the processes, recognizing the possible problems, implementing the practical solutions and checking the results to the post-project follow-up. In such cases the power of statistics can be fully utilized.

Even with projects in other sectors, such as health care, knowledge about the motivation and the expectations of the client are vital for the consultancy work to be effective. A deep understanding of the background is needed so that the suggestions are useful and realizable.

Although the above-mentioned type of participation in projects would of course be useful in the case of non-commercial units as well, usually the consultants can afford to allocate only a limited amount of time to the clients. So there is little hope for total involvement, which needs months or sometimes even years of continuous work. This lack of time results in the situation that even the formulation of the project and agreement on the outputs are not always done.

As statistical theories and methods are expanding, it is almost impossible for a person to be familiar with the most recent trends in more than one field of application. So specialization is needed (especially in the case of a commercial unit): either the unit itself concentrates on a specific subject (industrial statistics in the case of IBIS UvA; see below) or the specializations of the senior members of staff define the limits of the consultancy. For a non-commercial unit, there is more flexibility in choosing projects; one reason for this may be a much looser time schedule which allows time for research.

Table 3.III.1 gives a summary of the above mentioned issues.

Table 3.III.1 Comparison of commercial and non-commercial units.

Aspect	Commercial units	Non-commercial units
Financial status	Self-supporting	Supported by the host university
Pricing	Fixed (based on working days)	Flexible
Clients	Both internal and external	Mostly internal
Skills needed	Professional, managerial and communicational	Professional
Responsibility	Legal	Scientific
Selection of projects	Not always possible	Important
Method of work	Written contracts, policy statements	Informal agreements
Operational structure	Independent management	Possibly within the department
Project areas	Usually specialised	May be more diverse
Typical project depth	Total involvement	Clinic-type

3.III.3 Opportunities and problems when running a commercial consultancy unit

Let us investigate the main points related to running a commercial consultancy unit in a university department.

3.III.3.1 Research

Even in the case of commercial units, the university background should never be forgotten. Consultancy projects will provide the consultants with challenging research problems; for a wide range of examples, see Does (1999).

It is emphasised even in some earlier papers (see Marquardt, 1979) that the level of statistics applied in industry and science has become highly sophisticated. Often suitable procedures for non-standard situations can only be developed by a PhD-level consultant. It is the task of the head of the unit to encourage the publication of the results. Usually, it is not a major problem to get permission from clients to publish the results achieved in a scientific journal. However, it is surprising that these results are not always accepted as an achievement comparable to the (not always really applicable) methodological papers. The promotion and extension of the use of statistical methods to other disciplines are important goals of applied statistics (cf. Carter *et al.*, 1986). One should be aware of the danger of accepting any client only because of financial considerations: if there are only trivial problems involved in the consultancy projects, the consultant might feel overqualified for the work.

3.III.3.2 Teaching

The teaching aspect of consultancy requires that the consultant should possess the ability to communicate the statistical knowledge to the clients, who might have a very different statistical background. This ability is also important when delivering service courses to students of other faculties. The wide range of projects that staff members work on provide excellent examples for illustrating teaching.

On the other hand, a busy consultant might not have enough time to deal with the problems of individual students. It might also cause tensions if the consultant is the supervisor both of his PhD student's thesis and his consultancy work (cf. Boen, 1982).

3.III.3.3 Computing

A recent activity, which is heavily related to modern applied methods in statistics, is the use of computers. There are completely different attitudes towards this area. Some university members of staff are leading figures in non-commercial software development (R being the most widespread example, but there are other examples such as the BUGS package for carrying out Markov chain Monte Carlo analysis). These non-commercial packages are often developed through non-commercial consultancy units. On the other hand, commercial units are more time-conscious, so even if they develop their own procedures, they are rarely published. If they put more effort into their software development, it is more often for an (at least partially) commercialised product, as in the case of some teaching or simulation software (such as SIMUL8). Even in this case the large amount of time spent on the development and marketing of such products may not be effective enough to ensure large sales. So best practice seems to be to relate the software development to the projects, and any sales will be an added bonus.

3.III.3.4 Career

If the applied research of the consultant is recognised by the university, then career opportunities are no worse than for other university staff members who have mainly published papers in (theoretical) statistical journals. In addition, a consultant has extra options, being an entrepreneur as well as a scientist (through the 'total involvement' in the projects, and because practical experience has been acquired in team work and managerial skills – see Marquardt, 1979). This ensures that he is a strong candidate for management positions in industry or research institutes.

This is of benefit for the consultant himself, but it causes a major problem for the head of the unit: he must be able to keep the experienced employees (at least for a while), as too high a level of turnover of consultants causes problems in operating the unit. One way to avoid this is to give the consultants competitive salaries which are on a level with statisticians who work for private consultancy companies. This – together with the above-mentioned research opportunities and the relative freedom of working at a university – will hopefully be enough to solve the problem. However, this might cause tension between the lower-paid university staff and the consultants of the commercial unit, which can be resolved in two ways: first, the host department should also gain from the existence of the unit (usually some percentage of the income is handed over); second, participation in the unit's work should be made possible for other members of staff as well.

Table 3.III.2 gives a summary of this section.

Table 3.III.2 Opportunities and possible problems related to a commercial consultancy unit.

Aspect	Opportunities	Problems/dangers
Research	Real-life problems; possible (joint) publications also in non-statistical journals	Limited time is left for research; less scientific recognition
Teaching	Real-life examples for courses	Less time for individual work with (under)graduates
Computing	Case-related products, possible earnings	Too much time spent on development
Career	Entrepreneurial, communicational skills learned	Slower procedure of promotion

3.III.4 Units in europe

In this section we give some brief examples from Portugal and Great Britain (information by personal communication or from the internet) and present our own experience in two European countries (the Netherlands and Hungary) with rather different economic and scientific backgrounds.

An interesting mixture of commercial and non-commercial activities is carried out by GEPSI, a Portuguese research group, with main activities in the area of process and systems engineering. This mixture is partly due to the interdisciplinary nature of their main fields of interest, which range from engineering to chemistry, with statistics not playing a leading role. This group of leading scientists and active young members of staff have been extremely successful in applying for research grants – together with companies from the chemical industry and, in particular, the paper industry.

Another successful unit is the Medical Statistics group at Lancaster University. Pharmaceutical companies are often important clients of consultancy groups, as their statisticians are not always in a position to provide the most appropriate solutions to the practical problems arising at their companies.

Another British university, Reading, offers a wide range of training especially in agricultural research institutes, to animal health clients, public sector bodies, environmental organisations and developing world projects. The last is a main speciality of the unit. By using external funds to support projects in the Third World, the university is unique and its mission is to provide knowledge where it is most sparse and worthy of appreciation.

3.III.4.1 The Netherlands

The Institute for Business and Industrial Statistics of the University of Amsterdam (IBIS UvA) has been operational since 1 May 1994. It was able to start as a commercial statistical consultancy with a two-year start-up grant from the Faculty of Mathematics and Computer Science. From the outset the unit was sufficiently successful that the grant was not necessary.

The services that are provided deal with implementing statistical process control (see Does *et al.*, 1999) and related quantitative quality programs such as Six Sigma (see Van den Heuvel *et al.*, 1997; Hahn *et al.*, 1999), quality improvement (as a part of total quality management) courses and general statistical consultancy. Currently, there are seven enthusiastic consultants employed (four full-time senior consultants and three part-time senior consultants) three of whom are also professors in the Department of Mathematics. To support and constantly improve the consulting activities, IBIS UvA aims to:

- contribute to scientific research in business and industrial statistics on an international level (for overviews, see Does and Roes, 1996; Does, 1997, 1999);
- promote the application of industrial statistics in all relevant parts of society.

IBIS UvA has developed a comprehensive package of training courses and workshops. These involve statistical process control, Six Sigma, measurement system evaluation, design of experiments, statistically robust design, Taguchi methods, and Shainin – Bhote techniques. These workshops are strongly based on the broad and hands-on expertise available within IBIS UvA.

The customers of IBIS UvA cover a wide range of products and services: from low-volume to mass production; from industry to health care. The IBIS UvA approach has helped its customers to achieve lower costs, higher productivity and better quality.

From 1994 until 1997 IBIS UvA was part of the Department of Mathematics. The host university charges 5 % of the consultancy unit turnover and a fixed price of €3500 per consultant for housing. An extra 7 % of the turnover is paid for use of the faculty infrastructure. The total profit in the period from 1994 to 1997 was about €250 000, which was allocated into a fund. This fund can be used for sabbatical leave and for initiating research in industrial statistics.

In 1998 IBIS UvA became a private company, wholly owned by a holding company which controls all the university's commercial activities. The profits of IBIS UvA are now divided into three equal parts: one going to the holding company, one to the Department of Mathematics and one to IBIS UvA. The current situation allows IBIS UvA to keep the consultants' salaries at a level comparable to those of staff in other professional consultancy bureaux. This really was necessary in order to retain the employees. During the last five years the unit has made a profit in excess of €1 million.

3.III.4.2 Hungary

At Eötvös Loránd University, Budapest, in a traditionally theoretical department in probability and statistics, applied statistics in teaching became more important in the 1980s, when new insurance companies needed young mathematicians. Soon after this, however, there were heavy cuts in the universities' budgets and the number of staff fell by nearly 20 %, and even the daily life of the faculty became impossible without the aid of several grants (research projects or other development funds).

The support of the European Union was obtained in the form of a Joint European Project grant during the years 1995 to 1997 for gaining the knowledge and information needed to be able to create a successful consultancy unit. The Universities of Amsterdam (the Netherlands), Dortmund (Germany), Göteborg (Sweden), Lisbon (Portugal) and Sheffield (Great Britain) were involved in this project. At most of these universities there is still a statistical consultancy unit; see Zempléni (1997) for an overview.

The above-mentioned economic need motivated the choice for a commercial unit rather than a non-commercial one. The members of staff are university graduates and students of the department, so a balance between consultancy and other duties needed to be found – it was achieved within the first few years. Accounting and legal advice is given by the university, enabling the members of the unit to concentrate on the professional part of their work. Over 20 % of the project value has to be paid to the university and an additional 6 % to the host department. This service cannot be cheap. Thus the quality of the work and the reputation of the university have to be a factor when winning clients. This has been the case and, since its formal establishment in 1998, there have been several consultancy contracts fulfilled with a total value of nearly €30 000.

An additional gain for the unit was the successful participation in different national calls for applied projects (in the area of hydrology, for example) as well as the signing of major research contracts with insurance companies (risk analysis). The unit has been able to finance the studies of some PhD students, an important factor when the state is not in a position to give grants even to the best students. However, the status of the unit has not been changed, it has no full-time employees, and the projects are tackled by members of staff and the graduate students.

Another feature is the role of the unit in teaching. The courses taught by the active members of the unit are more application-oriented, with more emphasis on the problems they face during the solution of real-life problems. In addition, courses are held at different companies mostly not as the unit's own projects, but with unit members as teachers for other – mostly non-statistical – consultancy companies.

3.III.5 Conclusion

In most cases there is a need for a university-based statistical consultancy unit which is able to encourage valuable applied statistical research and can provide students with practical training. There is an option for such a unit to be commercial or non-commercial, which depends mostly on the university's willingness to finance this consultancy activity.

If the commercial route is chosen, then one can expect more problems with financial and legal administration, and one should be aware of the danger of being overrun by high budgetary expectations. The example of IBIS UvA has shown that it is possible to overcome these difficulties and to accumulate not only a substantial profit but also scientific recognition.

In the developing economies of central and eastern Europe one cannot expect such a quick expansion, but the need for advanced statistical applications is increasing, so it is worth being prepared for the challenges.

Acknowledgement

An earlier version of this chapter was presented at the Second Annual Meeting of ENBIS, and was supported by funding from the 'Growth' programme of the

European Community and was prepared in collaboration by member organisations of the Thematic Network - Pro-ENBIS EC contract number G6RT-CT-2001-05059.

References

Bacon, D. (1999) Expanding the use of statistical methods in industry through university-industry partnerships. In *Bulletin of the ISI: Proceedings of the 52nd Session (Helsinki)*, Vol. LVIII, Book 1, pp. 173–176. Voorburg, Netherlands: International Statistical Institute.

Boen, J.R. (1982) A self-supporting university statistical consulting center. *American Statistician*, **36**, 321–325.

Carter, R.L., Scheaffer, R.L. and Marks, R.G. (1986) The role of consulting units in statistics departments. *American Statistician*, **40**, 260–264.

Does, R.J.M.M. (1997) Industrial statistics at the University of Amsterdam. In A. Zempléni (ed.), *Proceedings of the Workshop on Statistics at Universities: Its Impact for Society*, pp. 57–65. Eötvös University Press, Budapest.

Does, R.J.M.M. (1999) Real-life problems and industrial statistics. In *Bulletin of the ISI: Proceedings of the 52nd Session (Helsinki)*, Vol. LVIII, Book 2, pp. 391–394. Voorburg, Netherlands: International Statistical Institute.

Does, R.J.M.M. and Roes, K.C.B. (1996) Industrial statistics and its recent contributions to total quality in the Netherlands. *Statistica Neerlandica*, **50**, 27–51.

Does, R.J.M.M. and Zempléni, A. (2001) Establishing a statistical consulting unit at universities. *Quantitave Methods*, **67**, 51–64.

Does, R.J.M.M., Roes, K.C.B and Trip, A. (1999) *Statistical Process Control in Industry*. Kluwer Academic, Dordrecht, Netherlands.

Hahn, G.J., Hill, W.J., Hoerl, R.W. and Zinkgraf, S.A. (1999) The impact of Six Sigma improvement – a glimpse into the future of statistics. *American Statistician*, **53**(3), 208–215.

Khamis, H.J. and Mann, B.L. (1994) Outreach at a university statistical consulting center. *American Statistician*, **48**, 204–207.

Kirk, R.E. (1991) Statistical consulting in a university: dealing with people and other challenges. *American Statistician*, **45**, 28–34.

Marquardt, D.W. (1979) Statistical Consulting in Industry. *American Statistician*, **33**, 102–107.

Minton, P.D. and Freund, R.J. (1977) Organisation for the conduct of statistical activities in colleges and universities. *American Statistician*, **31**, 113–117.

Snee, R.D. (1996) Nonstatistical skills that can help statisticians be more effective. *ASQC Statistics Division Newsletter*, **16**(5), 12–17.

Van den Heuvel, E.R., Oudshoorn, C.G.M. and Does, R.J.M.M. (1997) The Six Sigma quality program. In A. Zempléni (ed.), *Proceedings of the Workshop on Statistics at Universities: Its Impact for Society*, pp. 83–92. Eötvös University Press, Budapest.

Zempléni, A. (1997), Statistical consultancy at universities – special features: consulting units. In A. Zempléni (ed.), *Proceedings of the Workshop on Statistics at Universities: Its Impact for Society*, pp. 9–25. Eötvös University Press, Budapest.

3.IV

Consultancy? . . . What's in it for me?

Roland Caulcutt

3.IV.1 Introduction

It is unreasonable to expect that someone emerging from university, having spent three or four years studying mathematics and statistics, will be able immediately and effectively to function as a statistical consultant. Further development is required.

My colleagues and I have run many short courses entitled 'Consultancy Skills for Statisticians'. The course participants are relatively young statisticians, most of whom appear to have a very deep understanding of mathematical statistics. Many of them work as internal consultants in large organisations such as the Office for National Statistics, pharmaceutical companies and the National Health Service. Other participants work as external consultants with similar large organisations. If you have ever worked in large organisations you will probably agree that your prospects for advancement could be threatened if you did not fit into the culture of the organisation. In particular, it would be very unwise to disagree with your boss, to speak openly about your personal objectives or to show any sign of fear. In our consultancy skills course we deliberately focus on conflict, choice, objectives, truth and fear. We also discuss listening, learning, personality and control.

We start each course by asking the participants: 'What do your clients hope to *gain* by your involvement in their projects?' All participants agree that the client must hope to gain something, but many feel that it is not always clear just what benefit the client anticipates. Nonetheless, after group discussions, they produce a list of possible benefits, which often includes:

- Data analysis
- Confirmation of a preconceived conclusion
- A *p*-value
- A solution to a business problem
- A confidence interval
- A hypothesis test

- A design for an experiment
- A sample size, preferably a small sample size
- Analysis of variance
- Explanation of a computer print out
- A discussion of process changes.

This list certainly implies that the client needs *technical* advice. We do agree. Most clients do need technical advice, but we suggest that technical assistance alone may fail to address all of the client's needs. Furthermore, we suggest that a statistical consultant who focuses solely on providing technical advice may have dissatisfied clients.

We also ask our course members: 'What do your clients *fear* that they might get as a result of your involvement in their projects?' The participants appear to find this second question even more difficult than the first. Perhaps this extra difficulty should not surprise us, as the clients are unlikely to reveal their fears so readily as their hopes. Certainly our delegates' second lists tend to be shorter. They often include:

- Being made to look foolish
- Criticism of what they have done so far
- Being misunderstood
- Being treated with disrespect
- Boredom
- Having to visit the statistics department
- A reminder of previous bad experiences with statisticians.

This list differs somewhat from the first. It touches on the client's feelings. It indicates that many statisticians are aware of the possibility that the client's feelings might influence the course of the consultancy interaction. However, many course participants are reluctant to admit that their *own* feelings may also affect the outcome of the consultancy project. This becomes clear when they respond to the third and fourth questions: 'What do *you* hope to gain from participation in the consultancy project?' and 'What do *you* fear that you may get from your participation in the consultancy project?' These two questions are discussed with obvious reluctance by many delegates, some even claiming that they do not have any hopes or fears in the consultancy interaction. They are just doing a job.

We spend much of the course demonstrating that the feelings of the statistician and the client can have a profound effect on the consultancy interaction. This serves as a foundation for increasing the effectiveness of the statistical consultant.

3.IV.2 Fundamentals

Psychology is a vast subject. Professional psychologists tend to specialize and are understandably reluctant to offer a simple model that might explain the subtle interaction between consultant and client. However, we believe that such a model is necessary if we are to explore the consultancy interaction in sufficient depth to offer any hope of achieving lasting improvement in performance. With this objective in mind we offer the following:

- The behaviour of the client is driven by a desire to achieve his/her *personal objectives*.
- The behaviour of the statistician is driven by a desire to achieve his/her personal objectives.
- Both statistician and client have many objectives, all of which cannot be achieved as some objectives will be in *conflict* with other objectives, competing for scarce resources, including time and/or money.
- Both statistician and client must *prioritise* in choosing which objective(s) to pursue at any time. These *choices* will depend on the strength of the *needs* that the objectives are intended to satisfy.
- The behaviour of the statistician, or the client, will also be influenced by their values and their mental models. (Values tell us what behaviour is acceptable. Mental models, or paradigms, or theories, tell us what behaviour is likely to be successful.)

The above list constitutes a simple model that, we hope, will help to explain why statisticians and clients behave in the ways that they do. The model is not perfect, of course. We should not expect perfection. We should judge a model by its usefulness, and many of our course members do find it useful, when they attempt to understand the interaction that takes place during role-plays, which are an important feature of the course.

Perhaps you are reluctant to accept this model. Perhaps you have realised what the model implies. It implies that both statistician and client are, at any and every moment, asking: 'What's in it for me?' But, before you reject the model you should consider the support it has been given, at least in part, by many writers from very varied backgrounds. See, for example, Peck (1990), Dalai Lama (1999) and Dawkins (2006).

The 'what's-in-it-for-me' question may be an unmentionable in the culture in which you work. Nonetheless, we believe it is important for the statistician to ask 'What's in it for me?', 'What's in it for the client?' and 'What's in it for the other stakeholders?' throughout the consultancy project. Furthermore, we believe that the statistician will be better able to answer these questions if he/she understands the *needs* of all the stakeholders.

3.IV.3 Needs

Perhaps the name most closely associated with the study of human needs is Abraham Maslow. He gave us the well known hierarchy of needs:

- Physiological needs
- Safety needs
- Social needs
- Self-esteem needs
- Self-actualization needs

The first two needs are physical needs, the last three are psychological needs. Maslow (1970) suggested that a lower need will become important only when all of the 'higher' needs have been satisfied. Thus you would be unlikely to worry about your self-esteem if your life were threatened by food poisoning or malnutrition. During our training we choose to assume that the physical needs of course members will be largely satisfied, thereby allowing us to focus on the psychological needs.

For a relatively simple, but very useful, model of human needs we turn to the writings of Will Schutz (1984, 1994), who suggested that a person's fundamental psychological need is for self-esteem. He defined self-esteem as 'How you feel about how you see yourself'. We find that the term 'self-esteem' is frequently misunderstood so we define it even more simply as 'feeling good'. Thus a person with high self-esteem feels good and a person with low self-esteem feels bad. Schutz lists the characteristics of people with *low* self-esteem as:

- Lacking motivation and quitting easily
- Feeling helpless, and a victim of luck and other people
- Feeling like a person nobody would notice, or be interested in
- Unable to be relied upon to do a good job or make good decisions
- Feeling dull, bored, tired and lifeless
- Not liking themselves. Others would not like them if they knew them
- Blaming other people and circumstances for their bad feelings
- Lacking stamina and the ability to finish a job.

A statistical consultant who has low self-esteem should become more effective if his/her self-esteem can be raised. There is no shortage of books that offer advice on how this raising of self-esteem can be achieved, but many of these books are shallow or trivial and of doubtful value. To effectively overcome the debilitating influence of lower self-esteem we need to dig a little deeper to uncover the three components of the need for self-esteem. Schutz suggests that they are:

- A need to feel significant, important or worthy of attention.
- A need to feel competent, capable or dominant.
- A need to feel *likeable* or trustworthy.

Schutz suggests that a person can 'feel good' if, and only if, *all* three needs are satisfied. Many of our course members find this model of psychological needs very useful, especially whilst they are struggling to explain the behaviours they observe during consultancy role-plays. However, the model becomes even more powerful when we add to it the behaviours typically used to increase the probability that the needs will be satisfied. These behaviours are:

- Inclusion – either the need to include others, or the need to be included by others, or both.
- Control – either the need to control others, or the need to be controlled by others, or both.
- Openness – either the need to be open with others, or the need to have others be open with you, or both.

3.IV.4 Behaviours

We suggest that a statistical consultant can work more effectively if he/she attempts to assess behaviour in terms of inclusion, control and openness. (This would apply equally to the behaviour of the client, the consultant or some third party.) So what specific behaviours can be classed as inclusion behaviours and therefore regarded as indicating a need to feel significant by gaining attention or prominence? We suggest you consider the following:

- Weeping
- Shouting
- Arm-waving gestures
- Phoning
- Emailing
- Bringing news
- Making gifts
- Using names

- Eye contact
- Smiling
- Reporting symptoms of illness

You may see some of the above behaviours as good and some as bad. We suggest that all of them can be effective if used appropriately. Schutz would maintain that it is *rigidity* of behaviour that is ineffective in the consultancy interaction, or any other situation. For example, a client who has a great, and unsatisfied, need to feel significant might waste much time phoning or emailing other team members, when such contact is not appropriate.

Control behaviours are equally, if not more, important in the organisational environment and in the consultancy interaction. The statistical consultant could benefit from being aware of the following behaviours which may be driven by a need to feel competent:

- Interrupting other speakers
- Keeping people waiting
- Changing agendas
- Restricting the flow of information
- Demanding to know how, when, where, what and who
- Taking responsibility
- Avoiding responsibility
- Creating urgency

Each of the above behaviours may be entirely appropriate in certain circumstances. But, again, it is *rigid* use of control behaviours that leads to inefficiency and disruption. If the client is demanding to know every detail of the consultant's analysis, this may result in much wasted effort and considerable frustration. A client who frequently changes the times of meetings or arrives late at every meeting may, thereby, feel more competent but will not increase the effectiveness of the team. On the other hand, a client who expects the consultant to take over some of his managerial responsibilities may also put the project in jeopardy.

Do you discuss inclusion behaviours and control behaviours in the organisation where you work? I suspect that it is unlikely. It is even less likely that you discuss openness behaviours. These are closely related to what Schutz calls 'levels of truth'. He lists seven levels of truth which can help you to assess how willing you are to openly reveal to yourself and others the feelings that underpin your behaviour.

Schutz also lists the corresponding levels of listening, which are highly pertinent to the consultancy interaction. They are:

- Level 1 – I keep myself unaware of your presence.
- Level 2 – I am aware of you but deliberately do not listen.
- Level 3 – I listen only enough to prepare my defence.
- Level 4 – I listen enough to prepare my counterattack.
- Level 5 – I interrupt you because I wish to speak.
- Level 6 – I want to understand what you are saying.
- Level 7 – I want to understand what you are saying and I want to understand what you are feeling.

Clearly, it is highly desirable for both the consultant and the client to operate at level 6 or 7, at least part of the time. In our courses we demonstrate the additional benefits to be gained by raising listening from level 6 to level 7. We do not suggest that this is easily achieved. Course members readily agree, and some admit to using much lower levels of listening when under stress or in the home environment.

3.IV.5 Theory and practice

This discussion of motivation, needs and behaviour may appear rather theoretical. To put it into practice requires repeated attempts to use the models we offer to explain your behaviour and the behaviour of others. If you find that the models help to explain what has already occurred, your confidence may grow in the ability of the models to predict future behaviour.

Schutz (1984) offers you the opportunity to measure the strength of your needs (significance, competence and likeability) and the strength of your behaviours (inclusion, control and openness). Completing the questionnaires within the book may help your understanding, change your focus and may even prove to be a life-changing experience.

Block (2000), another very 'practical' book, offers consultants specific advice on the behaviour that is likely to be effective in any consultancy project. I have seen no evidence that Block ever read Schutz, or that Schutz ever read Block, but their recommendations are certainly compatible. For example, Block offers a 12-phase roadmap for the consultancy project and suggests that, in each phase, the client is asking 'Is this a consultant I can trust?'. He further suggests that, throughout each stage, the consultant should be asking 'Am I being authentic?' and 'Am I completing the business of this phase?'.

3.IV.6 Summary

I have suggested that both consultant and client are asking 'What's in it for me?'. If your input to a consultancy project does not offer benefits to the other participants, why would they want you to participate?

So, what benefits do clients seek? Technical assistance, of course, but clients also have many psychological needs that can impact upon the consultancy interaction. Schutz advises that these needs include significance, competence and likeability. None of these needs is visible, of course, but you may be able to assess the client's needs by observing the client's behaviour, focusing on inclusion, control and openness.

The consultant should also consider his/her own needs and behaviours, which will also impact on the efficient progress of the project. Critically important is the need of the consultant to appear authentic. Perhaps some consultants do not realise that this is a nagging worry of many clients.

References

Block, P. (2000) *Flawless Consulting: A Guide to Getting Your Expertise Used*. Jossey-Bass, San Francisco.

Dalai Lama (1999) *The Art of Happiness*. Coronet, London.

Dawkins, R. (2006) *The God Delusion*. Bantam Press, London.

Maslow, A.H. (1970) *Motivation and Personality*. Harper & Row, New York.

Peck, M.S. (1990) *The Road Less Travelled*. Arrow, London.

Schutz, W. (1984) *The Truth Option: A Practical Technology for Human Affairs*. Ten Speed Press, Berleley, CA.

Schutz, W. (1994) *The Human Element: Productivity, Self-Esteem, and the Bottom Line*. Jossey-Bass, San Francisco.

4

The statistical efficiency conjecture

Ron S Kenett, Anne De Frenne, Xavier Tort-Martorell and Chris McCollin

4.1 The quality ladder

Modern industrial organisations in manufacturing and services are subject to increasing competitive pressures and rising customer expectations. Management teams on all five continents are striving to satisfy and delight their customers while simultaneously improving efficiencies and cutting costs. In tackling this complex management challenge, an increasing number of organisations have shown that the apparent conflict between high productivity and high quality can be resolved through improvements in work processes and quality of design.

In this chapter we attempt to demonstrate the impact of statistical methods on process and product improvements and the competitive position of organisations. We describe a systematic approach to the evaluation of benefits from process improvement and quality by design (QbD) that can be implemented within and across organisations. We then formulate and validate the *statistical efficiency conjecture* that links management maturity with the impact level of problem solving and improvements driven by statistical methods.

The different approaches to the management of industrial organisations can be summarised and classified using a four-step *quality ladder* (Kenett and Zacks, 1998). The four approaches are: (1) fire fighting; (2) inspection; (3) process control; and (4) and quality by design. To each management approach there corresponds a

Statistical Practice in Business and Industry Edited by S.Y. Coleman, T. Greenfield, D.J. Stewardson and D.C. Montgomery

particular set of statistical methods, and the quality ladder maps each management approach to appropriate statistical methods.

Managers mainly involved in reactive fire fighting need to be exposed to basic statistical thinking. Their challenge is to evolve their organisation from typical data accumulation to data analysis so that numbers are turned into information and knowledge. Managers who attempt to contain quality and inefficiency problems through inspection and 100 % control can simplify their work by using sampling techniques. More proactive managers, who invest in process control and process improvement, are well aware of the advantages of control chart and process control procedures. At the top of the quality ladder is the approach where up-front investments are secured to run experiments designed to optimise product and process specifications. At that level of maturity, robust designs are run for example on simulation platforms, reliability engineering is performed routinely and reliability estimates are compared with field returns data to monitor the actual performance of products and improve the organisation's predictive capability (Kenett and Zacks, 1998; Bates *et al.*, 2006).

Efficient implementation of statistical methods requires a proper match between management approach and statistical tools. In this chapter we demonstrate, by means of case studies, the benefits achieved by organisations from process and quality improvement initiatives. The underlying theory behind the approach is that organisations that increase the maturity of their management system, moving from fire fighting to quality by design, enjoy increased benefits and significant improvements in their competitive positions.

In August 2002 the Food and Drug Administration (FDA) announced the Pharmaceutical current Good Manufacturing Practices (cGMP) for the 21st Century Initiative. In that announcement the FDA explained its intention to integrate quality systems and risk management approaches into existing programmes with the goal of encouraging industry to adopt modern and innovative manufacturing technologies. The cGMP initiative was spurred by the fact that since 1978, when the last major revision of the cGMP regulations was published, there have been many advances in manufacturing science and in the understanding of quality systems. This initiative resulted in three new guidance documents from the International Conference on Harmonisation – pharmaceutical development (ICH Q8), quality risk management (ICH Q9) and pharmaceutical quality systems (ICH Q10) – with the new vision that ensuring product quality requires "a harmonised pharmaceutical quality system applicable across the life cycle of the product emphasizing an integrated approach to quality risk management and science". This new approach is encouraging the implementation of QbD and hence, de facto, encouraging the pharmaceutical industry to move up the quality ladder (Nasr, 2007).

Figure 4.1 presents the quality ladder in graphical form. The right-hand side shows the management approach and the left-hand side shows the matching statistical techniques. In the next sections we discuss improvement projects and Six Sigma initiatives which help organisations go up the quality ladder.

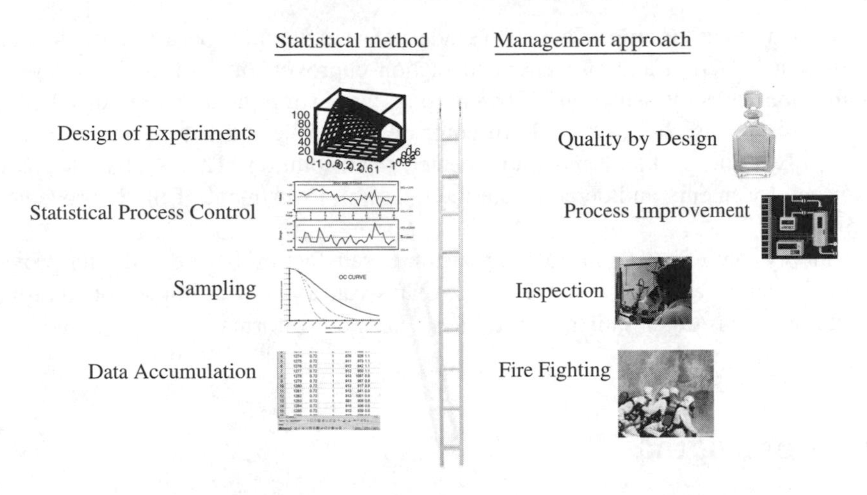

Figure 4.1 The quality ladder.

4.2 Improvement projects

Since the 1960s, many organisations have tackled the improvement of quality by means of focused improvement projects. A project-by-project improvement strategy relies on successful employee participation in project identification, analysis and implementation. Many factors influence a project's success; these are often overlooked and, in many cases, may be unknown. A project-by-project quality improvement programme must be supported by quality principles, analysis techniques, effective leaders and facilitators, and extensive training. While quality improvement teams can vary in size (even during the duration of the project), a typical team size is five to seven (Juran, 1986, 1989; Aubrey and Felkins, 1988; Godfrey and Kenett, 2006).

The proper selection and formulation of a problem is critical for effective quality improvement. The link between problem/project selection and team member selection must be direct and relevant. The regular management team, a quality council, or the team itself can select the problem or project.

While the end result of a quality improvement project remains the primary concern from a management perspective, time is critical for the success of quality improvement teams. The typical time for project completion is 3–5 months, which is about 32 person-hours per team member (2 hours per week for 16 weeks).

Team factors are critical to success. Managers should be involved in the project selection process and the outcome of the project. The team size and the length of the project are both important in ensuring that group dynamics and motivation are effectively employed. A structured improvement plan is essential, including a list of steps to be followed, with clear definitions, milestones and the tools to be used, many of them statistical.

Typical areas of improvement include defect reduction, performance to standard, cost reduction and customer satisfaction improvement. Some teams develop quality measures by using basic flowcharting, work simplification, and data collection tools. Teams should be able to compare data collected before and after they have solved the problem and implemented their solution. This enables teams to track improvements and demonstrate the return on investment of the improvement project.

Surveys are effective in tracking customer satisfaction before and after project solution implementation. With surveys, teams can assess the impact of improvements using customer satisfaction ratings. For more information on customer surveys, see Kenett (2006).

4.3 Six sigma

The Six Sigma business improvement strategy was introduced by Motorola in the 1980s and adopted successfully by General Electric and other large corporations in the 1990s. The key focus of all Six Sigma programmes is to optimise overall business results by balancing cost, quality, features and availability considerations for products and their production in a best business strategy. Six Sigma programmes combine the application of statistical and non-statistical methods to achieve overall business improvements. In that sense, Six Sigma is a more strategic and more aggressive initiative than simple improvement projects. Effective Six Sigma projects are orchestrated at the executive level and consist of a *define–measure–analyse–improve–control* (DMAIC) roadmap. Moreover, executives support Six Sigma as a business strategy when there are demonstrated bottom-line benefits.

The term *measure* to describe a phase in the Six Sigma DMAIC strategy can be misleading. Within Six Sigma training, the measure phase encompasses more than just measurements. It typically includes the tracking of key process output variables (KPOVs) over time, quantifying the process capability of these variables, gaining insight into improvement opportunities through the application of cause-and-effect diagrams and failure mode and effects analysis, and quantifying the effectiveness of current measurement systems. These activities help define how a KPOV is performing, and how it could relate to its primary upstream causes, the key process input variables (KPIVs). The measure step is repeated at the successive business, operations and process levels of the organisation, measuring baseline information for KPOVs in order to provide an accurate picture of the overall variation that the customer sees. Measurements provide estimates of the process capability relative to specifications and drive investigations of all significant sources of variation (including measurement system analyses). This information is then used throughout the *analyse, improve* and *control* phases to help make overall business improvements.

Several unique Six Sigma metrics, such as sigma quality level, yield and cost of poor quality (COPQ), have been developed to help quantify and reduce the *hidden factory* (hidden production costs). Six Sigma relies heavily on the ability to access information. When the most important information is hard to find, access or understand, the result is extra effort that increases both the hidden factory and the COPQ (Juran, 1989).

Many organisations set a dollar threshold as they begin to prioritise Six Sigma projects. Successful projects thereby provide returns that will pay for up-front investments in Six Sigma training and full-time Black Belt team leaders. The definition of what types of savings are considered (hard or soft) is also critical and drives specific behaviours to enable low hanging fruit to be identified first. In many organisations, soft or indirect benefits fall into categories such as cost avoidance related to regulatory or legal compliance or benefits related to improving employee morale or efficiency. Such benefits cannot be directly tied to operating margins or incremental revenues through a specific measurement. For more information on Six Sigma, see Chapter 2 of this book.

4.4 Practical statistical efficiency

Fisher (1922) states that 'the object of statistical method is the reduction of data'. He then identifies 'three problems which arise in the reduction of data'. These are:

- *specification* – choosing the right mathematical model for a population;
- *estimation* – methods to calculate, from a sample, estimates of the parameters of the hypothetical population;
- *distribution* – properties of statistics derived from samples.

Later, Mallows (1998) added a 'zeroth problem' – considering the relevance of the observed data, and other data that might be observed, to the substantive problem.

Building on a suggestion by Godfrey (1988, 1989), Kenett *et al.* (2003) define *practical statistical efficiency* (PSE) using an eight-term formula and demonstrated its applicability using five case studies. PSE is an important addition to the statistical consultancy toolkit; it enhances the ability of practising statisticians to show the extent of their contribution to the resolution of real-life problems. The original reason for introducing PSE was the pervasive observation that the application of statistical methods is, in many cases, an exercise in using statistical tools rather than a focused contribution to specific problems. PSE is calculated by a multiplication formula:

$$PSE = V\{D\} \times V\{M\} \times V\{P\} \times V\{PS\} \times P\{S\} \times P\{I\} \times T\{I\} \times E\{R\}$$

where $V\{D\}$ is the value of the data actually collected, $V\{M\}$ the value of the statistical method employed, $V\{P\}$ the value of the problem to be solved, $V\{PS\}$ the value of the problem actually solved, $P\{S\}$ the probability level that the problem actually gets solved, $P\{I\}$ the probability level that the solution is actually implemented, $T\{I\}$ the time the solution stays implemented, and $E\{R\}$ the expected number of replications.

A straightforward approach to evaluating PSE is to use a scale from 1 for 'not very good' to 5 for 'excellent'. This method of scoring can be applied uniformly for each PSE component. Some of the PSE components can be also assessed quantitatively: $P\{S\}$ and $P\{I\}$ are probability levels, $T\{I\}$ can be measured in months, and $V\{P\}$ and $V\{PS\}$ can be evaluated in monetary terms. $V\{PS\}$ is the value of the problem actually solved, as a fraction of the problem to be solved. If this is evaluated qualitatively, a large portion would be scored 4 or 5, a small portion 1 or 2.$V\{D\}$, the value of the data actually collected, is related to Fisher's zeroth problem presented in Mallows (1998). Whether PSE terms are evaluated quantitatively or qualitatively, PSE is a conceptual measure rather than a numerically precise one. A more elaborate approach to PSE evaluation can include differential weighting of the PSE components and/or non-linear assessments. In the last section we described a maturity ladder that maps the management approach with statistical techniques. The idea is that proper mapping of the two is critical for achieving high PSE.

As organisations move up the quality ladder, more useful data is collected, more significant projects get solved, and solutions developed locally are replicated throughout the organisation. We therefore postulate that increasing an organisation's maturity by going up the quality ladder results in higher PSEs and increased benefits. Let L denote the organisational level of management maturity on the quality ladder, where $L = 1$ stands for fire fighting, $L = 2$ inspection, $L = 3$ process improvement, and $L = 4$ quality by design. Let PSE be the practical statistical efficiency of a specific project, such that $1 \leq PSE \leq 5^8$, and $E\{PSE\}$ be the expected value of PSE over all projects. Our goal in this chapter, and related work, is to prove the following *statistical efficiency conjecture* on PSE improvement, which links expected practical statistical efficiency with the maturity of an organisation on the quality ladder.

Conjecture. Conditioned on the right variable, $E\{PSE\}$ is an increasing function of L.

Possible conditioning variables include company size, business area, ownership structure and market types. In the next section we provide a set of case studies to demonstrate this conjecture empirically.

4.5 Case studies

The 21 case studies described in the appendix to this chapter are taken from a mix of companies at different maturity levels on the quality ladder. The PSE was evaluated for each case study. Figure 4.2 plots PSE values against maturity level. Interestingly, the PSE scores do not seem to strongly reflect the maturity level of the company. The implication is that the PSE is affected by several variables, not just maturity level.

We stratify the data by company size and type of activity to explain the variability of PSE. As can be seen in Figure 4.3, PSE tends to increase with company size and, from Figure 4.4, companies with international activity tend to have higher PSE.

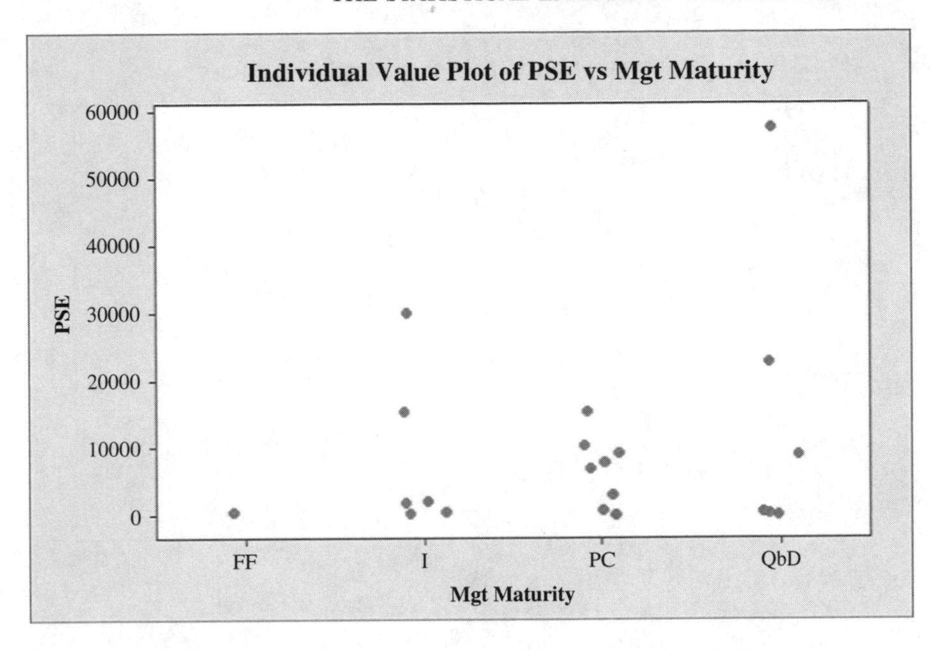

Figure 4.2 PSE versus maturity level (FF = fire fighting, I = inspection, P = process control, QbD = quality by design).

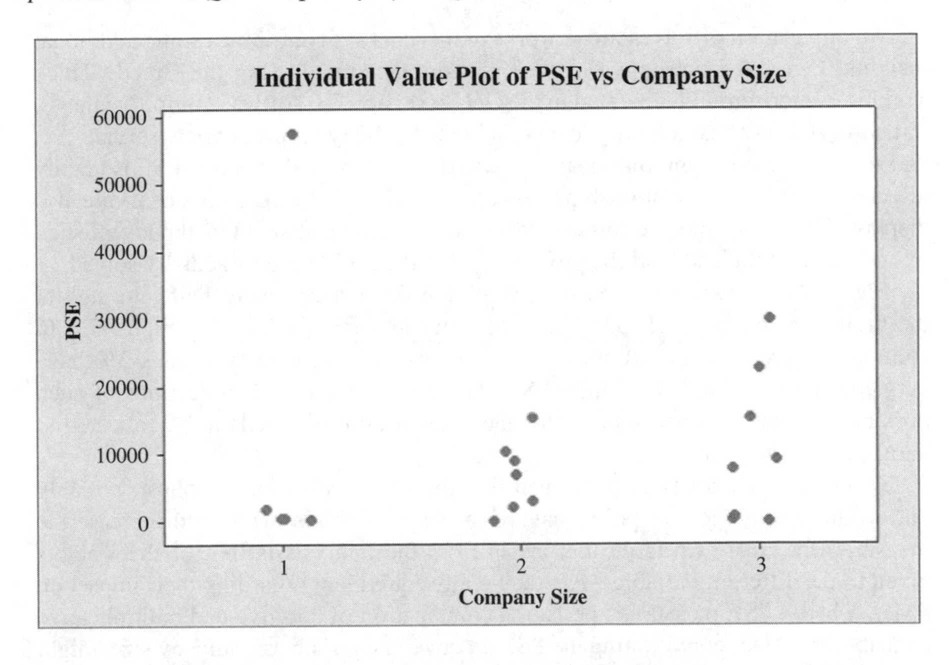

Figure 4.3 PSE by company size (1 = SME, 2 = large, 3 = very large).

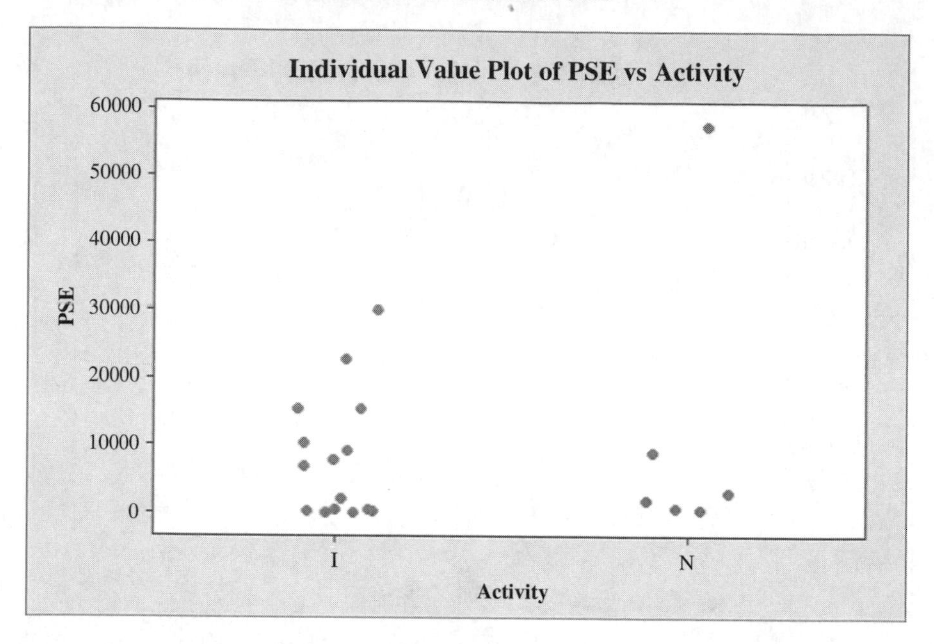

Figure 4.4 PSE by type of activity (I = international, N = national).

An application of Bayesian networks to this data set contributes some additional insights. Figure 4.5 presents a learned network generated using the Greedy Thick Thinning algorithm implemented in the GeNIe version 2.0 software (http://genie.sis.pitt.edu). It shows that maturity level is affected by the type (production, service, ...) and scope of activity (international versus national). It also shows that PSE is mainly affected by maturity. Relationships between the individual PSE components are also displayed. For example, we can see that the value of the statistical method is affected by the value of the data and the probability that the problem gets actually solved.

Figure 4.6 shows the fitted data conditioned on a company being located at the highest maturity level: 33 % have the very low PSE, 17 % very high PSE. In contrast, Figure 4.7 shows companies at the inspection maturity level: 42 % have very low PSE and 8 % very high PSE. These are only initial indications of such possible relationships and more data, under better control, needs to be collected to validate such patterns.

A repeatability analysis confirmed that the PSE evaluation is robust, possibly due to the use of the five-point scale. Expansion of this scale would increase the measurement error. An issue that arises from the analysis is that of the weights given to the different variables. Are they really equivalent regarding their impact on PSE? A better PSE measure is perhaps a combination of additive and multiplicative contributors. Also, conditioning the PSE on covariates, such as company size, might improve the picture. The point is that, even with our very rough maturity assessment and qualitative PSE, the right conditioning variable is unknown.

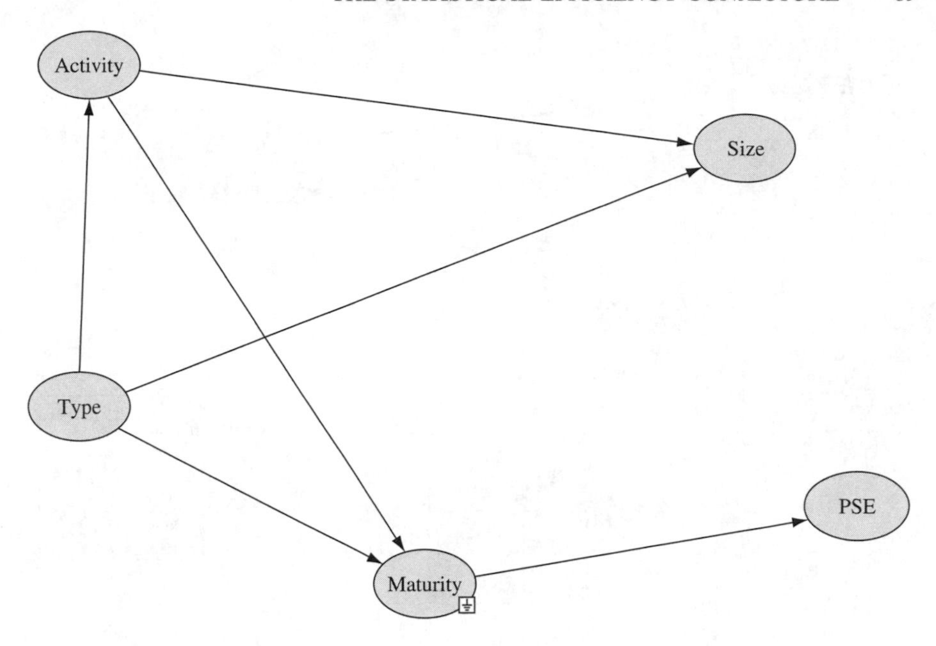

Figure 4.5 Bayesian network of PSE data.

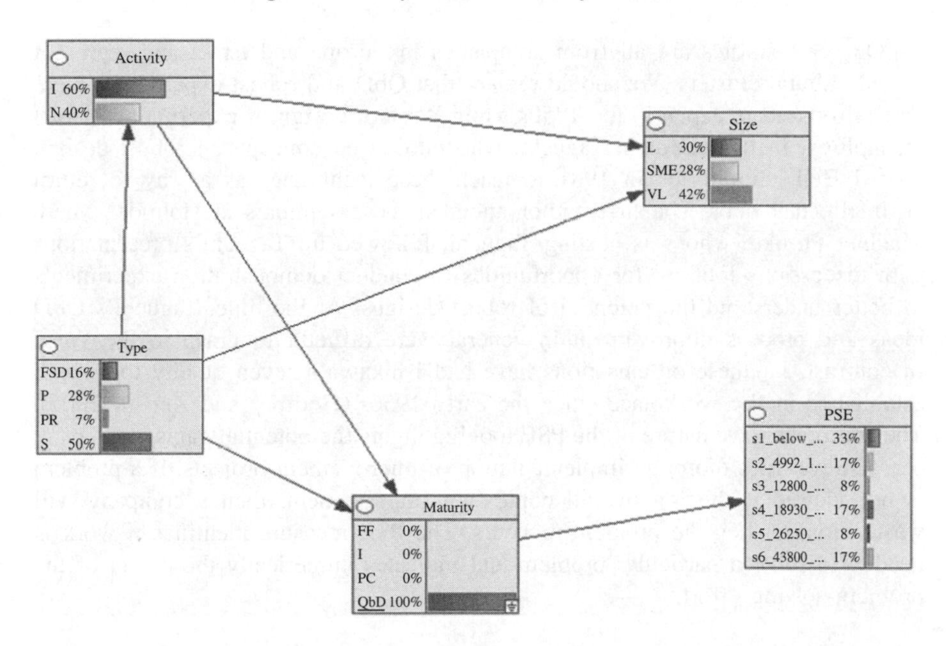

Figure 4.6 Bayesian network of PSE data with values for QbD maturity level.

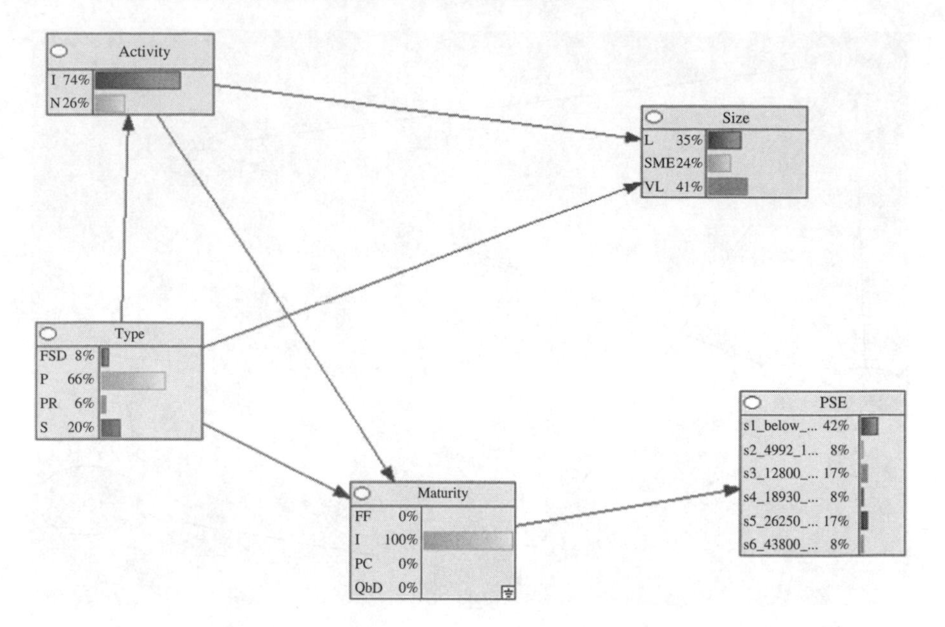

Figure 4.7 Bayesian network of PSE data with values for inspection maturity level.

Our case studies are all from companies in Europe and Israel and represent varied cultural clusters. We should remark that QbD and robust experiments were first introduced in Japan in the 1950s using Western design of experiments (DoE) technology. In fact, Genichi Taguchi, who introduced concepts of robust design, visited Bell laboratories in 1981 to teach these techniques as a way to return an intellectual debt. The first author attended these seminars at Holmdel where Madhav Phadke, who was hosting Taguchi, followed up Taguchi's presentations with discussions looking for opportunities to conduct demonstration experiments to better understand the potential of robust designs. At the time, Taguchi's QbD ideas and process improvement in general were difficult to grasp in the West. In contrast, Japanese organisations have had Ishikawa's seven quality tools well established in the workplace since the early 1950s (Godfrey and Kenett, 2007). The comprehensive nature of the PSE tool highlights the potential gains a company can achieve from thorough implementation of improvement projects. If a problem is not viewed within an overall context of improvement, then a company will waste money when the problem reoccurs. The PSE measure identifies if work is needed to solve a particular problem and indicates numerically the extent of the problem-solving effort.

4.6 Conclusions

This chapter is a first attempt to show how going up the quality ladder, from process improvement to QbD, can increase practical statistical efficiency. Our study of 21 case studies demonstrates how such an investigation can be carried out. Obviously more data is needed to prove or disprove the conjecture summarised in Table 4.1. If we disprove the conjecture we should have to expand the PSE formula by including some factors not currently considered. Such a research effort would help statisticians properly evaluate the impact of their work and demonstrate to others the added value they bring to their customers. The idea is to link PSE evaluations with the maturity level of organisation on the quality ladder. If the conjecture linking the two holds, statisticians interested in achieving a high impact on organisations will also have to become proactive in getting organisations to go up the quality ladder – thus also becoming 'management consultants'. In reviewing the careers of W. Edwards Deming, J.M. Juran, Brian Joiner and others, we can see this common denominator (see Joiner, 1985, 1994; Deming, 1986; Juran, 1986, 1989; Godfrey and Kenett, 2007; Kenett and Thyregod, 2006). They all began as statisticians, with significant contributions to statistical methodology, and developed into world-class management consultants.

Table 4.1 Characteristics of the quality ladder maturity levels.

Quality ladder	Fire fighting	Inspection	Process control	Quality by design
Statistical tool	None	Sampling	Statistical CS	DoE
Time	Short	Short	Medium	Long
Action	Correction	Correction	Correction Prevention	Prevention
Benefits	Very low	Low	High	Very high
Customer satisfaction	Very low	Low	High	Very high
Organisation maturity	Very low	Low	High	Very high
PSE measure	Very low	Low	High	Very high

References

Aubrey, C.A. and Felkins, P.K. (1988) *Teamwork: Involving People in Quality and Productivity Improvement*. Quality Press, Milwaukee, WI.

Bates, R., Kenett, R., Steinberg, D. and Wynn, H. (2006) Achieving robust design from computer simulations. *Journal of Quality Technology and Quantitative Management*, **3**(2), 161–177.

Deming, W.E. (1986) *Out of the Crisis*. MIT Press, Cambridge, MA.

Fisher, R.A. (1922) On the mathematical foundations of theoretical statistics. *Philosophical Transactions of the Royal Society, Series A*, **222**, 309–368.

Godfrey, A.B. (1988) Statistics, quality and the bottom line: Part 1. *ASQC Statistics Division Newsletter*, **9**(2), 211–213.

Godfrey, A.B. (1989) Statistics, quality and the bottom line: Part 2. *ASQC Statistics Division Newsletter*, **10**(1), 14–17.

Godfrey, A.B. and Kenett, R.S. (2007) Joseph M. Juran: A perspective on past contributions and future impact. *Quality and Reliability Engineering International*, **23**, 653–663.

Joiner, B.L. (1985) The key role of statisticians in the transformation of North American industry. *American Statistician*, **39**(3), 233–234.

Joiner, B.L. (1994) *Fourth Generation Management: The New Business Consciousness*. McGraw Hill, New York.

Juran, J.M. (1986) *The quality trilogy: A universal approach to managing for quality*. In *Proceedings of the ASQC 40th Annual Quality Congress in Anaheim, California*.

Juran, J.M. (1989) *Juran on Leadership for Quality*. Free Press, New York.

Kenett, R.S. (2006) On the planning and design of sample surveys. *Journal of Applied Statistics*, **33**(4), 405–415.

Kenett, R.S. and Zacks, S. (1988) *Modern Industrial Statistics: Design and Control of Quality and Reliability*. Duxbury Press, San Francisco.

Kenett, R.S. and Thyregod, P. (2006) Aspects of statistical consulting not taught by academia. *Statistica Neerlandica*, **60**(3), 396–412.

Kenett, R.S., Coleman, S.Y. and Stewardson, D. (2003) Statistical efficiency: The practical perspective. *Quality and Reliability Engineering International*, **19**, 265–272.

Mallows, C. (1998) The zeroth problem. *American Statistician*, **52**, 1–9.

Nasr, M. (2007) Quality by design (QbD) – a modern system approach to pharmaceutical development and manufacturing – FDA perspective. FDA Quality Initiatives Workshop, North Bethesda, MD, 28 February.

Appendix 4A

Case 1 (with instructions)

Problem name: Missile Storage Failure Data

Date of project: 1977 Reporter: C. McCollin
I (International)
S (Service)
VL (Very large company size)
QbD (Very high management maturity)

Scope of activity (S): International, national
Type of business (T): Prototype, full-scale development, production, service
Company size (CS): SME, large, very large
Management maturity (MM): Fire fighting, inspection, process improvement, QbD

Problem description:
Shipborne missiles kept in storage on a land facility tested periodically for operational status.

Context:
Data collected to monitor dormant failure rates

Data collected:
Number of failures N, time of operation during test t_1, time of non-operation t_2

Data analysis:
Standard regression using mainframe computer package: $N = \lambda_1 t_1 + \lambda_2 t_2$

Conclusions and recommendations:
Negative failure rates!! That more work was required on improving software routines or upgrading software (if possible).

Impact:
Critical analysis of available off-the-shelf software for in-company analysis.
The number that reflects the evaluation of the PSE criteria on a scale from 1 to 5 is underlined.
5 – Very high value
4 – High value
3 – Average value
2 – Low value
1 – Very low value

	Criteria	Value –				Level +
1	$V\{D\}$ = value of the data actually collected	1	2	3	$\underline{4}$	5
2	$V\{M\}$ = value of the statistical method employed	1	$\underline{2}$	3	4	5
3	$V\{P\}$ = value of the problem to be solved	1	2	$\underline{3}$	4	5
4	$V\{PS\}$ = value of the problem actually solved	$\underline{1}$	2	3	4	5
5	$P\{S\}$ = probability that the problem actually gets solved	$\underline{1}$	2	3	4	5
6	$P\{I\}$ = probability the solution is actually implemented	$\underline{1}$	2	3	4	5
7	$T\{I\}$ = time the solution stays implemented	$\underline{1}$	2	3	4	5
8	$E\{R\}$ = expected number of replications	$\underline{1}$	2	3	4	5

$PSE = 24$

Case 2

Problem name: Electricity Generation Company Valve Failure Data

Date: 1983 Reporter: C. McCollin
N (National)
S (Service)
VL (Very large company size)
PI (Medium management maturity)

Problem description:
Analysis of valve failure data for a complete generating facility for reliability calculations

Context:
Data collected for monitoring possible plant safety shutdown issues

Data collected:
Type of valve, time to failure

Data analysis:
Weibull process estimates

Conclusions and recommendations:
Reliability of many valves was decaying and working valves were being maintained although they had not failed (possibly leading to the reliability decay). Available maintenance procedures were based on Weibull distribution analysis software (hence maintenance intervals based on incorrect statistical assumption (independent and identical distribution)).
Set of procedures written on how to analyse repairable systems data including searching for trend, seasonality.

Impact: None.

	Criteria	Value –			Level +	
1	$V\{D\}$ = value of the data actually collected	1	2	3	$\underline{4}$	5
2	$V\{M\}$ = value of the statistical method employed	1	2	3	$\underline{4}$	5
3	$V\{P\}$ = value of the problem to be solved	1	2	3	$\underline{4}$	5
4	$V\{PS\}$ = value of the problem actually solved	1	$\underline{2}$	3	4	5
5	$P\{S\}$ = probability that the problem actually gets solved	$\underline{1}$	2	3	4	5
6	$P\{I\}$ = probability the solution is actually implemented	$\underline{1}$	2	3	4	5
7	$T\{I\}$ = time the solution stays implemented	$\underline{1}$	2	3	4	5
8	$E\{R\}$ = expected number of replications	1	2	3	4	$\underline{5}$

$PSE = 640$

Case 3

Problem name: Military System Reliability Prediction

Date: 1985 Reporter: C. McCollin
I (International)

S (Service)
VL (Very large company size)
QbD (High management maturity)

Problem description:
Provide mean time between failure (MTBF) estimates to customer as part of contract

Context:
Customer requirement for data to be provided for life cycle cost estimates

Data collected:
Calendar time to failure, duty cycle, number of relevant failures

Data analysis:
MTBF calculation

Conclusions and recommendations:
Information forwarded to customer to use for spares provisioning.

Impact:
Spares provisioning.

	Criteria	Value –				Level +
1	$V\{D\}$ = value of the data actually collected	1	2	$\underline{3}$	4	5
2	$V\{M\}$ = value of the statistical method employed	1	$\underline{2}$	3	4	5
3	$V\{P\}$ = value of the problem to be solved	1	2	3	4	$\underline{5}$
4	$V\{PS\}$ = value of the problem actually solved	1	2	3	4	$\underline{5}$
5	$P\{S\}$ = probability that the problem actually gets solved	1	2	3	4	$\underline{5}$
6	$P\{I\}$ = probability the solution is actually implemented	1	2	3	4	$\underline{5}$
7	$T\{I\}$ = time the solution stays implemented	1	2	$\underline{3}$	4	5
8	$E\{R\}$ = expected number of replications	1	$\underline{2}$	3	4	5

$PSE = 22\,500$

Case 4

Problem name: Military System Failure Analysis

Date: 1985 Reporter: C. McCollin
I (International)
S (Service)

VL (Very large company size)
QbyD (High management maturity)

Problem description:
Customer declares system is unreliable based on in-service field data. Requires some action to be taken.

Context:
Customer upset

Data collected:
About 50 failure reports

Data analysis:
Classification of reports into how many systems failed (2!!), types of failure (mainly corrosion, rust) – led to conclusion that these two systems had been dropped in the sea, picked up at a later date and then investigated for failures.

Conclusions and recommendations:
Specification and testing tender costs forwarded to customer for next-update systems if they are required to work after being in such an environment. No response from customer.

Impact:
Customer did not criticize system again.

	Criteria	Value –				Level +
1	$V\{D\}$ = value of the data actually collected	1	2	3	4	5
2	$V\{M\}$ = value of the statistical method employed	1	2	3	4	5
3	$V\{P\}$ = value of the problem to be solved	1	2	3	4	5
4	$V\{PS\}$ = value of the problem actually solved	1	2	3	4	5
5	$P\{S\}$ = probability that the problem actually gets solved	1	2	3	4	5
6	$P\{I\}$ = probability the solution is actually implemented	1	2	3	4	5
7	$T\{I\}$ = time the solution stays implemented	1	2	3	4	5
8	$E\{R\}$ = expected number of replications	1	2	3	4	5

$PSE = 500$

Case 5

Problem name: Military System Warranty Analysis

Date: 1985 Reporter: C. McCollin

I (International)
P (Production type)
VL (Very large company size)
Inspection (Low management maturity)

Problem description:
Customer queries why delivered systems are all failing within the warranty period. No future contracts with this customer's country or near neighbours until problem resolved.

Context:
Country could not fly their military aircraft.

Data collected:
About 80 warranty reports

Data analysis:
Classification of failures into whether they were design, manufacturing, testing, systematic or one-off failures (one-off failures may be used for MTBF calculations).

Conclusions and recommendations:
Only one one-off failure classified, hence MTBF was good enough for system if design, manufacturing and test had been satisfactory. Production manager carpeted, project manager sidelined, design engineer who helped reporter analyse the design failures promoted. Ban lifted by countries.

Impact:
New procedures on systems management for project leaders.

	Criteria	Value −				Level +
1	$V\{D\}$ = value of the data actually collected	1	2	$\underline{3}$	4	5
2	$V\{M\}$ = value of the statistical method employed	$\underline{1}$	2	3	4	5
3	$V\{P\}$ = value of the problem to be solved	1	2	3	4	$\underline{5}$
4	$V\{PS\}$ = value of the problem actually solved	1	2	3	4	$\underline{5}$
5	$P\{S\}$ = probability that the problem actually gets solved	1	2	3	4	$\underline{5}$
6	$P\{I\}$ = probability the solution is actually implemented	1	2	3	4	$\underline{5}$
7	$T\{I\}$ = time the solution stays implemented	1	2	3	$\underline{4}$	5
8	$E\{R\}$ = expected number of replications	1	2	3	$\underline{4}$	5

$PSE = 30\,000$

Case 6

Problem name: Commercial Software Failure Analysis

Date: 1991 Reporter: C. McCollin
I (International)
FSD (Full-scale development type)
SME (SME size)
QbD (High management maturity)

Problem description:
Analysis of a very large data set (100 000 records), search for structure including change points. Why are there failures after delivery?

Context:
Data arose from major software producer as part of the Alvey Software Reliability Modelling project.

Data collected:
Date of failure, number of faults per failure per source code per product version, number of repairs per failure per source code per product version, in-house/in service, programmer

Data analysis:
Sorting and ordering routines written to identify missing or corrupt data, exploratory data analysis, time series (ARIMA), proportional hazards modelling, proportional odds modelling, logistic regression, discriminant analysis, principal components analysis, hazard estimation, proportional intensity modelling

Conclusions and recommendations:
Main findings were that number of failures was dependent on the day of the week, statistical structure is a function of development effort (no effort = random structure, effort = reliability growth).

Impact:
Programmers carpeted by project manager for not working to contract. Data could not answer why there are failures after delivery.

	Criteria	Value				Level
		−				+
1	$V\{D\}$ = value of the data actually collected	1	2	3	4	$\underline{5}$
2	$V\{M\}$ = value of the statistical method employed	1	2	3	$\underline{4}$	5
3	$V\{P\}$ = value of the problem to be solved	$\underline{1}$	2	3	4	5
4	$V\{PS\}$ = value of the problem actually solved	1	$\underline{2}$	3	4	5
5	$P\{S\}$ = probability that the problem actually gets solved	$\underline{1}$	2	3	4	5
6	$P\{I\}$ = probability the solution is actually implemented	$\underline{1}$	2	3	4	5
7	$T\{I\}$ = time the solution stays implemented	$\underline{1}$	2	3	4	5
8	$E\{R\}$ = expected number of replications	1	2	3	4	$\underline{5}$

$PSE = 200$

Case 7

Problem name: Availability Analysis

Date: 1995 Reporter: C. McCollin
N (National)
PR (Prototype type)
SME (SME size)
FF (Lowest management maturity)

Problem description:
Carry out an availability analysis to show customer that requirement was being met

Context:
Small company did not know how to meet customer requirement.

Data collected:
Date of delivery, number and types of failures

Data analysis:
Availability analysis

Conclusions and recommendations:
Requirement were met with a possible saving of not manufacturing unrequired units.

Impact:
None. Company carried on developing original solution at extra costs

	Criteria	Value −				Level +
1	$V\{D\}$ = value of the data actually collected	1	2	3	<u>4</u>	5
2	$V\{M\}$ = value of the statistical method employed	1	2	3	<u>4</u>	5
3	$V\{P\}$ = value of the problem to be solved	1	<u>2</u>	3	4	5
4	$V\{PS\}$ = value of the problem actually solved	1	2	3	4	<u>5</u>
5	$P\{S\}$ = probability that the problem actually gets solved	<u>1</u>	2	3	4	5
6	$P\{I\}$ = probability the solution is actually implemented	<u>1</u>	2	3	4	5
7	$T\{I\}$ = time the solution stays implemented	<u>1</u>	2	3	4	5
8	$E\{R\}$ = expected number of replications	1	2	<u>3</u>	4	5

$PSE = 480$

Case 8

Problem name: Commercial switch reliability analysis

Date: 1994 Reporter: C. McCollin
I (International)
P (Production type)
SME (SME size)
Inspection (Low management maturity)

Problem description:
Carry out exploratory reliability analysis on switch life data

Context:
Data made available by company for analysis. Previous Weibull analyses made available.

Data collected:
Operating time to failure, initializing current on delivery, electrical specification documents

Data analysis:
Proportional hazards modelling

Conclusions and recommendations:
Time to failure was found to be dependent on initializing current. Recommendation by author to company to determine why variable was significant (would require failure mode and effects analysis and DoE).

Impact:
None. Company carried on using available techniques. No requirement by standards or customers to avert this problem.

	Criteria	Value –				Level +
1	$V\{D\}$ = value of the data actually collected	1	2	3	<u>4</u>	5
2	$V\{M\}$ = value of the statistical method employed	1	2	3	<u>4</u>	5
3	$V\{P\}$ = value of the problem to be solved	1	2	3	4	<u>5</u>
4	$V\{PS\}$ = value of the problem actually solved	1	<u>2</u>	3	4	5
5	$P\{S\}$ = probability that the problem actually gets solved	<u>1</u>	2	3	4	5
6	$P\{I\}$ = probability the solution is actually implemented	<u>1</u>	2	3	4	5
7	$T\{I\}$ = time the solution stays implemented	<u>1</u>	2	3	4	5
8	$E\{R\}$ = expected number of replications	1	2	<u>3</u>	4	5

$PSE = 480$

Case 9

Problem name: Electrical cut-off protection switch test data analysis

Date: 1998 Reporter: C. McCollin
I (International)
P (Production type)
VL (Very large company size)
Inspection (Low management maturity)

Problem description:
Carry out exploratory reliability analysis on electrical switch test data

Context:
Teaching company course on reliability techniques

Data collected:
Operating time to failure, stress level, batch number, number of failures

Data analysis:
Proportional hazards modelling, Weibull analysis

Conclusions and recommendations:
First 50 hours of test under normal stress was having no effect on the time to failure of the unit.

Impact:
None. Company did not want to approach customer with change to test conditions for fear of losing contract. If implemented, cost savings of thousands of pounds.

	Criteria	Value –			Level +	
1	$V\{D\}$ = value of the data actually collected	1	2	3	<u>4</u>	5
2	$V\{M\}$ = value of the statistical method employed	1	2	<u>3</u>	4	5
3	$V\{P\}$ = value of the problem to be solved	<u>1</u>	2	3	4	5
4	$V\{PS\}$ = value of the problem actually solved	1	2	3	<u>4</u>	5
5	$P\{S\}$ = probability that the problem actually gets solved	<u>1</u>	2	3	4	5
6	$P\{I\}$ = probability the solution is actually implemented	<u>1</u>	2	3	4	5
7	$T\{I\}$ = time the solution stays implemented	<u>1</u>	2	3	4	5
8	$E\{R\}$ = expected number of replications	1	2	3	<u>4</u>	5

$PSE = 192$

Case 10

Problem name: Reduce number of defects in a car assembly chain

Date: 1993 Reporter: A. De Frenne
I (international)
P (Production type)
VL (Very large company size)
Process improvement (High management maturity)

Problem description:
Reduce the number of defects on the car assembly chain

Context:
SPC was implemented in the company without or with very light worker training. Method was misused.

Data collected:
Check sheets, control charts on one month at different stages

Data analysis:
Control chart limits, cause and effect diagram

Conclusions and recommendations:
The quality improvement of one assembly stage was measured by the next assembly stage (client). Better communication between them reduced number of defects by 40 %.

Impact:
Reduction of defects due to a careful follow-up with control chart

	Criteria	Value –				Level +
1	$V\{\boldsymbol{D}\}$ = value of the data actually collected	1	2	$\underline{3}$	4	5
2	$V\{\boldsymbol{M}\}$ = value of the statistical method employed	1	2	$\underline{3}$	4	5
3	$V\{\boldsymbol{P}\}$ = value of the problem to be solved	1	2	$\underline{3}$	4	5
4	$V\{\boldsymbol{PS}\}$ = value of the problem actually solved	1	2	$\underline{3}$	4	5
5	$P\{\boldsymbol{S}\}$ = probability that the problem actually gets solved	1	2	3	$\underline{4}$	5
6	$P\{\boldsymbol{I}\}$ = probability the solution is actually implemented	1	2	3	$\underline{4}$	5
7	$T\{\boldsymbol{I}\}$ = time the solution stays implemented	1	$\underline{2}$	3	4	5
8	$E\{\boldsymbol{R}\}$ = expected number of replications	1	2	$\underline{3}$	4	5

$PSE = 7776$

Case 11

Problem name: Production start-up of new product

Date: 1997 Reporter: A. De Frenne
I (International)
P (Prototype type)
VL (Very large company size)
Process improvement (High management maturity)

Problem description:
Tune the production equipment to produce within legal specifications

Context:
Weight and pressure of a prototype aerosol were out of control. It has to be fixed within legal limits.

Data collected:
Weight and pressure measurements for one week's production.

Data analysis:
Control chart and histogram

Conclusions and recommendations:
The tuning of the equipment was not fine enough. Each modification of production setting introduced too much or not enough product, rarely at target value. Better training of worker on machine tuning solved the problem.

Impact:
Product better on target and within specification limits.

	Criteria	Value			Level	
		–				+
1	$V\{D\}$ = value of the data actually collected	1	2	3	$\underline{4}$	5
2	$V\{M\}$ = value of the statistical method employed	1	2	$\underline{3}$	4	5
3	$V\{P\}$ = value of the problem to be solved	1	2	3	$\underline{4}$	5
4	$V\{PS\}$ = value of the problem actually solved	1	2	3	$\underline{4}$	5
5	$P\{S\}$ = probability that the problem actually gets solved	1	2	3	$\underline{4}$	5
6	$P\{I\}$ = probability the solution is actually implemented	1	2	3	$\underline{4}$	5
7	$T\{I\}$ = time the solution stays implemented	1	2	$\underline{3}$	4	5
8	$E\{R\}$ = expected number of replications	$\underline{1}$	2	3	4	5

$PSE = 9216$

Case 12

Problem name: Defect reduction in precision smelting

Date: 2000 Reporter: A. De Frenne
I (International)
P (Production type)
L (Large company size)
Inspection (Low management maturity)

Problem description:
Headquarters asked to use DoE to reduce the percentage of defects in precision smelting pieces

Context:
60 % of production was defective for some specific pieces

Data collected:
History of production for 3 years

Data analysis:
Cause and effect diagram, identification of key characteristics at each step of the process, DoE

Conclusions and recommendations:
DoE identified the significant key characteristics of the process. The optimal combination was completely reversed from the current settings. Simulation programs confirmed the optimal result.

Impact:
Defect reduction from 60 % to 19 %.

	Criteria	Value –				Level +
1	$V\{D\}$ = value of the data actually collected	1	2	<u>3</u>	4	5
2	$V\{M\}$ = value of the statistical method employed	1	2	3	<u>4</u>	5
3	$V\{P\}$ = value of the problem to be solved	1	2	3	<u>4</u>	5
4	$V\{PS\}$ = value of the problem actually solved	1	2	3	<u>4</u>	5
5	$P\{S\}$ = probability that the problem actually gets solved	1	2	3	<u>4</u>	5
6	$P\{I\}$ = probability the solution is actually implemented	1	2	3	<u>4</u>	5
7	$T\{I\}$ = time the solution stays implemented	1	2	3	4	<u>5</u>
8	$E\{R\}$ = expected number of replications	<u>1</u>	2	3	4	5

$PSE = 15\,360$

Case 13

Problem name: Peeling of coating on aircraft wing

Date: 2002 Reporter: A. De Frenne
N (National)
P (Production type)
L (Large company size)
QbD (Very high management maturity)

Problem description:
Coating of aircraft piece of wings was peeling without reproducibility. Understand the source of the trouble.

Context:
About 30 % of pieces were peeling and for scrap.

Data collected:
History of production conditions for 5 runs.

Data analysis:
Cause-and-effect diagram, identification of key characteristics, DoE

Conclusions and recommendations:
None of the parameters from the DoE was found significant. However, well-controlled DoE allowed identification of the key characteristic due to the logic of the production organisation. Problem was fixed.

Impact:
Defect reduction from 30 % to 0 %

	Criteria	Value –				Level +
1	$V\{D\}$ = value of the data actually collected	1	<u>2</u>	3	4	5
2	$V\{M\}$ = value of the statistical method employed	1	2	3	<u>4</u>	5
3	$V\{P\}$ = value of the problem to be solved	1	2	<u>3</u>	4	5
4	$V\{PS\}$ = value of the problem actually solved	1	2	<u>3</u>	4	5
5	$P\{S\}$ = probability that the problem actually gets solved	1	2	3	4	<u>5</u>
6	$P\{I\}$ = probability the solution is actually implemented	1	2	3	4	<u>5</u>
7	$T\{I\}$ = time the solution stays implemented	1	2	3	4	<u>5</u>
8	$E\{R\}$ = expected number of replications	<u>1</u>	2	3	4	5

$PSE = 9000$

Case 14

Problem name: Production improvement in catalytic converter substratum

Date: 2002 Reporter: A. De Frenne
I (International)
P (Production type)
L (Large company size)
Process control (High management maturity)

Problem description:
The final product should be circular. Identify sources of improvement when product is fresh before cooking in the oven.

Context:
Product size changes dramatically during cooking in the oven. Very tight specifications are imposed on production of fresh product.

Data collected:
History of product measurements by laser for 1 month

Data analysis:
Control chart of product dimensions

Conclusions and recommendations:
Control charts show no potential improvement at the fresh product production. Improvement could be realised up-front on raw material and with better control of the oven.

Impact:
Production changes to be implemented but were not accepted by management.

	Criteria	Value			Level	
		−				+
1	$V\{D\}$ = value of the data actually collected	1	2	3	$\underline{4}$	5
2	$V\{M\}$ = value of the statistical method employed	1	2	$\underline{3}$	4	5
3	$V\{P\}$ = value of the problem to be solved	1	2	$\underline{3}$	4	5
4	$V\{PS\}$ = value of the problem actually solved	$\underline{1}$	2	3	4	5
5	$P\{S\}$ = probability that the problem actually gets solved	$\underline{1}$	2	3	4	5
6	$P\{I\}$ = probability the solution is actually implemented	$\underline{1}$	2	3	4	5
7	$T\{I\}$ = time the solution stays implemented	$\underline{1}$	2	3	4	5
8	$E\{R\}$ = expected number of replications	$\underline{1}$	2	3	4	5

$PSE = 36$

Case 15

Problem name: Reduce porosity in Aluminium injection process for making wheels

Date: 1990 Reporter: X. Tort-Martorell
I (International)
P (Production type)
L (Large company size)
Process improvement (High management maturity)

Problem description:
Reduce porosity and possibly other types of defects (mainly adhered material and filling problems) in car aluminium wheels produced by a gravity injection process

Context:
The company asked for outside help in using DoE to improve the process. Outside help included training in DoE of several plant engineers and using the case as a learning-by-doing experience.

Data collected:
A 2^5 design with three responses, followed by a 2^3 design and several confirmatory experiments

Data analysis:
Calculation and interpretation of effects, interaction plots

Conclusions and recommendations:
Three factors were found important (two of them interacting, one had a quadratic effect). Better operating conditions were identified.

Impact:
The rejection rate drop from 20–25 % to 5–10 %. In spite of the improvement the factory closed 2 years later.

	Criteria	Value –			Level +	
1	$V\{D\}$ = value of the data actually collected	1	2	3	<u>4</u>	5
2	$V\{M\}$ = value of the statistical method employed	1	2	3	4	<u>5</u>
3	$V\{P\}$ = value of the problem to be solved	1	2	3	<u>4</u>	5
4	$V\{PS\}$ = value of the problem actually solved	1	2	3	<u>4</u>	5
5	$P\{S\}$ = probability that the problem actually gets solved	1	2	3	<u>4</u>	5
6	$P\{I\}$ = probability the solution is actually implemented	1	2	3	<u>4</u>	5
7	$T\{I\}$ = time the solution stays implemented	1	<u>2</u>	3	4	5
8	$E\{R\}$ = expected number of replications	<u>1</u>	2	3	4	5

$PSE = 10\,240$

Case 16

Problem name: Designing a test for comparing slot machines

Date: 1996 Reporter: X. Tort-Martorell
I (International)
FSD (Full-scale development)
L (Large company size)
Process improvement (High management maturity)

Problem description:
After slot machines are designed and prototypes produced, they are tested in real conditions. The aim is to devise a testing methodology (location, time, type of data gathered, test to conduct, ...) to forecast whether the slot machine is going to be liked by customers and be a money-maker.

Context:
The new machines are tested by comparing them to existing good machines. There are important factors to be taking into account, the most important one being the novelty effect (during an initial period machines are always money-makers).

Data collected:
Income produced by different types of machines trough time taking into account other important factors. Data related to trials of the test procedure designed.

Data analysis:
Graphical methods, DoE, statistical tests

Conclusions and recommendations:
The conclusion of the project was a test procedure described in a step-by-step protocol that included a MINITAB macro to analyse the data coming from the test.

Impact:
Introduction of better prototypes to the market

	Criteria	Value –				Level +
1	***V*{*D*}** = value of the data actually collected	1	2	<u>3</u>	4	5
2	***V*{*M*}** = value of the statistical method employed	1	2	<u>3</u>	4	5
3	***V*{*P*}** = value of the problem to be solved	1	2	3	<u>4</u>	5
4	***V*{*PS*}** = value of the problem actually solved	1	2	3	<u>4</u>	5
5	***P*{*S*}** = probability that the problem actually gets solved	1	2	3	<u>4</u>	5
6	***P*{*I*}** = probability the solution is actually implemented	1	2	3	<u>4</u>	5
7	***T*{*I*}** = time the solution stays implemented	1	2	<u>3</u>	4	5
8	***E*{*R*}** = expected number of replications	<u>1</u>	2	3	4	5

$PSE = 6912$

Case 17

Problem name: Reduction of label defects in a cava bottling process

Company name: Codorniu
Date: 1997 Reporter: X. Tort-Martorell
N (National)
P (Production type)
L (Large company size)
Process improvement (High management maturity)

Problem description:
Cava bottles have six labels. The correct position and alignment of these is crucial in customer-perceived quality.

Context:
There was no clear classification of defects and specifications.

Data collected:
Production of a clear defect classification, description and specifications for each type of defect. Data collected on 2500 bottles of different cava types.

Data analysis:
Pareto analysis, cause-and-effect diagram, identification of key causes (stratification, correlation). Design of a control procedure based on statistical process control.

Conclusions and recommendations:
Definition of types of defects and ways to measure them were fundamental and, through the data collected, identify process improvements. The control procedure allowed for easy monitoring.

Impact:
Important defect reductions. Wide variability, depending on the production line and type of defect, ranging from 4–30 % for some types to 3–5 % in others.

	Criteria	Value				Level
		−				+
1	$V\{D\}$ = value of the data actually collected	1	2	$\underline{3}$	4	5
2	$V\{M\}$ = value of the statistical method employed	1	$\underline{2}$	3	4	5
3	$V\{P\}$ = value of the problem to be solved	1	$\underline{2}$	3	4	5
4	$V\{PS\}$ = value of the problem actually solved	1	$\underline{2}$	3	4	5
5	$P\{S\}$ = probability that the problem actually gets solved	1	2	3	$\underline{4}$	5
6	$P\{I\}$ = probability the solution is actually implemented	1	2	3	$\underline{4}$	5
7	$T\{I\}$ = time the solution stays implemented	1	2	3	$\underline{4}$	5
8	$E\{R\}$ = expected number of replications	1	$\underline{2}$	3	4	5

$PSE = 3072$

Case 18

Problem name: Monitoring the cleaning process in a municipality

Date: 1998 Reporter: X. Tort-Martorell
N (National)
S (Public service type)
S (SME size)
Inspection (Low management maturity)

Problem description:
Cleaning services for municipal dependencies and the road and park network are outsourced. Citizens complain about dirtiness and the municipal authorities have

the conviction that the money spent should be enough to accomplish a satisfactory cleaning level. They would like an inspection procedure to produce a cleanness index and check on the subcontractor.

Context:
The municipality has 60 000 inhabitants and occupies 30 km^2

Data collected:
Cleanness attributes for different types of dependencies

Data analysis:
Sampling, operating characteristic curves, control charts. Definition of inspection procedures and training for inspectors. Definition of weights for different types of dirtiness according to citizens' perceptions.

Conclusions and recommendations:
The procedure designed was included in the tender for the new cleaning contract incorporating economic penalties.

Impact:
Difficult to assess due to the lack of previous data

		Value				Level
	Criteria	−				+
1	***V*{*D*}** = value of the data actually collected	1	2	3	4	5
2	***V*{*M*}** = value of the statistical method employed	1	2	3	4	5
3	***V*{*P*}** = value of the problem to be solved	1	2	3	4	5
4	***V*{*PS*}** = value of the problem actually solved	1	2	3	4	5
5	***P*{*S*}** = probability that the problem actually gets solved	1	2	3	4	5
6	***P*{*I*}** = probability the solution is actually implemented	1	2	3	4	5
7	***T*{*I*}** = time the solution stays implemented	1	2	3	4	5
8	***E*{*R*}** = expected number of replications	1	2	3	4	5

$PSE = 1728$

Case 19

Problem name: Welding Process Optimisation

Company name: Alstom Transport
Date: 2001 Reporter: X. Tort-Martorell
I (International)
P (Production type)

VL (Very large company size)
Process control (High management maturity)

Problem description:
Optimise the parameters of a welding process conducted by a new robot in the train chassis. Both the materials to be welded and the welding conditions were quite specific.

Context:
Factory engineers knew a lot about welding but little about experimental design. They had conducted some experiments and started production that was good in general, but showing some contradictions and unexpected results.

Data collected:
A 2^{8-3} design assigned factors to columns to allow for a favourable confounding pattern. A lot of work in preparing the plates to be welded and the procedures to measure the responses.

Data analysis:
Calibration, measurement system analysis, regression analysis, calculation and interpretation of effects

Conclusions and recommendations:
More important than the knowledge and improvements gained from the conclusions of the design was the knowledge gained from the preparation (understanding that the robot-programmed intensity was different than the real intensity, that intensity varied during the welding process, understanding the causes of sparks).

Impact:
Faster welding with increased traction and shearing

	Criteria	Value				Level
		−				+
1	$V\{\boldsymbol{D}\}$ = value of the data actually collected	1	2	3	$\underline{4}$	5
2	$V\{\boldsymbol{M}\}$ = value of the statistical method employed	1	2	3	4	$\underline{5}$
3	$V\{\boldsymbol{P}\}$ = value of the problem to be solved	1	2	3	$\underline{4}$	5
4	$V\{\boldsymbol{PS}\}$ = value of the problem actually solved	1	2	$\underline{3}$	4	5
5	$P\{\boldsymbol{S}\}$ = probability that the problem actually gets solved	1	2	3	$\underline{4}$	5
6	$P\{\boldsymbol{I}\}$ = probability the solution is actually implemented	1	2	3	$\underline{4}$	5
7	$T\{\boldsymbol{I}\}$ = time the solution stays implemented	1	2	3	$\underline{4}$	5
8	$E\{\boldsymbol{R}\}$ = expected number of replications	$\underline{1}$	2	3	4	5

$PSE = 15\,360$

Case 20

Problem name: Scale-Up Optimisation

Company name: 'Sofist'
Date: 2004 Reporter: R. Kenett
I (International)
P (Production type)
L (Large company size)
Inspection (Low management maturity)

Problem description:
Optimise the parameters of a chemical compound during scale-up pilot experiments using a simulation package to simulator the production process

Context:
Engineers of the pilot plan were requested to optimise a chemical compound used in the pharmaceutical industry. The materials involved in producing the compound are of very high cost, so that the cost of physical experimentation is prohibitively high. The pilot unit manager decided to invest in a sophisticated simulation package to help reduce experimentation cost and speed up the scale-up process.

Data collected:
A space-filling Latin hypercube design assigned factor levels to experimental runs. Experiments were conducted using a special-purpose simulation package for scale-up processes. Kriging models were used to determine the response surface and derive optimal set-up conditions.

Data analysis:
Stochastic emulators, radial base kriging models, multi-objective optimisation methods

Conclusions and recommendations:
The simulation results first had to be calibrated using a small set of simple experimental runs. Calibration consisted of comparing simulation results to physical experiments results. Following calibration, a full-scale simulation experiment was conducted and optimisation of economic indicators was performed.

Impact:
Significantly improved parameter set-up points, as measured by the simulation predictions of the economic indicators. The results, however, were ignored by management who preferred to rely on simple physical experiments.

	Criteria	Value –				Level +
1	$V\{D\}$ = value of the data actually collected	1	2	3	**4**	5
2	$V\{M\}$ = value of the statistical method employed	1	2	3	4	**5**
3	$V\{P\}$ = value of the problem to be solved	1	2	3	4	**5**
4	$V\{PS\}$ = value of the problem actually solved	1	2	3	**4**	5
5	$P\{S\}$ = probability that the problem actually gets solved	1	2	3	**4**	5
6	$P\{I\}$ = probability the solution is actually implemented	**1**	2	3	4	5
7	$T\{I\}$ = time the solution stays implemented	**1**	2	3	4	5
8	$E\{R\}$ = expected number of replications	**1**	2	3	4	5

$PSE = 1600$

Case 21

Problem name: Employee Suggestion System

Company name: 'Basil'
Date: 2006 Reporter: R. Kenett
N (National)
S (Service)
SME (SME company size)
Process control (High management maturity)

Problem description:
The results of the Basil employee survey indicated several weaknesses, among them the participation of the workforce through suggestions and other bottom-up communication channels.

Context:
Basil conducted an employee satisfaction survey, in which every employee participated. This outstanding result reflected the efficient planning and execution of the survey and the motivation of the employees to have their voice heard. The company set up a steering committee to manage focused initiatives to address the weaknesses. The committee was essentially made up of top management and chaired by the chief executive. An improvement team, with a mission to improve the flow of employee suggestions, was launched by the steering committee, along with several other teams directed at other issues. A team leader was assigned and a monthly progress review by the steering committee was conducted. The team was commissioned to work with the DMAIC roadmap.

Data collected:

The employee suggestion system improvement team reviewed minutes and reports covering a period of 12 months to identify where, how and if employee suggestions were identified and implemented. The data was organised in tabular form.

Data analysis:

Pareto charts, trend charts and P charts for comparing proportion of adopted suggestions by various classifications. Following the diagnostic analysis, a full intranet-based suggestion was designed and deployed.

Conclusions and recommendations:

The team was able to prove that a great majority of the employee suggestions recorded in various minutes of meetings and reports were not implemented. It was also clear that no formal recognition or systematic tracking of these contributions was carried out.

Impact:

The new suggestion system generated many ideas that resulted in significant savings and improved motivation of the workforce. This positive experience enhanced participation of the workforce in several additional areas.

		Value				Level
	Criteria	−				+
1	$\boldsymbol{V\{D\}}$ = value of the data actually collected	1	2	3	$\underline{4}$	5
2	$\boldsymbol{V\{M\}}$ = value of the statistical method employed	1	2	3	4	$\underline{5}$
3	$\boldsymbol{V\{P\}}$ = value of the problem to be solved	1	2	3	$\underline{4}$	5
4	$\boldsymbol{V\{PS\}}$ = value of the problem actually solved	1	2	$\underline{3}$	4	5
5	$\boldsymbol{P\{S\}}$ = probability that the problem actually gets solved	1	2	3	$\underline{4}$	5
6	$\boldsymbol{P\{I\}}$ = probability the solution is actually implemented	1	2	3	$\underline{4}$	5
7	$\boldsymbol{T\{I\}}$ = time the solution stays implemented	1	2	3	$\underline{4}$	5
8	$\boldsymbol{E\{R\}}$ = expected number of replications	1	2	$\underline{3}$	4	5

$PSE = 57\,600$

5

Management statistics

Irena Ograjenšek and Ron S Kenett

5.1 Introduction

Management statistics is the basis for informed decision-making processes in every organisation. It provides a systematic approach to measuring organisational performance. As such, it is fundamental to the pursuit of business excellence which, in turn, enables businesses to achieve their overriding goal of staying in business (Dransfield *et al.*, 1999).

Measurement of organisational performance relates to the impact of products and/or services produced by an organisation on the environment in which the organisation operates. The environment provides the organisation with essential inputs that allow it to function, and absorbs its outputs. The extent to which an organisation's inputs, processes, and outputs are in harmony with the environment determines its sustainability. For long-term sustainability, an organisation needs to achieve two simultaneous outcomes: *satisfied stakeholders* (owners, employees, customers, and community) on the one hand, and its *own economic health* on the other. Organisational performance can be measured at three levels: strategic, tactical, and operational. In the framework of this chapter, we will have a detailed look at each of these.

In order to manage organisational performance successfully, managers should have at least an elementary knowledge of the following four fields in line with Deming's *Concept of Profound Knowledge* (Swift, 1995): systems and optimisation theory; statistical theory; managerial techniques and procedures; and psychology. However, given the specific nature of their work and the level at which the organisation's performance is measured, managers should strive to deepen their

Statistical Practice in Business and Industry Edited by S.Y. Coleman, T. Greenfield, D.J. Stewardson and D.C. Montgomery

understanding of individual fields in order to improve their own performance and the performance of their employees.

5.2 Types of indicators

Three different types of measures are needed in order to devise an approach to measurement of organisational performance which can be used at strategic, tactical, and operational levels within an organisation:

- indicators of past performance (lag indicators);
- indicators of current performance (real-time indicators);
- indicators of future performance (lead indicators).

Typical *lag indicators* of business success are profitability, sales, shareholder value, customer satisfaction, product portfolio, product and/or service quality, brand associations, and employee performance. Most frequently used lag indicators are traditional accounting indicators. Unfortunately they do not allow the confident prediction of future success, but tend to be suitable only for control after the event (troubleshooting instead of managing improvement). Using the quarterly financial statement indicators to manage improvement is, for Dransfield *et al.* (1999), like steering a car along the road by looking in the rear-view mirror.

Significant *lead indicators* are the result of:

- customer analysis (segments, motivations, unmet needs);
- competitor analysis (identity, strategic groups, performance, image, objectives, strategies, culture, cost, structure, strengths and weaknesses);
- market analysis (size, projected growth, entry barriers, cost structure, distribution systems, trends, key success factors);
- environmental analysis (technological, governmental, economic, cultural, demographic, technological).

As pointed out by Aaker (2001), these are all elements external to organisation. Their analysis should be purposeful, focusing on the identification of opportunities, threats, trends, uncertainties, and choices. The danger of becoming excessively descriptive should be recognised and avoided. Readily available industry measures (such as production, sales and stock indices, as well as employment indicators) regularly published by national statistical offices and other government bodies should also be included and interpreted in this framework.

Monitoring of lag and lead indicators is crucial at the strategic and tactical level, while *real-time* (process) *indicators* are important at the operational level.

5.3 Measurement of organisational performance at the strategic level

5.3.1 Key concerns

At the strategic level, typical concerns boil down to the direction the organisation, facing the challenges and problems presented by the outside world, should take. Examples can be verbalised in questions such as:

- How should we respond to the terrorist threat to business?
- How should we respond to EU enlargement?
- Should we attempt to penetrate the Asian market?
- Should we review our employment policy?

Decision-makers at this level deal with typical 'what if'-type problems which are generally of a qualitative nature. These are best dealt with in the framework of the PEST and the SWOT analysis.

5.3.2 PEST analysis

Political, economic, social, and technological (PEST) analysis is concerned with a framework of macro-environmental factors used when scanning the environment of a given organisation. In combination with SWOT analysis, it is applied to gain an understanding of business and environmental factors such as market growth or decline, market potential, business position, and direction of operations.

Political factors under scrutiny are tax policy, employment laws, environmental regulations, trade restrictions and tariffs, and political stability. Prominent among the *economic* factors are economic growth, interest rates, exchange rates, and inflation rate. *Social* factors usually include population growth rate, age distribution, career attitudes, and perceptions of safety. *Technological* factors include ecological and environmental aspects and are used to determine entry barriers, minimum efficient production levels, and outcomes of outsourcing decisions. The PEST factors can be classified as opportunities or threats in the framework of SWOT analysis, which is discussed next.

5.3.3 SWOT analysis

SWOT analysis is a strategic planning tool used to evaluate organisational strengths, weaknesses, opportunities, and threats from the viewpoint of a business venture, project, or any other situation that demands decision-making. The technique is credited to Albert Humphrey, who led a research project at Stanford University in the 1960s and 1970s using data from Fortune 500 companies.

Strengths are defined as internal attributes that can help the organisation to achieve the set objectives (such as a well-educated workforce). *Weaknesses* are internal attributes which are harmful to achieving the set objectives (such as obsolete machinery). *Opportunities* are external conditions helping the organisation to achieve the set objectives (such as a low fuel price, development of a new market niche). *Threats*, on the other hand, are external conditions which are harmful to achieving the set objectives (for example the persistent terrorist threat in air traffic).

SWOT analysis is best performed in matrix form, combining external opportunities and threats with internal strengths and weaknesses to form the appropriate strategy as shown in Figure 5.1. In practice, SWOT analysis is usually applied in combination with techniques of creative thinking such as brainstorming, mind-mapping, and/or Delphi technique. These help identify the organisation's strengths, weaknesses, opportunities and threats.

SWOT analysis		**Internal perspective**	
		Strengths	**Weaknesses**
External perspective	**Opportunities**	*SO strategies:* Build on organization's strengths to make use of opportunities.	*WO strategies:* Eliminate weaknesses to enable new opportunities.
	Threats	*ST strategies:* Use organization's strengths to divert threats.	*WT strategies:* Develop strategies to avoid weaknesses that could be exploited by threats.

Figure 5.1 The SWOT matrix.

5.3.4 Statistical skills required by management

Ograjenšek (2002) defines *quantitative literacy* in terms of three components:

- *statistical literacy* (ability to select, use, and interpret results of the proper statistical method to solve a given problem);
- *computer literacy* (ability to use the proper statistical software to solve a given problem); and
- *web (on-line) literacy* (ability to find and access data and metadata on-line).

At the strategic level, it is web literacy, rather than the statistical literacy, of managers that is the most important component of quantitative literacy, since dealing with 'what if'-type problems often calls for thorough research based on secondary statistical data (such as demographic and economic projections prepared by national statistical offices or central banks), and a growing number of secondary statistical resources are accessible on-line.

5.4 Measurement of organisational performance at the tactical level

5.4.1 Key concerns

The key concern of tactical management is the proper implementation of strategic goals. Typical problems encountered in the process include definition of desired leadership characteristics and employee competences, preparation of short-term plans, and establishment of an efficient performance management system based on the so-called *key performance indicators* (KPIs).

Examples of KPIs include (Dransfield *et al.*, 1999):

- *owner-related measures* – profitability, revenue, costs, market share, relevant efficiency measures (for example cycle time);
- *customer-related measures* – on-time delivery, complaints, response time;
- *employee-related measures* – indicators of absenteeism, safety, labour turnover, competencies;
- *community-related measures* – safety, outcomes of environmental audits.

There are three complementary approaches to setting up a system of evidence-based KPIs in an organisation. We initially present the integrated model used to map out cause-and-effect relationships, go on to discuss the balanced scorecard that provides management with a navigation panel, and conclude with a brief introduction to economic value added (EVA), an economic model measuring the creation of wealth.

5.4.2 The integrated model

The *integrated model* is designed to map out cause-and-effect relationships to help improve management decisions (Kenett, 2004; Godfrey and Kenett, 2007).

Sears, Roebuck and Co. implemented the first integrated model. It was called the employee-customer-profit model (Rucci *et al.*, 1998). The cause-and-effect chain links three strategic initiatives of Sears – that is, to be:

- a compelling place to work;
- a compelling place to shop; and
- a compelling place to invest.

To push forward these initiatives Sears' management looked for answers to three basic questions:

- how employees felt about working at Sears;
- how employee behaviour affected the customer shopping experience; and
- how the customer shopping experience affected profits.

The model presented in Figure 5.2 reflects detailed answers to these questions and identifies the drivers to employee retention, customer retention, customer recommendation, and profits. Sears have been able to map out these variables and determine that, for them, a 0.5 increase in employee attitude causes a 1.3 unit increase in customer satisfaction, which creates a 0.5 % increase in revenue growth.

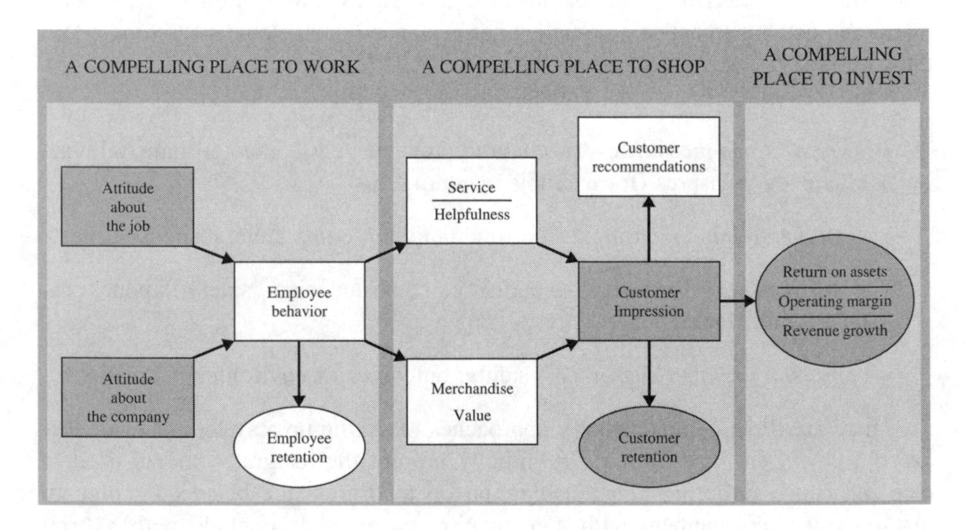

Figure 5.2 The integrated model (Rucci et al., 1998).

As another example, we show data from a company specializing in home office delivery (HOD) of water bottles in Figure 5.3. We established that increase in employees' satisfaction with their immediate supervisor, by branch, is directly related to customer satisfaction at that branch. In the six branches investigated, higher employee satisfaction correlates well with higher customer satisfaction. We can almost exactly predict customer satisfaction on the basis of employee satisfaction level.

Yet another example of an integrated model is the national customer satisfaction barometer, which is a measure for evaluating and comparing customer satisfaction across products, companies, and industries. Sweden was the first to successfully implement the barometer in 1989 (Fornell, 1992). Following the Swedish prototype, the American Customer Satisfaction Index (ACSI) was developed in a joint project of the American Quality Foundation and the University of Michigan Business School (Fornell *et al.*, 1996; Anderson and Fornell, 2000).

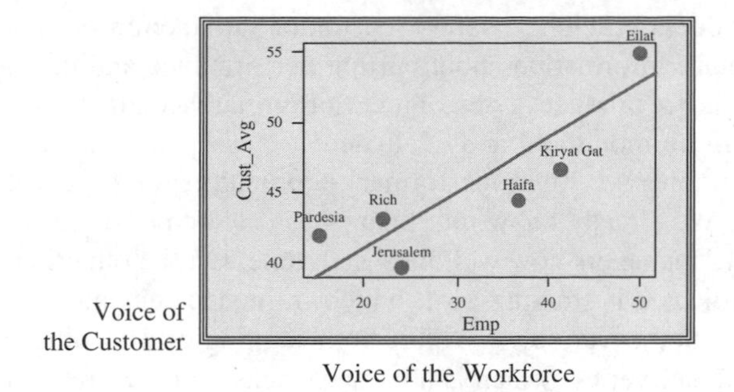

Figure 5.3 The relationship between employee and customer satisfaction (Kenett, 2004).

The ACSI is a structural equation model with six endogenous variables measuring perceived and expected quality, perceived value, satisfaction level, customer complaints, and customer retention (see Figure 5.4). In order to assess the customer satisfaction level a partial least squares (PLS) analysis is conducted of exogenous data gathered through telephone surveys. Similar projects have also been developed and implemented in Asia, Europe, and South America (among them the European Customer Satisfaction Index, introduced in 1999).

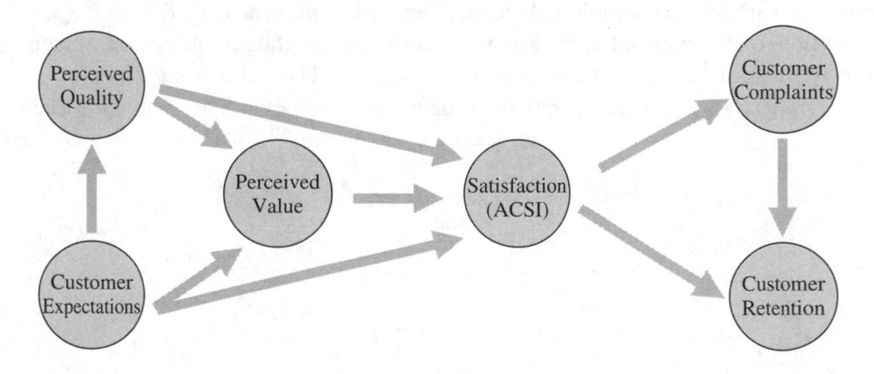

Figure 5.4 The American Customer Satisfaction Structural Equation Model.

5.4.3 Balanced scorecard

To translate vision and strategy into objectives, Kaplan and Norton (1992) created the *balanced scorecard*. This is meant to help managers keep their finger on the pulse of the business. Each organisation will emphasise different measures, depending on their strategy. Management is, in effect, translating its strategy into objectives that can be measured.

The balanced scorecard helps organisations identify and track a number of financial and non-financial measures to provide a broader view of the business. Data analysis is not limited to accounting data. A company may decide to use indicators of process efficiency, safety, customer satisfaction or employee morale. These may capture information about current performance and indicate future success. The aim is to produce a set of measures matched to the business so that performance can be monitored and evaluated.

Leading variables are future performance indicators, and lagging variables are historic results. We already know that financial measurements are typically lagging variables, telling managers how well they had done. On the other hand, an example of a leading indicator is training cost, which influences customer satisfaction and repeat business. Some variables exhibit both lagging and leading characteristics, such as 'on-time deliveries' which is a lagging measure of operational performance and a leading indicator of customer satisfaction.

The balanced scorecard usually includes four broad categories: financial performance, customers, internal processes, as well as learning and growth. Typically, each category will include two to five measures. If the business strategy is to increase market share and reduce operating costs, the measures may include market share and cost per unit.

Another business may choose financial indicators that focus on price and margin, willingly forgoing market share for a higher-priced niche product. These measures should be set after the strategy is in place. A list of objectives, targets, measurements and initiatives comes with each variable. The saying 'we manage what we measure' holds true. One reason why the balanced scorecard works is that it raises awareness. A sample balanced scorecard is presented in Figure 5.5.

Although the balanced scorecard is typically used in the private sector, examples of its application in the public sector are also known. The public sector management of Charlotte (North Carolina) has been using this framework to manage the city.

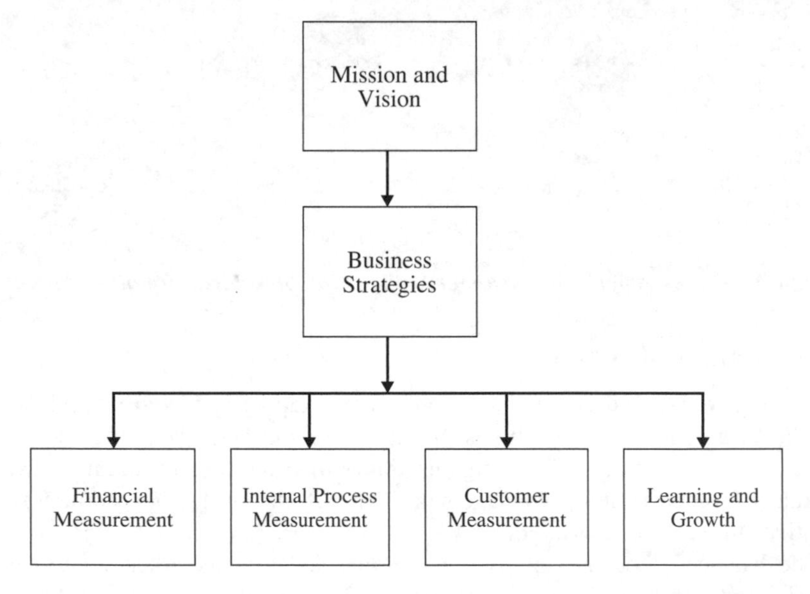

Figure 5.5 A sample balanced scorecard (Gordon and Gordon, 1998).

Their scorecard includes performance indicators that track improvements in community safety, quality of transportation, economic development, and effectiveness of government restructuring. Other more local measures of fire prevention, garbage collection, and sidewalk maintenance are used for individual departmental scorecards. This illustrates the flexibility of the scorecard and re-emphasises the point that the scorecard measures can be determined only after the overall strategy is accepted and understood.

The balanced scorecard, if properly implemented, is an excellent management framework that can help managers to track different factors which influence performance. But it lacks a single focus for accountability. Management needs one overriding goal to summarise the interaction between the variables and, ultimately, determine success. Shareholders entrust management with the implementation of strategy but their primary concern is earning an adequate return on their investment. EVA, a measure discussed in the next subchapter, tells us whether managers have correctly balanced the scorecard measures.

5.4.4 Economic value added

Economic value added is the one measure used to monitor the overall value creation in a business. There are many value drivers that need to be managed, but there can only be one measure that demonstrates success. A single measure is needed as the ultimate reference of performance to help managers balance conflicting objectives – and that measure is EVA.

The EVA measure was created by Stern Stewart & Co. (Stewart, 1991) to address the challenges companies faced in the area of financial performance measurement. By measuring profits after subtracting the expected return to shareholders, EVA indicates economic profitability. As shown in empirical studies (Boulos *et al.*, 2001), EVA tracks share prices much more accurately than earnings, earnings per share, return on equity or other accounting metrics. Creating sustainable improvements in EVA is the same as increasing shareholder wealth.

EVA is not a new concept. Economists have known about the residual income framework for years, but businesses have only recently begun to make the switch from managing for earnings to managing for value. EVA has facilitated this process by providing practical applications that operating managers can use. One of the great benefits of the balanced scorecard is that it illuminates the objectives which drive the strategy. In the same manner, EVA provides a common language across the organisation. Typically, different metrics are used for different processes. For example, sales growth and market share are discussed when strategy is formulated, net present value or internal rate of return (IRR) are calculated when capital investments are considered, earnings and earnings per share are considered when an acquisition is contemplated, yet trading profit is used when bonuses are determined. Which do you manage? Managers who use IRR to get a project approved do not need to worry about their actual performance because IRR cannot be used to measure performance in the middle of a project. Managers who are rewarded against a budget know that budget negotiation skills can be more lucrative than actually delivering results.

EVA can be the single metric used in all of these management processes. When decisions are made, performance is measured and compensation is determined. By using the same measurement you get accountability. EVA thus simplifies the lives not only of tactical but also of operating managers, who barely understand the interaction between the multiple existing measures.

5.4.5 Managerial commitment as the basis of an efficient performance management system

At the tactical level, managers' statistical and computer literacy are the most important components of quantitative literacy. This is similar to the operational level, yet with a different focus.

All three complementary approaches to setting up evidence-based KPIs in an organisation begin with new measurement techniques. At the outset, it is important to understand the barriers that can prevent a project from becoming successful. Creating more or better information is not necessarily going to lead to better decisions. In fact, adding another measurement to the existing pile of measurements will actually complicate matters. Managers must be motivated to act upon new information. When managers are not rewarded for making the right decision, the right decision will rarely be made.

Another critical factor is achieving a balance between simplicity and accuracy. An overly complex model may be more accurate, but the project could collapse under the weight of mountains of data that hide the relevant facts. And, as an overly complex model may be too difficult for managers to understand, an excessively simplistic model might lead to wrong decisions.

Senior management commitment is imperative for the project team to receive the support necessary to deliver results. Without this commitment, the project will be viewed as another 'flavour of the month' or, worse, another 'bean-counter' initiative. The desired result is not to get better information, but to increase the value of the organisation. The only way these frameworks will help achieve this goal is by using the new information to make the right decisions. People, not information, add value to the organisation, so the project must positively affect management behaviour to be successful.

Long-term, sustainable increases in performance will come when an organisation's culture is transformed from one focused on doing one's job to one centred on value creation. The implementation of a measurement system alone will not create sustainable changes. Freedom to act upon the information can only be delegated when there is accountability for the results. A compensation plan that properly rewards the desired actions is needed to expect true 'owner-like' behaviour.

The integrated model, the balanced scorecard and EVA are complementary tools that can help a company achieve greater success in the current dynamic and competitive business environment. The integrated model can help managers understand the cause-and-effect relationships of their decisions. The balanced scorecard broadens the view of performance to include financial and non-financial indicators of both a leading and lagging nature. EVA provides a link between decisions, performance measures, and rewards, which draws attention of managers to creating value.

These frameworks help managers focus on performing better. Regardless of whether one of these tools has already been implemented, the other two should be considered for implementation. Managers perform best when they have the information, decision frameworks, performance measures, and rewards that motivate them to behave like owners. It is important to have a good strategy, but it is just as important to have managers who are motivated to execute the strategy and produce results.

5.5 Measurement of organisational performance at the operational level

5.5.1 Key concerns

Operational management is directly concerned with processes in which people within an organisation interact to create value-added products and services (Dransfield *et al.*, 1999). Examples of dilemmas at this level are illustrated in the following list:

- How could we improve the capacity of this product line?
- How many temporary workers should be hired during the summer?
- Should this batch be reworked?
- Why is this process suddenly out of control?

An understanding of variation is the key prerequisite for measurement of organisational performance at the operational level. All processes in the company are functions of five elements: materials, people, machines or equipment, methods or procedures, and environment (Beauregard *et al.*, 1992: 15). There is variation in each element; each element thus contributes to variation in the process output as shown in the following formula (which should be understood as a simple illustration, not a statement about linear nature of links among the elements):

$$V_{\text{mat}} + V_{\text{p}} + V_{\text{mach}} + V_{\text{met}} + V_{\text{env}} = V_{\text{po}}.$$

where

V_{mat} = variation in materials,
V_{p} = variation in people (such as their knowledge, abilities, willingness to work),
V_{mach} = variation in machines (equipment),
V_{met} = variation in methods (procedures),
V_{env} = variation in environment,
V_{po} = variation in process output.

Since the variation of process inputs inevitably contributes to variation in the process output, analysis is necessary to determine whether process adjustments have to be made or not.

The roots of variation are either in common or special causes. *Common* causes of variation are reflected in the so-called chronic or historic variation, which seems to be an inherent part of the process. It can only change if the process itself is changed. This type of variation can be reduced in the process of continuous quality improvement based on the Deming (PDCA) cycle which is designed as a continuous four-step problem-solving approach and which we discuss in more detail in the following section.

Special causes of variation are due to acute or short-term influences that are not normally part of the process as it was designed or intended to operate. Statistical process control (SPC) is the tool to use in order to reduce the influence of special causes. For Juran (1989: 28) finding and eliminating them is similar to fire fighting, since there is no permanence in achieved solutions.

Historically, quality management was concerned with finding and eliminating the special causes of variation. Juran (1989: 29) talks about the *original zone of quality control* as opposed to the *new zone of quality control*, in which finding and eliminating the common causes of variation presents an opportunity for continuous quality improvement.

5.5.2 Key tools at the operational level

Measurement of organisational performance at the operational level follows the essence of the Deming cycle: it is a continuous revision and improvement of existing standard operating procedures, as well as replacement of existing standards with new, improved ones. Its continuity is viewed as the best warranty against bureaucratization of processes and their components. The four steps of the cycle are shown in Figure 5.6. They can be briefly described as follows:

- *Plan*. An area for improvement is chosen. Data are collected, the current state of affairs is analysed, and problems are identified. Action plans for change are prepared with the goal of process improvement in a pre-specified time framework.

- *Do*. The action plan is executed and changes are implemented. Detailed records of executed steps, gaps between goals, and actual outcomes of planned activities, as well as possible changes of goals, are usually prepared in order to be reviewed in the next step of the cycle.

- *Check*. The results of the previous step are studied in detail. Gaps between planned goals and actual outcomes are given special attention.

- *Act*. A proper reaction to both wanted and unwanted results of the previous steps is the essence of this phase. If the results are satisfactory, companies are prompted to standardise and institutionalise the procedures in which they were achieved, thus making them known and available to all employees.

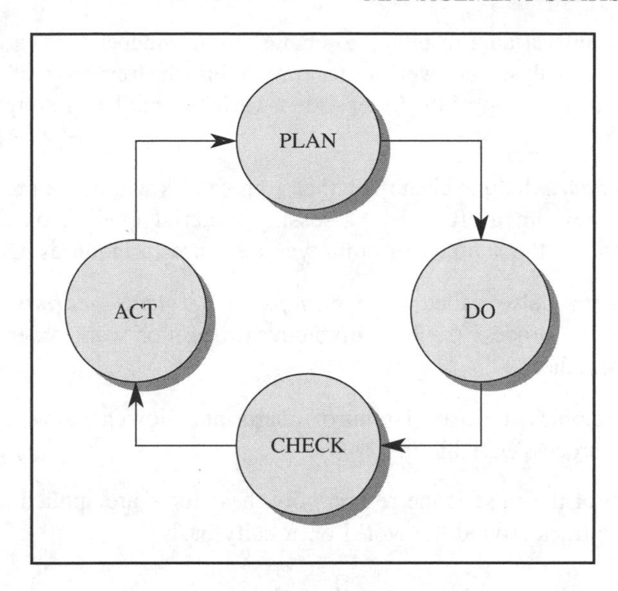

Figure 5.6 Deming (PDCA) cycle.

Where the results are unsatisfactory, procedures must be revised and either improved or discarded.

The use of statistical methods plays a very important role in all four steps of the Deming cycle, be it in the analysis of factors influencing the current state of affairs, forecasts of future developments, or comparisons of planned goals and actual outcomes.

In the framework of the Deming cycle, Ishikawa (1985) popularised a set of seven simple tools for employees to use (usually without any in-depth statistical training):

- *Check sheets* are the simplest and easiest-to-use data collection sheets which often serve as memory joggers to make an operation or a process mistake-proof. Maintenance personnel use them when performing preventive maintenance to ensure they do not overlook a step. Interviewers use them to determine whether a candidate meets all requirements for a job.

- *Pareto charts* help graphically separate the few most important causes of a problem from the many trivial ones. In the analysis of order errors or invoice errors Pareto diagrams can be made by type of error, by cause, or by employee.

- *Histograms* display attribute and variable data in order to discover data patterns. In a truck rental company, for example, accidents can be analysed by cause, by truck type, by truck age, by time of the day, by length of rental, and so on.

- *Scatter plots* depict a relationship between two variables. They are useful for determining if one variable is dependent on another, for showing how a dependent variable responds to change in an independent variable (cause-and-effect analysis) as well as for predicting the response of a dependent variable to a setting of the independent variable that has not previously been measured.

- *Control charts* help evaluate whether a process is within the preset statistical limits. They are useful when assessing material quality, on-time delivery performance, the supplier or employee response to inquiries and similar.

- *Flowcharts* – also called *service maps* or *service blueprints* – show how steps in any process (such as invoicing process or work order process) are linked together.

- *SPC* combines the use of control charts and flowcharts with the goal of reduced process variability.

The simplicity of their use is the reason why these tools are applied in production and service facilities around the world on a daily basis.

5.5.3 Key skills at the operational level

According to Fortuna (1992: 20–21) unfamiliarity with the concept of variation can have serious consequences:

- management failing to understand past performance and embarking on a fruitless search for a definitive cause of a single observed occurrence;

- management accepting wrong conclusions about past performance and trends, which can lead to flawed plans for the future;

- management acting or reacting to trends that are perceived but do not actually exist, or management failing to act when the data signal a systematic shift;

- management rewarding and punishing people for system faults – things over which they have no control because the inherent variation in the system is independent of individual performance.

For all these reasons, Snee (1999: 143) argues that an understanding of variation should be regarded as the core competence of statisticians, one giving them competitive advantage over other experts. Thyregod and Conradsen (1999: 145) add that, apart from statistical thinking, statisticians also need to understand business and management issues so that they can address problems in their proper context and prioritise the necessary efforts in accordance with the order of effectiveness and the principle of diminishing returns.

It would seem that at this level a knot should be tied between managerial competences and statistical as well as computational skills – the way it is done in the framework of the Six Sigma approach where the proactive manager with profound statistical skills strives to anticipate and avoid problems in the never-ending quest to improve operations, and, ultimately, the bottom line.

5.6 The future of management statistics

What future for management statistics? The need for evidence-based management has never been greater. The institutional framework for consistent use of management statistics – a quality management system – seems to be in place in most organisations. Why, then, do managers not apply statistical tools in the decision-making processes more frequently?

Although Czarnecki (1999) claims that the quality movement represents an impetus for measurement, it could be ascertained that even ISO-certified companies focus primarily on the philosophy inherent in total quality management, without really embracing the statistical methodology. In other words, concepts such as quality circles, or Deming's fourteen points, have been widely accepted and used. Statistical aspects of quality management, on the other hand, have usually remained the neglected component of certified quality management systems, due to measurement problems and the quantitative illiteracy of employees. Of the two problems, quantitative illiteracy is doubtless the more problematic to overcome. According to Gunter (1998),

> ISO-mania is symptomatic of an ever more pervasive decline in the quality profession . . . the retreat from quantitative methodology to soft quality . . . [which] emphasises human relations, personnel organisation, communications procedures, meeting conduct, and the like. I do not wish to dismiss this stuff as useless; some is both necessary and important. However, without the core quantitative disciplines that actually measure and effect real improvement, it ends up being mere fluff and distraction – all form and no substance. . . . I believe that the flight from quantitative methodology in general – and statistical methodology in particular – is basically a capitulation. . . . It is a lot easier to send everyone through a class on how to hold better meetings or improve interdepartment communication than it is to learn SPC or experimental design. So by dismissing quantitative methods as inadequate and ineffective, one avoids having to expend that effort.

The fact is that most MBA courses on quality management do not incorporate statistical aspects, which probably makes them easier to sell. The negative attitude of (business) students towards statistics is well documented (Gordon, 1995; Sowey, 1998; Francis, 2002; Martin, 2003; Ograjenšek and Bavdaž Kveder, 2003) and countless discussions on how to overcome it have not yet managed to produce any long-term satisfactory results.

Because older versions of the ISO 9001 standard address the issue of statistical analysis in two very general, very brief paragraphs (see Box 5.1 for the 1994 version), and the necessary links to International Statistical Standards were first published in 1999 (ISO/TR 10017:1999), such setting of priorities comes as no surprise.

In the 2000 version of the standard (ISO 9001: 2000) a more process-based structure was introduced together with an increased focus on continuous quality improvement and customer orientation. This version emphasises an evidence-based

Box 5.1 Clause 4.20 of ISO 9001: 1994.

4.20 STATISTICAL TECHNIQUES

4.20.1 Identification of Need

The supplier shall identify the need for statistical techniques required for establishing, controlling and verifying process capability and product characteristics.

4.20.2 Procedures

The supplier shall establish and maintain documented procedures to implement and control the application of the statistical techniques identified in 4.20.1.

approach to decision-making, and accordingly deals with issues of measurement, analysis and improvement in much more detail in its clause 8 (see Box 5.2 for the general requirements in clause 8.1).

Box 5.2 Clause 8.1 of ISO 9001: 2000.

8 MEASUREMENT, ANALYSIS AND IMPROVEMENT

8.1 General

The organisation shall plan and implement the monitoring, measurement, analysis and improvement processes needed

a) to demonstrate conformity of the product,

b) to ensure conformity of the quality management system,

c) to continually improve the effectiveness of the quality management system.

This shall include the determination of applicable methods, including statistical techniques, and the extent of their use.

Standardisation, ISO. These standards can be obtained from any ISO member and from the Website of ISO Central Seretariat at the following address: www.iso.org. Copyright remains with ISO. Reproduced with permission from ISO.

Apart from the general requirements provided in clause 8.1, clause 8 also includes guidelines on measurement of customer satisfaction, internal audits, monitoring and measurement of processes, as well as monitoring and measurement of products. Furthermore, the clause addresses the control of measurement systems, non-conforming products, and the issue of data analysis. Despite the heroic effort in the update of the guidance document ISO/TR 10017: 2003 to present a list of ISO 9001: 2000 clauses, clause by clause, and an identification of need for quantitative data and appropriate statistical techniques associated with the implementation of the clauses, it is, however, questionable whether this important revision of the standard and the guidance document can result in rapid positive changes of attitude towards the use of the statistical toolbox in companies with ISO-certified quality management systems.

This doubt is further reinforced by awareness that statistical methods have always been used more frequently in those companies whose production vitally depends on statistical trials (for example in chemical, pharmaceutical or food industries). Overall, however, during the whole total quality management and ISO certification boom, the promotion of the use of statistical methods was systematically neglected both at the company level and at the level of economy as a whole – and has been to this day, although solutions are proposed from time to time.

European statisticians carry their share of responsibility for such a state of affairs. It is therefore of the utmost importance that they recognize the need to establish a professional network. Such a network, which should connect theoretical and applied statisticians and statistical practitioners who, in their professional environments, have lacked interaction with and stimulation from the like-minded professionals, was founded in December 2000 as the European Network for Business and Industrial Statistics (ENBIS). Its mission and goals are presented in Box 5.3.

Box 5.3 ENBIS mission and goals.

The mission of ENBIS is [to]:

- Foster and facilitate the application and understanding of statistical methods to the benefit of European business and industry,
- Provide a forum for the dynamic exchange of ideas and facilitate networking among statistical practitioners (a statistical practitioner is any person using statistical methods whether formally trained or not),
- Nurture interactions and professional development of statistical practitioners regionally and internationally.

Additionally, ENBIS also strives:

- To promote the widespread use of sound science-driven, applied statistical methods in European business and industry,
- To emphasise multidisciplinary problem solving involving statistics,
- To facilitate the rapid transfer of statistical methods and related technologies to and from business and industry,
- To link academic teaching and research in statistics with industrial and business practice,
- To facilitate and sponsor continuing professional development,
- To keep its membership up to date in the field of statistics and related technologies,
- To seek collaborative agreements with related organisations.

Source: http://www.enbis.org/index.php?id=9 (accessed September 2007).

Functioning as a web-based society, ENBIS connects members who feel that statistics is vital for economic and technical development, and, consequently, improved competitiveness of European companies. Therefore, the ENBIS initiative has to be hailed as an important step in giving the quantitative aspects of management – and consequently management statistics – the attention they deserve.

References

Aaker, D.A. (2001) *Strategic Market Management*, 6th edition. John Wiley & Sons, Inc., New York.

Anderson, E.W. and Fornell, C. (2000) Foundations of the American Customer Satisfaction Index. *Total Quality Management*, **11**(7), 869–882.

Beauregard, M.R., Mikulak, R.J. and Olson, B.A. (1992) *A Practical Guide to Statistical Quality Improvement. Opening up the Statistical Toolbox*. Van Nostrand Reinhold, New York.

Boulos, F., Haspeslagh, P. and Noda, T. (2001) Getting the value out of value-based management: Findings from a global survey on best practices. *Harvard Business Review*, **79**(7), 65–73.

Czarnecki, M.T. (1999) *Managing by Measuring. How to Improve Your Organisation's Performance through Effective Benchmarking*. AMACOM, American Management Association, New York.

Dransfield, S.B., Fisher, N.I. and Vogel, N.J. (1999) Using statistics and statistical thinking to improve organisational performance. *International Statistical Review*, **67**(2), 99–150.

Fornell, C. (1992) A national customer satisfaction barometer: The Swedish experience. *Journal of Marketing*, **56**(1), 6–21.

Fornell, C., Johnson, M.D., Anderson, E.W., Cha, J. and Bryant, B.E. (1996) The American Customer Satisfaction Index: Nature, purpose and findings. *Journal of Marketing*, **60**(4), 7–18.

Fortuna, R.M. (1992) The quality imperative. In E.C. Huge (ed.), *Total Quality: A Manager's Guide for 1990s*, pp. 3–25. Kogan Page, London.

Francis, G. (2002) Choosing to study independently – when is it a good idea? In L. Pereira-Mendoza (ed.), *Proceedings of the Fifth International Conference on Teaching of Statistics*. Voorburg, Netherlands: International Statistical Institute.

Godfrey, A.B. and Kenett, R.S. (2007) Joseph M. Juran: A perspective on past contributions and future impact. *Quality and Reliability International*, **23**, 653–663.

Gordon, D. and Gordon, T. (1998) Measuring excellence: A case study in the use of the balanced scorecard. *Control*, May, 24–25.

Gordon, S. (1995) A theoretical approach to understanding learners of statistics. *Journal of Statistics Education*, **3**(3).

Gunter, B. (1998) Farewell fusillade. *Quality Progress*, **31**(4), 111–119.

Ishikawa, K. (1985) *What is Total Quality Control? The Japanese Way*. Prentice Hall, Englewood Cliffs, NJ.

Juran, J.M. (1989) *Juran on Leadership for Quality. An Executive Handbook*. Free Press, New York.

Kaplan, R.S. and Norton, D.P. (1992) The balanced scorecard: Measures that drive performance. *Harvard Business Review*, **70**(1), 71–79.

Kenett, R.S. (2004) The integrated model, customer satisfaction surveys and Six Sigma. In *Proceedings of the First International Six Sigma Conference*, CAMT, Wrocław, Poland.

Martin, M. (2003) 'It's like... you know': The use of analogies and heuristics in teaching introductory statistical methods. *Journal of Statistics Education*, **11**(2).

Ograjenšek, I. (2002) Business statistics and service excellence: Applicability of statistical methods to continuous quality improvement of service processes. Doctoral dissertation, Faculty of Economics, University of Ljubljana.

Ograjenšek, I. and Bavdaž Kveder, M. (2003) Student acceptance of ITT-supported teaching and internal course administration: Case of business statistics. In *Statistics Education and the Internet, Proceedings of the IASE Satellite Conference on Statistics Education, Berlin, August 11–12, 2003*. Voorburg, Netherlands: International Statistical Institute and International Association for Statistical Education.

Rucci, A.J., Kirn, S.P. and Quinn, R.T. (1998) The employee-customer-profit chain at Sears. *Harvard Business Review*, **76**(1), 82–97.

Snee, R.D. (1999) Statisticians must develop data-based management and improvement systems as well as create measurement systems. *International Statistical Review*, **2**, 139–144.

Sowey, E. (1998) Statistical vistas: Perspectives on purpose and structure. *Journal of Statistics Education*, **6**(2).

Stewart, B. (1991) *The Quest for Value*. HarperCollins, New York.

Swift, J.A. (1995) *Introduction to Modern Statistical Quality Control and Management*. St. Lucie Press, Delray Beach, FL.

Thyregod, P. and Conradsen, K. (1999) Discussion. *International Statistical Review*, **2**, 144–146.

6

Service quality

Irena Ograjenšek

6.1 Introduction

Every manufactured good has service elements such as storage, marketing, distribution, insurance, customer support, and advertising. Every service has manufactured tangible elements: banks provide statements; airlines provide tickets; restaurants serve foods. There is no clear distinction, but there is a *tangibility spectrum*. Introduced by Shostack (1977), it shows the proportions of tangible to intangible elements for different products (see Figure 6.1).

Services usually have more intangible elements than manufactured goods. Other important defining characteristics of services are as follows:

- *Inseparability*. Production and consumption usually take place at the same time – a haircut cannot be stored for future use.
- *Variability*. A restaurant can spoil us with a superb meal one day and disappoint us with mediocre fare a week later.
- *Heterogeneity*. Services range from simple to complex, from high-contact to low-contact, from fully customised to fully standardised, from personal to business services, and so on.

The enormous heterogeneity of services makes it impossible to define quality improvement routines applicable to all service industries. Routines used in financial or airline global players are not comparable with those used in small local service companies.

Also, most modern management texts describe quality improvement in terms of manufactured goods. Sometimes they dedicate one chapter to services. I shall

Statistical Practice in Business and Industry Edited by S.Y. Coleman, T. Greenfield,
D.J. Stewardson and D.C. Montgomery

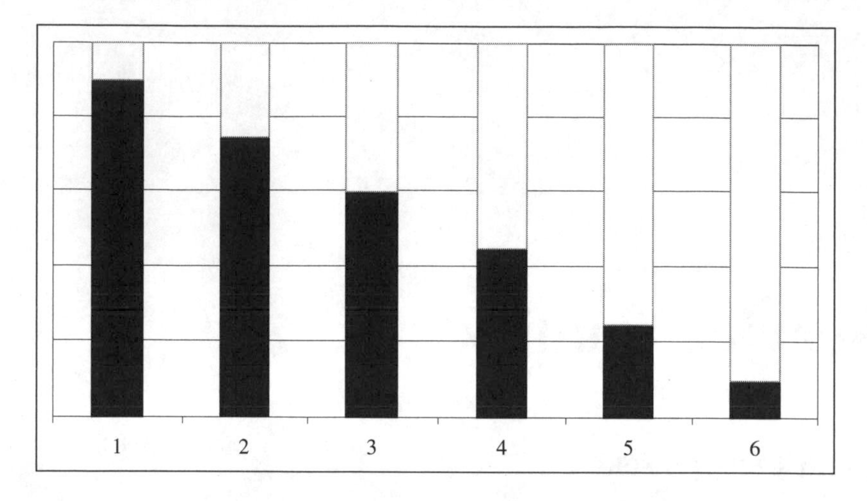

Figure 6.1 Tangibility spectrum. The shaded part of the column represents the tangible component and the white part the intangible component for the following products: (1) salt, (2) cars, (3) fast food, (4) on-line banking, (5) education, and (6) surgery (adapted from Shostack (1977: 77) and Kasper et al. (1999: 15)).

focus exclusively on services but will show where services have learned lessons from manufacturing.

6.2 Use of statistical methods to improve service quality: an overview of approaches

There are two distinct approaches to the use of statistical methods in quality improvement of services. These are the *production* approach, which uses traditional statistical quality control tools, and the *marketing* approach, which builds on the social sciences toolbox. Further differences are summarised in Table 6.1.

The production approach has limitations in services because objective direct real-time measures of key service attributes can rarely be obtained. Instead, managers tend to act upon subjective results of customer surveys, even though they may be too late or too unreliable to be of any value for quality improvement efforts.

As shown in Table 6.2, the production and marketing approach can be characterised using the four approaches to defining quality described by Garvin (1984):

- *Manufacturing-based approach.* Quality is conformance to predefined requirements, where divergence from requirements is a decrease in quality.
- *Product-based approach.* Quality is reflected in the presence or absence of product attributes. The customer compares attributes of products from the same product group and ranks them according to their perceived quality.

Table 6.1 Comparison of production and marketing approaches.

Criteria of comparison	Production approach	Marketing approach
Theoretical background	Classical texts on statistical quality control	Texts on service management and marketing
Predominant focus	Internal (back office – design and production – technical quality)	External (front office – delivery – perceived quality)
Standards	Objective (precisely determinable)	Subjective (customer expectations, customer intentions)
Statistical toolbox	Traditional SQC toolbox	Social sciences toolbox
Mode of application	Ex ante, real time and ex post	Rarely in real time, usually ex post
Goal of application	Design quality into service processes to prevent negative service experiences	Use customer feedback to discover areas in need of quality improvement
Quality of tangible components	Measured directly with an array of different tools	Measured indirectly through expert and customer observations
Quality of intangible components	No tools available	Measured indirectly through expert and customer observations
Resulting measures	Direct process measures	Indirect customer perception measures
Time frame of measurement	Measures available relatively quickly	Measures available with considerable delay
Cost of measurement	Measures obtained at a low cost	Measures obtained at a high cost (due to survey design, implementation and evaluation)
Frequency of application	Very high (hourly, daily ...)	Very low (annually or biannually)
Required level of employee quantitative literacy	Includes some of the simplest methods applicable with a minimum of educational effort	Includes some of the most complex analytical methods – proper training necessary to avoid the danger of simplification through pull-down menus in statistical software packages

Table 6.2 Production and marketing approach to improve service quality characterised by Garvin's approaches to define quality.

Garvin's approaches to defining quality	Production approach	Marketing approach
Manufacturing-based	☺	
Product-based		☺
User-based		☺
Value-based	☺	☺

- *User-based approach.* Customers determine quality according to products' fitness for use. Products that best satisfy customer needs or wants are those with the highest quality. This approach is highly subjective (fitness for use of the same product can be perceived differently by different customers).
- *Value-based approach.* Quality is assessed by comparing sacrifices (such as cost or time) and benefits. A quality product conforms to requirements at an acceptable cost (internal focus, characteristic for production approach) or to performance at an acceptable price (external focus, characteristic for marketing approach).

Garvin also identifies a fifth, *transcendent* or *philosophical* approach: quality cannot be defined precisely, it can be recognised only through experience. Such experience presents a basis for the formation of customer quality expectations and is embedded in Garvin's approaches to defining quality that characterise the marketing approach to improving service quality.

6.3 Production approach to improve service quality

6.3.1 Quality standards in the framework of the production approach

The production approach focuses on optimisation of service processes from the management viewpoint. Quality standards in this framework can be expressed:

- **In non-monetary terms.** Examples include:
  - the normal wine or coffee temperature in the restaurant,
  - the prescribed pizza oven temperature,
  - the health and hygiene rules for workers,
  - the tolerance level for the monthly number of customer complaints,

  - the tolerable length of waiting time,
  - the number of customers served in a given time frame.

- **In monetary terms.** Usually, upper tolerance levels for different categories of quality costs are used.

These quality standards can be applied in observational and inferential studies.

6.3.2 Toolbox of observational studies

Observational studies build on the application of the *basic statistical toolbox*, which includes a set of seven tools popularised by Ishikawa (1985) and listed in Chapter 5. Simplicity of use is the strength and common characteristic of these seven tools. They help companies see how often things happen, when and where they happen and in what different forms they may present themselves to the observer. They usually do not require employees to have high quantitative literacy. They do, however, provide an excellent basis for application of logical problem-solving, where data gathering and visual presentation act as catalysts for thinking and stimulate understanding of the observed phenomenon.

As illustrated in Box 6.1, application of informed observations to gather, organise and visually present data can be an effective way of finding patterns and links among causes and effects in the observed phenomenon.

Box 6.1 Use of observational studies in service industries.

The president of a small mortgage-lending bank facing serious competition from similar larger institutions had decided that to survive in the market for home mortgage loans, excellent service should be the bank's trademark. As part of a larger initiative, a selected team conducted a preliminary statistical study of transaction data. Team members found that the average time to complete a loan (the elapsed time from the moment a homeowner approached the bank until the loan was granted) was 24 days. This was similar to the completion times of competitors. A market analysis showed that a shorter completion time would give the bank a significant advantage over the competitors. The president nominated a team to work on reducing the waiting time for loan approvals.

After flowcharting the process on the basis of available transaction data and observing the processing of new loans for 3 months, the team found that enormous time savings could be achieved. A Pareto chart indicated that an overwhelmingly large part of the elapsed time was due to documents travelling between various offices since they had to be read and approved by so many different people. Further analysis showed that the steps in the process could be combined, and be done by one person. This could greatly reduce both the waiting time and the potential for errors. As a consequence, a set of standard operating procedures was prepared for the new process. After the system changes were implemented, a statistical study showed that the average time to obtain a home loan was reduced to 4 days, providing this bank with a significant advantage over its competitors.

Source: Ograjenšek and Bisgaard (1999: 6).

6.3.3 Toolbox of inferential studies: design of experiments

Inferential studies build on the application of the *extended statistical toolbox* with the goal of preventing flaws in services due to poor design. *Design of experiments* (DoE) should be used to determine how changes in service design influence service delivery and, ultimately, service quality. Examples include experiments with environmental factors influencing service quality such as light, noise, or humidity. Furthermore, it is possible to experiment with back and front office settings, opening hours, availability of phone operators, response time to on-line enquiries, employee supervision, and so on.

6.4 Marketing approach to improve service quality

6.4.1 Quality standards in the framework of the marketing approach

The marketing approach focuses on the customer assessment of their service experience. The emphasis is on the analysis of quality attributes and of overall quality. Willingness to pay merits less attention.

As in the production approach, quality standards in the framework of the marketing approach can be expressed in non-monetary or monetary terms (see Figure 6.2).

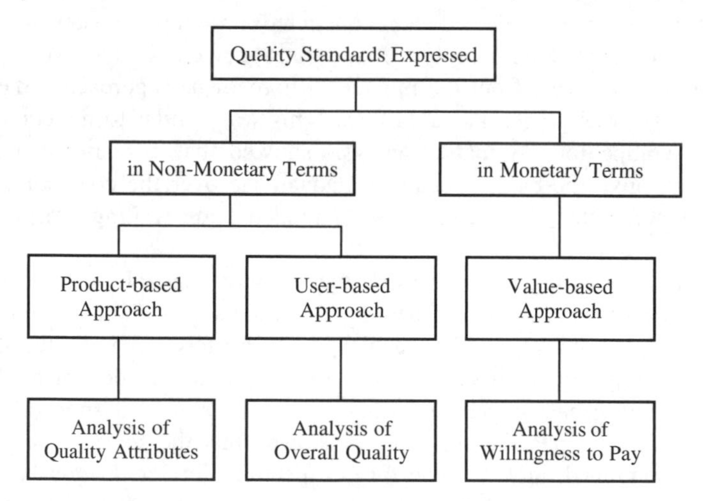

Figure 6.2 Quality standards and tools of the marketing approach.

When defining quality standards in non-monetary terms, there is little room to avoid the highly subjective *customer expectations* formed through:

- *search attributes* which are determined and evaluated before the service is delivered. These usually involve tangible aspects of a service, are physically visible, and more related to equipment than to the person delivering a service. Price is a typical search attribute in the service setting.
- *experience attributes* which are experienced either during or shortly after service delivery. These include characteristics such as taste or duration of well-being.
- *belief* (or *credence*) *attributes* which may be impossible to evaluate even a long time after service delivery. Examples include results of medical or car maintenance procedures. Few customers possess medical or mechanical skills sufficient to evaluate whether these services are or were necessary and performed in a proper manner.

The problem with many services is that they are high in experience and belief properties (which are difficult to evaluate from the customer point of view) and low in search properties (which are easiest to evaluate). To overcome this problem, companies usually try to *create* proper expectations that will be met with the actual service performance.

Customer intentions can also be used as subjective non-monetary quality standards. This is problematic when they reflect the ideal rather than actual buying behaviour of customers who act as quality evaluators.

6.4.2 Product-based approach: analysis of attributes

Service attributes can be analysed using either rating scales, penalty/reward analysis or the vignette method.

6.4.2.1 Use of rating scales in analysis of attributes: SERVQUAL

Service attributes are often analysed using *rating scales*. The most widely used and most carefully scrutinised rating scale is SERVQUAL (the name stands for SERVice QUALity). The scale is a result of a systematic ongoing study of service quality that began in 1983. The model defines quality as the difference between customer perceptions and expectations with regard to quality of delivered service. Respondents are asked to answer two sets of questions dealing with the same subject, one set at a general level (such as quality of service in financial institutions), and one for a company of interest (such as quality of service in bank XYZ). The first (general) set of questions are quality expectations (E_i) and the second (specific) set are quality perceptions (P_i), as shown in Table 6.3.

Respondents choose from a modified Likert scale which measures intensity of agreement or disagreement with any given statement. Response options range from

Table 6.3 The SERVQUAL questionnaire.

Quality dimension	Expectations (E_i)	Perceptions (P_i)
Tangibles	Excellent companies will have modern-looking equipment.	XYZ has modern-looking equipment.
	The physical facilities at excellent companies will be visually appealing.	XYZ's physical facilities are visually appealing.
	Employees of excellent companies will be neat in appearance.	XYZ's employees are neat in appearance.
	Materials associated with the service (such as pamphlets or statements) will be visually appealing in an excellent company.	Materials associated with the service (such as pamphlets or statements) are visually appealing at XYZ.
Reliability	When excellent companies promise to do something by a certain time, they will do so.	When XYZ promises to do something by a certain time, it does so.
	When customers have a problem, excellent companies will show a sincere interest in solving it.	When you have a problem, XYZ shows a sincere interest in solving it.
	Excellent companies will perform the service right first time.	XYZ performs its service right first time.
	Excellent companies will provide their services at the time they promise to do so.	XYZ provides its services at the time it promises to do so.
	Excellent companies will insist on error-free records.	XYZ insists on error-free records.
Responsiveness	Employees of excellent companies will tell customers exactly when services will be performed.	Employees of XYZ tell you exactly when the service will be performed.
	Employees of excellent companies will give prompt service to customers.	Employees of XYZ give you prompt service.
	Employees of excellent companies will always be willing to help customers.	Employees of XYZ are always willing to help you.

Table 6.3 (*Continued*)

Quality dimension	Expectations (E_i)	Perceptions (P_i)
	Employees of excellent companies will never be too busy to respond to customer requests.	Employees of XYZ are never too busy to respond to your requests.
Assurance	The behaviour of employees of excellent companies will instil confidence in customers.	The behaviour of XYZ's employees instils confidence in you.
	Customers of excellent companies will feel safe in their transactions.	You feel safe in your transactions with XYZ.
	Employees of excellent companies will be consistently courteous with customers.	Employees of XYZ are consistently courteous with you.
	Employees of excellent companies will have the knowledge to answer customer questions.	Employees of XYZ have the knowledge to answer your questions.
Empathy	Excellent companies will give customers individual attention.	XYZ gives you individual attention.
	Excellent companies will have operating hours convenient to all their customers.	XYZ has operating hours convenient to you.
	Excellent companies will have employees who give customers personal attention.	XYZ has employees who give you personal attention.
	Excellent companies will have the customers' best interests at heart.	XYZ has your best interests at heart.
	The employees of excellent companies will understand the specific needs of their customers.	Employees of XYZ understand your specific needs.

Adapted from Parasuraman *et al.* (1991a: 446–449).

strongly agree and lesser levels of agreement to neutral (neither agree nor disagree) and lesser levels of disagreement to strongly disagree.

For each of the 22 items (service attributes), a quality judgement can then be computed according to the following formula:

$$\text{Perception}(P_i) - \text{Expectation}(E_i) = \text{Quality}(Q_i)$$

From the formula, the following can be implied:

- If expectations exceed perceptions, quality is poor.
- If perceptions exceed expectations, quality is excellent.
- If customers have low expectations which are met, quality exists.

The SERVQUAL score (perceived service quality) is obtained by the following equation:

$$Q = \frac{1}{22} \sum_{i=1}^{22} (P_i - E_i).$$

The $P_i - E_i$ difference scores can be subjected to an iterative sequence of item-to-item correlation analyses, followed by a series of factor analyses to examine the dimensionality of the scale using the oblique rotation, which identifies the extent to which the extracted factors are correlated.

The original SERVQUAL scale had 97 items that resulted in ten service quality dimensions. The presently established 22-item scale shown in Table 6.3 has five quality dimensions defined as follows:

- *tangibles* – physical facilities, equipment, and appearance of personnel;
- *reliability* – ability to perform the promised service dependably and accurately;
- *responsiveness* – willingness to help customers and provide prompt service;
- *assurance* – knowledge and courtesy of employees as well as their ability to convey trust and confidence;
- *empathy* – individual care and attention that company provides its customers.

Although SERVQUAL has been widely used in business-to-business and business-to-customer settings, this does not mean the scale has not been subject to constant re-examination and criticism. The main objections to SERVQUAL are as follows:

- *Object of measurement*. It is not clear whether the scale measures service quality or customer satisfaction.

- *Length of the questionnaire.* The SERVQUAL questionnaire is too long. It could be shortened by elimination of expectation scores (see discussion under *Use of* $(P_i - E_i)$ *difference scores* for more details), elimination of certain items (those without the clear mode) and/or fusion of the interrelated dimensions of *reliability, responsiveness* and *assurance* into one dimension called *task-related receptiveness.*

- *Timing of questionnaire administration.* The main issue here is whether to distribute the questionnaire before or after the service experience. In other words, should expectations be solicited before the service experience or away from the actual point of service delivery and unrelated to an encounter? Some researchers compromise by collecting their data after the service experience at the actual point of service delivery. Consequently, they fear that this might have loaded their results towards performance, while those of other researchers might have been loaded towards expectations.

- *Use of the Likert scale.* The issues such as the number and labelling of points or the inclusion of a middle alternative in the scale are very important. SERVQUAL authors use a seven-point scale, while in many replication studies a five-point scale is adopted to increase response rate and response quality.

  Another problem is the equality of distances between points on the Likert scale as perceived by the respondent. It should be noted that one person's 'complete satisfaction' might be less than another's 'partial satisfaction'. Furthermore, once the respondents have marked the extreme point and want to express an even stronger opinion on the next item, this can no longer be reflected in the answer, since the maximum score has already been given.

  While some authors argue that none of these problems matter as long as the answers are normally distributed, others point out that in practice the majority of service quality surveys tend to result in highly skewed customer responses. This is why an average rating based on the arithmetic mean of the customer responses is likely to be a poor measure of central tendency, and may not be the best indicator of service quality.

- *Use of* $P_i - E_i$ *difference scores.* Ambigous definition of expectations in the SERVQUAL model seems to be a major problem. Increasing $P_i - E_i$ scores may not always correspond to increasing levels of perceived quality which impugns the theoretical validity of the SERVQUAL's perceived quality framework.

  Also questioned are the value and purpose of two separate data sets (perceptions and expectations). Some researchers suggest that it might be better not to use difference scores since the factor structure of the answers given to the questions about expectations and perceptions, and the resulting difference scores, are not always identical. Additionally, research shows that performance perception scores alone give a good quality indication. Therefore a modified SERVQUAL scale using only performance perceptions to measure service quality (called SERVice PERFormance or SERVPERF) has been proposed.

A three-component measurement model including perceptions, expectations and perceived importance ratings or weights of service attributes was also suggested. A rationale for this model is the following: while customers might expect excellent service on each attribute, all of the attributes involved may not be equally important to them. However, it seems that inclusion of perceived importance ratings does not substantially improve the explanatory power of the model.

- *Generalization of service quality dimensions*. The empirically identified five-factor structure cannot be found in all service industries. Only the existence of the *tangibles* dimension is confirmed in all replication studies, in which the number of distinct service quality dimensions otherwise varies from one to nine. It seems that the dimensionality of service quality might be determined by the type of service a researcher deals with.

- *The static nature of the model*. There exist a number of long-term service processes (such as education) where both perceptions and expectations (and consequently quality evaluations) change in time. For these service processes, a dynamic model of service quality should be developed.

As a consequence of these critiques, various alternative formulations of the service quality model have been suggested. The only one that, to date, seems to outperform SERVQUAL is SERVPERF, the performance-only measure of service quality.

6.4.2.2 Penalty/reward analysis

Penalty/reward analysis, developed by Brandt (1987, 1988), is used for identification of service attributes whose (lack of) presence can either diminish or increase service value in the eyes of potential customers.

Desirable attributes which are not part of a service cause customer dissatisfaction. The same can be stated for undesirable attributes which are part of a service. These are therefore labelled *penalty factors*.

Reward factors, on the other hand, are those attributes which allow a provider to exceed customer expectations. Sometimes referred to as *delighters*, they account for higher perceived quality, and consequently also higher customer satisfaction.

To identify service attributes as either penalty or reward factors, *penalty/reward contrast* (PRC) *analysis* is applied. PRC analysis is nothing more than a multiple regression analysis based on the use of dummy variables. For each attribute, perceived quality is compared to quality expectations, and the net result of the comparison statistically linked to total customer satisfaction. In other words, should expectations exceed perceptions, the attribute under scrutiny causes an average decrease in total customer satisfaction, and vice versa. It is therefore possible to compensate the negative effects of penalty factors with positive effects of reward factors.

6.4.2.3 Vignette method

The *vignette method* or *factorial survey approach* is a less well-known method developed by Rossi and Anderson (1982) and based on *conjoint analysis*. A basic idea in conjoint analysis is that any service can be broken down into a set of relevant attributes. By defining services as collections of attributes, it is possible to develop a number of *vignettes*, fictitious combinations of selected attributes (also referred to as *critical quality characteristics* which are usually identified beforehand, such as in focus groups) representing the service as a whole. Vignettes are then assessed as indicated in Figure 6.3. Afterwards, the results of the vignette method are analysed using a frequency table, with attributes acting as independent variables in the process.

VIGNETTE NUMBER ___		
Friendly receptionist		
Friendly room service		
Modest furniture		
No elevator		
How would you evaluate a hotel with these characteristics?		
(1) very good	(2) good	(3) satisfactory
(4) unsatisfactory	(5) bad	(6) very bad

Figure 6.3 Example of a vignette for a hotel (adapted from Haller, 1998: 115).

The vignette method is an efficient way to assess a large variety of service design options based on a simulation of the actual customer behaviour in purchase decision-making. It is often not necessary to evaluate all possible combinations of attributes, since some of them are a priori deemed unsuitable and rejected by designers. Frequently only prototypes are presented to evaluators.

One drawback of the vignette method is the requirement that the target segment be known in advance. The representativeness of the sample is of critical importance.

A further limitation of the method is the fact that it often oversimplifies real-life situations since only a limited number of attributes can be used in the analysis. Also, it is debatable whether *customer intentions*, expressed in the process of vignette evaluation, directly correspond to actual customer behaviour. Another factor not to be dismissed too lightly is *respondent fatigue*, especially if numerous vignettes have to be evaluated one after another.

With the advent of the internet, this method is frequently applied on-line in place of costly face-to-face interviews.

6.4.3 User-based approach: analysis of overall quality

6.4.3.1 Overview of methods

Methods used within the framework of the user-based approach are the critical incident technique, analysis of complaints, analysis of contacts and mystery shopping. Instead of important service quality *attributes*, these methods focus on *events*. They build on the surmise that overall quality evaluations are not based solely on the outcome of a service, but depend to a great extent on the process of service delivery. The overall service quality is therefore highly vulnerable to the *halo effect*: a single very positive or very negative experience in the process of service delivery decisively determines the overall perceived service quality.

6.4.3.2 Critical incident technique

The *critical incident technique* is a systematic procedure for recording events and behaviours that are observed to lead to success or failure on a specific task. It is based on a survey of customer experiences that were perceived either as extremely positive or extremely negative.

Using this method, data are collected through structured, open-ended questions, and the results are content-analysed. Respondents are asked to report specific events from the recent past (6–12 months). These accounts provide rich details of first-hand experiences in which customers have been satisfied or dissatisfied with service quality.

The discussion of factors which determine the critical incidents (moments of truth) is beyond the scope of this chapter. The same goes for discussion of the social competence of people involved in critical incidents. At one level, the buyer and seller are involved. At another, other customers (not excluding the buyer's family members) and other employees (not excluding the seller's supervisors) also play a role in the critical incident. Suffice it to say that certain quality improvements can be undertaken ex ante by working with employees, listening to them and their complaints, conducting special employee satisfaction surveys, empowering employees where feasible, giving them management support where needed by acknowledging that customer is not always right, and so on.

6.4.3.3 Analysis of complaints

Analysis of complaints is generally regarded as the basis of *complaint management*, which is widespread in service industries and essentially about recognizing *service failure* and making an effort to accomplish *service recovery*. The important issue is to use the opportunity that presents itself if the company is notified about a service failure (which is often not the case), since the company's response has the potential either to restore customer satisfaction and reinforce loyalty (this is service recovery, which has an important influence on quality assessment), or to exacerbate the situation and drive the customer to a competing firm.

Analysis of complaints is based on evaluation of unsolicited or solicited negative customer feedback (the latter obtained either via comment cards or in the form of customer surveys).

Unsolicited feedback is the most immediate source of customer service evaluation and one that quickly identifies reasons for customer dissatisfaction. Complaints received in person, by mail, phone, fax or on-line should be registered for future analysis. The same is also true for *comment cards* and results of *customer surveys*. Although only a limited amount of actual data is available from these often highly standardised sources, data can usually be grouped into usable classifications and analysed to show trends over a period of time. The basic statistical toolbox from the production approach can be used to great advantage in this setting.

There is one weakness to the analysis of complaints and comment cards: it reflects opinions of only those customers who decide to communicate with the company. To get a more unbiased evaluation of complaints, a customer survey should be carried out using a representative sample of customers.

6.4.3.4 Analysis of contacts

Analysis of contacts is a relatively new method based on *blueprinting*. As discussed in Chapter 5, a *service blueprint* shows each step in the service process and is therefore very similar to a flowchart. The only difference between the two concepts is the greater detail used in a service blueprint with regard to the graphical presentation of activities performed to provide a service. A service blueprint includes estimates of the time required for each activity, an indication of activities that necessitate customer contact, and an identification of activities where a service failure might occur.

Service blueprinting is usually applied within the framework of the operations management, for example to define activities that can be performed separately and those that can be co-processed to reduce waiting time which has a significant impact on the assessment of individual service quality attributes and/or overall quality. Apart from these benefits, the use of service blueprints is especially important with regard to the analysis of (customer) contacts.

The qualitative analysis of contacts attempts to determine how customers perceive the 'moments of truth' in service activities, which require their presence (and sometimes also their participation). Data for analysis is easiest to obtain with the help of specially trained call centre operators.

Given the fact that many customers tend to change their behaviour if under observation, it is usually better to let the customer report his or her feelings and perceptions ex post, using the *sequential incident method* (a step-by-step analysis of the service encounter). Often, this method is combined with either the critical incident technique or the analysis of complaints or both.

6.4.3.5 Mystery shopping

A qualitative tool of observation, *mystery shopping* has been growing increasingly popular in the service setting. The basic idea of this method is to look at the process(es) under scrutiny from the outside and measure their efficiency from a number of viewpoints:

- *The checker.* A specially trained employee of the company has the advantage of being familiar with the company's service standards. Airlines often use this type of analysis. The disadvantage of the use of checkers as mystery shoppers is the fact that they might be recognised as such, for they unconsciously act differently from average customers and may provoke the *observer effect* (when observed personnel deviate from their usual behaviour and present themselves at their best). An additional danger is *company blindness*. This occurs if checkers are only able to account for the company's internal evaluation criteria, while they are either completely ignoring, or ignorant of, both standards of competition and customer expectations.

- *The expert.* Specially trained outsiders can often be found carrying out mystery shopping in the field of gastronomy. Their tests are very highly regarded by the general public, for experts are (or should be) familiar with standards of the competition as well as general industry standards. The main impediment to expert tests is the fact that experts do not necessarily belong to the customer segment the company focuses on, and may have different expectations than the average customer.

- *The customer.* Evaluators selected to play the role of customers normally fit the socio-demographic or psychographic group profile for the customer segment(s) on which the company focuses. Banks and insurance companies as well as food and clothes retailers often use this type of mystery shopping. The main disadvantage of this approach is customers' lack of expertise. To offset it, companies usually organise preparatory training sessions.

All three approaches to mystery shopping are widely used in the airline sector for internal and external benchmarking. All of them are highly subjective. However, the level of subjectivity can be reduced to a certain extent by using as many evaluators as possible. Domino's Pizza, a US-based pizza company, used 800 customer-inspectors for each of their outlets over a period of 2 months, and then compared the results of their observations (Haller, 1998: 139). In this way, individual outliers levelled out and a more balanced picture emerged.

6.4.4 Value-based approach: analysis of willingness to pay

At the core of the willingness-to-pay concept lies a comparison between service benefits and the sacrifices necessary to acquire the service (such as cost, time consumption, physical or psychological effort). This trade-off is used as the basis for quality assessment. Given its nature, the concept is best applied ex ante (in the design phase).

There are three ways to measure service quality within the framework of this approach. In the first, each service benefit is assigned a utility U_i, which is then directly compared with the price p_i in the quotient $h_i = p_i/U_i$. This approach only makes sense if the customer is able to choose among several available quotients, looking for the smallest one in the process. A single isolated quotient is of no value.

In the second, price is viewed as one of the quality attributes. In the conjoint analysis framework, a series of two-attribute comparisons (one of them is always price) can then be conducted to discover existing customer preferences and assign them proper weights.

In the third, respondents are presented with a number of vignettes. Apart from asking them how they would evaluate a given combination of service attributes (see the discussion of the vignette method), they are also asked what is the highest price they would be prepared to pay for this particular combination of attributes. This information helps define customer tolerance limits for the cost–benefit (price/perceived quality) ratio.

The third possibility seems to be of the greatest practical value. Figure 6.4 shows the tolerance limits for a set of ten vignettes representing ten different hotels. Vignettes are evaluated by a sample of respondents. Perceived service quality is measured on a nine-point scale. The zero line represents the 'normal' price level for a 'normal' hotel. Positive quality deviations can be rewarded by a price up to 50 % higher. Negative quality deviations need to be compensated with a price up to 30 % lower, or the management faces the loss of customers.

6.5 The production and marketing approaches: Two sides of the same coin

Companies have traditionally managed services by manipulating engineering and operational attributes (internal quality measures) and observing market outcomes – customer satisfaction, customer behaviour and revenues (external quality measures). Given increased global competitiveness, this traditional approach needs a distinct spin. To constantly improve service quality, the external quality measures of the service process should be used as lead quality indicators. Bad external quality measures (such as low customer survey ratings and plummeting revenues) should

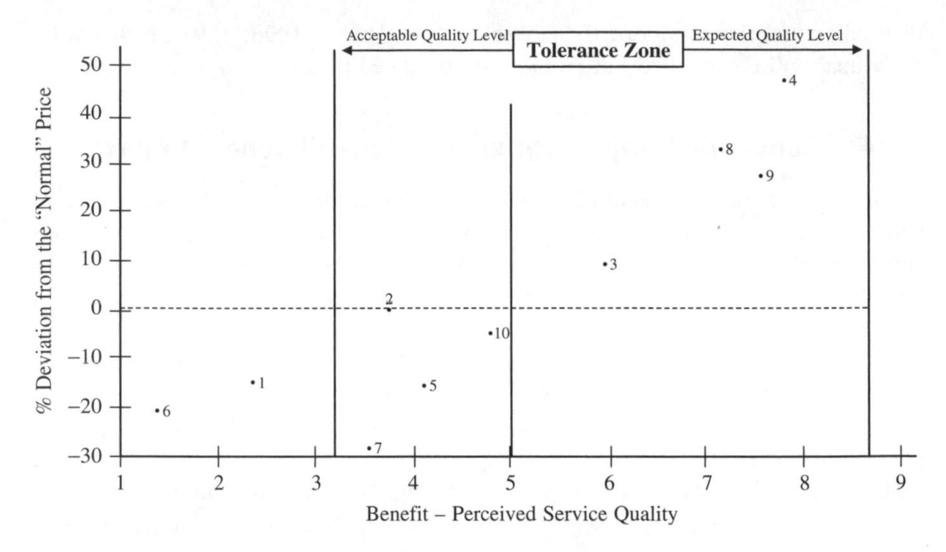

Figure 6.4 Graphical presentation of tolerance limits for a set of ten hotels.

be understood as clear signals that service processes need to be redesigned, and vice versa. For an illustrative example see Box 6.2.

Box 6.2 Combined use of production and marketing approaches to improve service quality.

The city of Madison, Wisconsin (USA), had serious problems with service garages for police cars, garbage trucks and other vehicles. A flowchart of the repair process and a Pareto chart were constructed from the data collected in a customer satisfaction survey. The Pareto chart showed that a major problem was the 'down time' – the time between the reporting of an instance of a defective vehicle and its delivery back to the customer. A two-month observational study of the process revealed that approximately 80 % of the down time was spent waiting (for parts, assembly, completion of paperwork). Further analysis showed that these problems were caused by factors such as lack of spare parts. This in turn was a consequence of the city's vehicle purchasing policy which involved buying at the lowest price. Therefore, the city's car park consisted of a large number of different types of vehicles, and different types of spare parts had to be kept in inventory or be ordered from outside. Another serious problem was the lack of a preventive maintenance policy, resulting in the transformation of minor problems into major and costly ones.

After the mechanics had flowcharted the repair process and performed a statistical analysis of data from the process, they presented their results to the Mayor. Upon evaluating the results, the Mayor changed the city's vehicle purchasing policy and the car park was standardised. The spare parts inventory was made more efficient, which reduced both waiting time and costs.

A preventive maintenance programme was also initiated. Shortly after implementation of these initiatives, significant returns on investment were observed. For example, an audit a year later showed that the average downtime had been reduced by 14 %, the number of vehicle breakdowns by 57 %, and maintenance costs by 23 %. In addition, the average cost of repair was reduced from $395 to $214.

Source: Bisgaard (2000: 300–301).

References

Bisgaard, S. (2000) The role of scientific method in quality management. *Total Quality Management*, **11**(3), 295–306.

Brandt, D.R. (1987) A procedure for identifying value-enhancing service components using customer satisfaction survey data. In C. Surprenant (ed.), *Add Value to Your Service: The Key to Success, 6th Annual Services Marketing Conference Proceedings*, pp. 61–64. American Marketing Association, Chicago.

Brandt, D.R. (1988) How service marketers can identify value-enhancing service elements. *Journal of Services Marketing*, **12**(3), 35–41.

Garvin, D.A. (1984) What does 'product quality' really mean? *Sloan Management Review*, **25**(Fall), 25–43.

Haller, S. (1998) *Beurteilung von Dienstleistungsqualität. Dynamische Betrachtung des Qualitätsurteils im Weiterbildungsbereich*. Deutscher Universitäts-Verlag and Gabler Verlag, Wiesbaden.

Ishikawa, K. (1985) *What is Total Quality Control? The Japanese Way*. Prentice Hall, Englewood Cliffs, NJ.

Kasper, H., van Helsdingen, P. and de Vries Jr., W. (1999) *Services Marketing Management. An International Perspective*. John Wiley & Sons, Ltd, Chichester.

Ograjenšek, I. and Bisgaard, S. (1999) Applying statistical tools to achieve service excellence. In *Collection of Papers of the International Conference on Quality Management and Economic Development (QMED)*, Portorož (Slovenia), 2–3 September.

Parasuraman, A., Berry, L.A. and Zeithaml, V.A. (1991a) Refinement and reassessment of the SERVQUAL scale. *Journal of Retailing*, **67**(4), 420–450.

Rossi, P.H and Anderson, A.B. (1982) The factorial survey approach: An introduction. In P.H. Rossi and S. Nock (eds), *Measuring Social Judgements: The Factorial Survey Approach*, pp. 15–67. Sage, Beverly Hills, CA.

Shostack, G.L. (1977) Breaking free from product marketing. *Journal of Marketing*, **41**(2), 73–80.

Further reading

Bateson, J.E.G. and Hoffman, K.D. (1999) *Managing Services Marketing*. Dryden Press, Fort Worth, TX.

Beauregard, M.R., Mikulak, R.J. and Olson, B.A. (1992) *A Practical Guide to Statistical Quality Improvement. Opening up the Statistical Toolbox*. Van Nostrand Reinhold, New York.

Brown, S.W. and Bond III, E.U. (1995) The internal market/external market framework and service quality: Toward theory in services marketing. *Journal of Marketing Management*, **11**(1–3), 25–39.

Carman, J.M. (1990) Consumer perceptions of service quality: An assessment of the SERVQUAL dimensions. *Journal of Retailing*, **66**(2), 33–55.

Cronin Jr., J.J. and Taylor, S.A. (1992) Measuring service quality: A reexamination and extension. *Journal of Marketing*, **56**(3), 55–68.

Cronin Jr., J.J. and Taylor, S.A. (1994) SERVPERF versus SERVQUAL: Reconciling performance-based and perceptions-minus-expectations measurement of service quality. *Journal of Marketing*, **58**(1), 125–131.

DeSarbo, W.S., Huff, L., Rolandelli, M.M. and Choi, J. (1994) On the measurement of perceived service quality. A conjoint analysis approach. In R.T. Rust and R.L. Oliver (eds), *Service Quality. New Directions in Theory and Practice*, pp. 201–222. Sage, Thousand Oaks, CA.

Dolan, R.J. (1990)*Conjoint analysis: A manager's guide. Harvard Business School Case Study 9-590-059*. Harvard Business School, Boston.

George, W.R. and Gibson, B.E. (1991) Blueprinting. A tool for managing quality in service. In S.W. Brown, E. Gummesson, B. Edvardsson and B. Gustavsson (eds), *Service Quality. Multidisciplinary and Multinational Perspectives*, pp. 73–91. Lexington Books, Lexington, MA.

Liljander, V. and Strandvik, T. (1992) Estimating zones of tolerance in perceived service quality and perceived service value. Working Paper No. 247, Swedish School of Economics and Business Administration, Helsinki.

Ograjenšek, I. (2002) Applying statistical tools to improve quality in the service sector. In A. Ferligoj and A. Mrvar (eds), *Developments in Social Science Methodology* (Metodološki zvezki 18), pp. 239–251. FDV, Ljubljana.

Ograjenšek, I. (2003) Use of customer data analysis in continuous quality improvement of service processes. In A. Mrvar (ed.), *Proceedings of the Seventh Young Statisticians Meeting* (Metodološki zvezki 21), pp. 51–69. FDV, Ljubljana.

Parasuraman, A., Zeithaml, V.A. and Berry, L.L. (1985) A conceptual model of service quality and its implications for future research. *Journal of Marketing*, **49**(Fall), 41–50.

Parasuraman, A., Zeithaml, V.A. and Berry, L.L. (1988) SERVQUAL: A multiple-item scale for measuring customer perceptions of service quality. *Journal of Retailing*, **64**(1), 12–40.

Parasuraman, A., Berry, L.L. and Zeithaml, V.A. (1991b) Understanding, measuring and improving service quality. Findings from a multiphase research program. In S.W. Brown, E. Gummesson, B. Edvardsson and B. Gustavsson (eds), *Service Quality. Multidisciplinary and Multinational Perspectives*, pp. 253–268. Lexington Books, Lexington, MA.

Parasuraman, A., Zeithaml, V.A. and Berry, L.L. (1994) Reassessment of expectations as a comparison standard in measuring service quality: Implications for further research. *Journal of Marketing*, **58**(1), 111–124.

Teas, R.K. (1994) Expectations as a comparison standard in measuring service quality: An assessment of a reassessment. *Journal of Marketing*, **58**(1), 132–139.

Teas, R.K. (1993) Consumer expectations and the measurement of perceived service quality. *Journal of Professional Services Marketing*, **8**(2), 33–54.

Teas, R.K. (1993a) Expectations, performance evaluation, and consumers' perceptions of quality. *Journal of Marketing*, **57**(4), 18–34.

Zeithaml, V.A. and Bitner, M.J. (1996) *Services Marketing*. McGraw-Hill, New York.

7

Design and analysis of industrial experiments

Timothy J Robinson

7.1 Introduction

To understand almost any system or process, purposeful changes are made to the independent variables and subsequent changes are noted on the dependent variable(s) of interest. The set of purposeful changes and the randomisation of those changes are what encompass the notion of the statistical design of experiments. George Box said: 'to see how a system functions when you have interfered with it, you have to interfere with it' (quoted in Ryan, 2007: 1). In this chapter, I shall briefly discuss the planning and running of industrial experiments and the associated analyses.

The process or system of interest to the industrial practitioner generally involves a combination of operations, machines, methods, people and other resources that transform some input into an output characterised by one or more *response* variables. Some of the inputs are *controllable factors* (denoted by $x_1, \ldots, x_k$) whereas others are *uncontrollable factors* (denoted by $z_1, \ldots, z_l$). Figure 7.1 helps to visualise a typical process.

The design of any experiment should take into consideration (1) the objectives of the experiment, (2) the extent to which extraneous factors (identifiable or non-identifiable) may influence the response, (3) the cost of the experiment and the sources of cost, (4) the number of factors being investigated, (5) the model with which the experimenter is interested in relating the response to the factors (both

Statistical Practice in Business and Industry Edited by S.Y. Coleman, T. Greenfield,
D.J. Stewardson and D.C. Montgomery

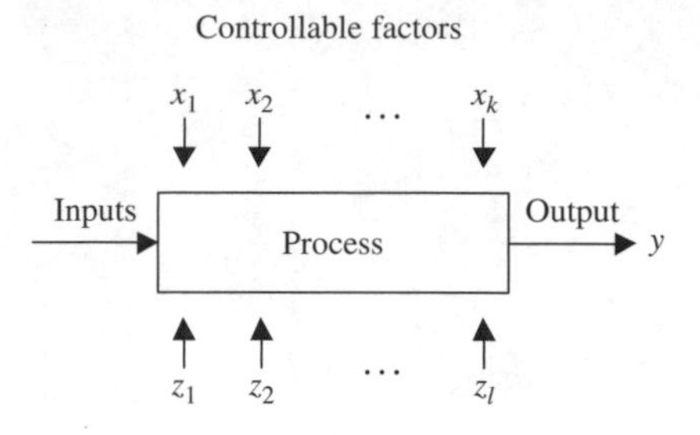

Figure 7.1 Schematic of industrial process.

controllable and uncontrollable), and (6) the extent to which sequential experimentation may be used.

To illustrate these points, consider an example involving a chemical process where we may be interested in the relationship between batch yield (y) and two controllable process variables, reaction time (x_1) and reaction temperature (x_2). The defined strategy of manipulating the factors is known as the *experimental design*. A specific combination of the levels of each factor is known as a *treatment combination*. In our example, we may use a 'high' or $+1$ level of x_1 and a 'low' or -1 level for x_2. Thus, ($x_1 = 1, x_2 = -1$) denotes one of the four possible treatment combinations in the design. If any of the factors is categorical in nature, an indicator or 'dummy' coding is used. For example, an important factor in this chemical process may be the presence or absence of a particular catalyst. Thus, a third factor, x_3, could be defined as $x_3 = 1$ if the catalyst is present and $x_3 = 0$ if it is not present. Each treatment combination that is carried out is known as an *experimental run*. Thus, the experimental *design* consists of a series of *experimental runs*. Generally the run order is assumed to be completely randomised. Randomisation should be used whenever possible so as to reduce the possibility of systematic effects on the response via extraneous factors. Certainly, the cost of the experiment may influence the randomisation scheme as certain factors may have levels that are difficult and/or costly to change. Experiments that involve these *hard to change* factors are often run using a split-plot randomisation scheme (to be discussed later).

Suppose the goal of the experimenter is to determine the factor combination (x_1, x_2) which results in the maximum batch yield, y. If the experimenter does not have a feel for the optimal locations of x_1 and x_2, it may be better to conduct a small initial experiment and then gradually move the design region to a more desirable location via sequential experimentation (to be discussed later). Before determining an appropriate experimental design, the experimenter needs to select a reasonable model for relating the response to the two independent factors. In general, the relationship between the response (y) and the independent design variables can be written as

$$y = f(\xi_1, \xi_2) + \varepsilon,$$

where the functional form of f is typically unknown, ε reflects the random error in our response, and ξ_1 and ξ_2 denote reaction time and reaction temperature in their natural units (such as minutes and degrees Celsius). It is often convenient to transform the natural variables to coded variables, x_1 and x_2, which are dimensionless and have mean zero. In this discussion, we will refer to the coded variable notation. Although f is unknown, f can generally be well approximated within a small region of x_1 and x_2 by a lower-order polynomial. In many cases, either a first-order or second-order model is used. When two design variables are used, the first-order model (often known as the *main effects* model) can be written as

$$y = \beta_0 + \beta_1 x_1 + \beta_2 x_2 + \varepsilon. \tag{7.1}$$

If interaction exists between the factors, the term $\beta_{12}x_1x_2$ can be added to the model above. Figures 7.2 and 7.3 show the main effects model ($\beta_0 = 50$, $\beta_1 = 8$,

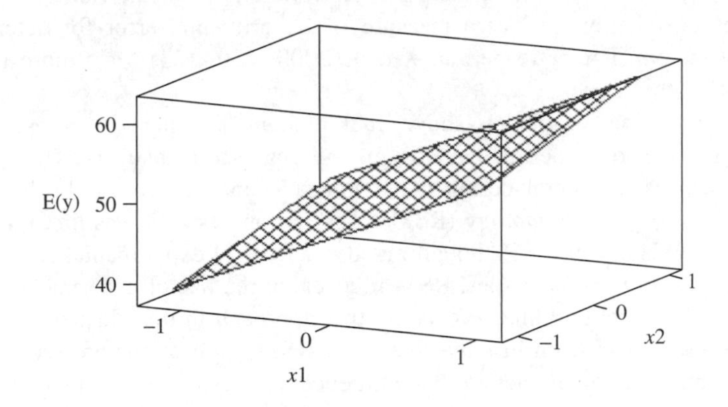

Figure 7.2 Plot of response surface for mean function $E(y) = 50 + 8x_1 + 3x_2$.

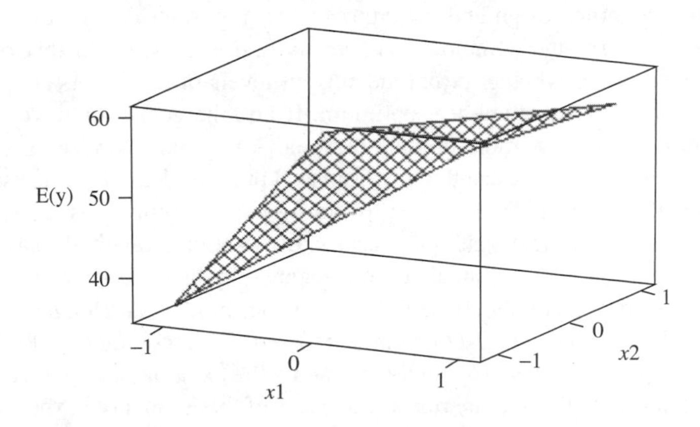

Figure 7.3 Plot of response surface for mean function $E(y) = 50 + 8x_1 + 3x_2 - 4x_1x_2$.

and $\beta_2 = 3$) and the main effects with interaction ($\beta_{12} = -4.0$) model, respectively. Notice that with the addition of an interaction term, the model allows for a twist in the response surface but no curvature is allowed for. In many situations, the curvature in the response is stronger than the twist which can occur in a first-order model with interaction so pure quadratic terms are required ($\beta_{11}x_1^2$ and/or $\beta_{22}x_2^2$). I shall describe these second-order models in more detail later.

Each unit to which a treatment combination is applied is known as an *experimental unit*, and each unit upon which we make a measurement is known as an *observational unit*. Experimental error variance reflects the variability among the experimental units. In the present example, the experimental unit and the observational unit are identical in that they are both a batch in the process. In many experiments, the observational unit may not be equivalent to the experimental unit. For instance, the observational unit may be a sample from the experimental unit. Often the variance among multiple observations from the same experimental unit is mistakenly used as a measure of experimental error for determining important factors. I refer readers to Kuehl (2000: 159–163) for a more in-depth discussion of this issue.

Upon observing the models above, it is evident that there is a close relation between the use of experimental design and regression analysis. The relationship between experimental design and regression analysis forms the foundation of *response surface methodology* (RSM). The notion of RSM was first introduced by Box and Wilson (1951). Although much of classical experimental design seeks to look at treatment comparisons, RSM focuses on the use of experimental design for prediction model building. I focus on this approach in this chapter.

It is important to keep in mind that most RSM applications are sequential. If there are many potential factors that influence the response, the first priority is to simplify the model by screening out the unimportant terms. The initial screening experiment is often referred to as *phase zero* of the study. After choosing an appropriate screening design and determining the important factors, the researcher progresses to *phase one*, commonly known as *region seeking*. In this phase, the researcher determines whether or not the current levels of the factors are producing values of the response that are near optimum. If not, the goal is to move across the response surface to other, more optimal regions. A popular optimisation algorithm known as steepest ascent/descent, to be discussed in more detail later, is used in this phase. Once the experimenter is close to the region of optimal response, the goal becomes to model the response more accurately within a relatively small region around the optimum. The optimal response generally occurs where the response surface exhibits curvature. As a result, the first-order model is no longer appropriate and a second- or higher-order polynomial is used. To accommodate higher-order models, the experimenter appropriately augments the existing design. Over the next several sections, I shall elaborate on various parts of this sequential experimentation process.

7.2 Two-level factorial designs

Most response surface studies begin with some form of a two-level *factorial* design. If we have k factors, the full two-level factorial design consists of every combination of the two levels for the k factors, and hence 2^k treatment combinations. As a result, these designs are often called 2^k *factorial designs*. These designs are particularly useful in the following areas: (1) phase zero of the study when the emphasis is on screening of factors; (2) phase one in which the experimenter uses the 2^k design to fit the first-order model used in steepest ascent/descent; and (3) as the building blocks for creating other designs when the experimenter is attempting to fit higher-order polynomials within the optimal design region.

The simplest design in the 2^k family is the 2^2 factorial design in which there are only two factors, each run at two levels. For example, Rao and Saxena (1993) studied the effect of moisture (A) and furnace temperature (B) on the composition of flue gases when pinewood is burned. The researchers studied moisture contents of 0 % and 22.2 % and furnace temperatures of 1100 K and 1500 K. The experiment was replicated twice and the data are shown in Table 7.1. The low and high levels of A and B are denoted '−' and '+' respectively. The four treatment combinations are represented by lower-case letters. More explicitly, a represents the combination of factor levels with A at the high level and B at the low level, b represents A at the low level and B at the high level and ab represents both factors being run at the high level. By convention, '(1)' is used to denote A and B each run at the low level. A geometrical representation of the design region is shown in Figure 7.4.

Table 7.1 Twice replicated 2^2 factorial design.

			Concentration of CO^2 in flue gas	
Treatment combination	A	B	Replicate 1	Replicate 2
(1)	−	−	20.3	20.4
a	+	−	13.6	14.8
b	−	+	15.0	15.1
ab	+	+	9.7	10.7

In two-level factorial designs we can estimate two types of effects: main effects and interaction effects between factors. A *main effect* is defined as the difference in the average response at the high level of a factor and the low level of a factor. If this effect is positive, the response increases as we go from the low level of the factor to the high level. Thus, in general for the 2^2 factorial designs, the main

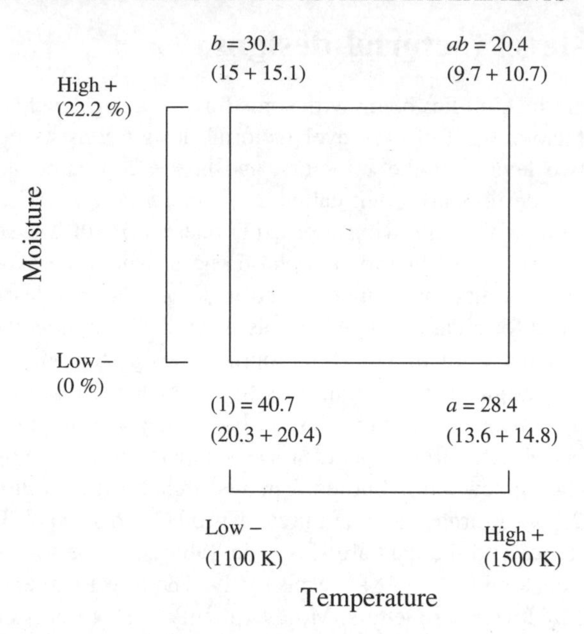

Figure 7.4 Geometric layout of 2^2 *design space.*

effects for factors *A* and *B* are defined as:

$$\begin{aligned} A &= \overline{y}_{A+} - \overline{y}_{A-} = \frac{1}{2n}[ab + a - b - (1)], \\ B &= \overline{y}_{B+} - \overline{y}_{B-} = \frac{1}{2n}[ab + b - a - (1)]. \end{aligned} \tag{7.2}$$

The interaction effect, *AB*, is the difference in the average of the right to left diagonal design points in Figure 7.4 (*ab* and (1)) minus the left to right diagonal design points (*b* and *a*), or

$$AB = \frac{1}{2n}[ab + (1) - a - b]. \tag{7.3}$$

For the pinewood data, the effects are $A = -5.5$, $B = -4.65$ and $AB = 0.65$. From equations (7.2) and (7.3) it is evident that each effect in a 2^2 design is a function of a *contrast* of design points. Each contrast represents a single degree of freedom. The sum of squares for any contrast is equal to the contrast squared divided by the number of observations making up each contrast. Thus we can write the sum of squares in the 2^2 design as

$$SS_A = \frac{[ab + a - b - (1)]^2}{4n},$$

$$SS_B = \frac{[ab + b - a - (1)]^2}{4n},$$

$$SS_{AB} = \frac{[ab + (1) - a - b]^2}{4n}.$$

In all 2^k factorial designs, the sums of squares are independent and thus additive due to the orthogonality of the effects.

A process model in RSM can be used to predict the response at any point in the design space. Considering the pinewood example, the first-order regression model with interaction is written

$$y = \beta_0 + \beta_1 x_1 + \beta_2 x_2 + \beta_{12} x_1 x_2 + \varepsilon.$$

A slope is simply the change in y divided by the change in x. As a result, the slope associated with X_1 (or factor A) is given by

$$\beta_1 = \frac{\overline{y}_{A+} - \overline{y}_{A-}}{2} = \frac{\text{Main effect for} A}{2}.$$

In general, any slope parameter in the 2^2 design is just the main effect of the particular factor divided by 2. The intercept, β_0, is the grand mean of the responses. For the pinewood example, the fitted regression model is then given by

$$\begin{aligned}\hat{y} &= 14.95 + \left(\frac{-5.5}{2}\right) x_1 + \left(\frac{-4.65}{2}\right) x_2 + \frac{0.65}{2} x_1 x_2 \\ &= 14.95 - 2.75 x_1 - 2.325 x_2 + 0.325 x_1 x_2.\end{aligned}$$

Typically the fitted regression model is found through ordinary least squares.

When there are three factors (A, B, and C), each at two levels, a 2^3 factorial design is appropriate. The design region is now represented by a cube in which each of the eight treatment combinations represents a vertex of the cube. Extending the notation discussed for 2^2 designs, the treatment combinations in standard order are (1), *a*, *b*, *ab*, *c*, *ac*, *bc*, and *abc*. The geometric representation of the 2^3 design is shown in Figure 7.5. There are seven degrees of freedom for the eight treatment combinations, of which three are for the main effects (A, B, C), three are for the two-factor interactions (*AB, AC, BC*) and one is for the three-factor interaction (*ABC*).

The methods of analysis and design construction discussed so far naturally extend to 2^k designs for $k > 2$. Clearly, as k becomes large, the number of treatment combinations increases very quickly. For instance, if $k = 6$ then 64 treatment combinations are needed for one replicate of a 2^6 design. As is often the case, the resources available for experimentation are not enough for more than one replicate of a full factorial design, especially when there are many factors. When the full regression model is fitted (all main effects and all possible interactions), there are no degrees of freedom available for variance estimation. In this scenario there are

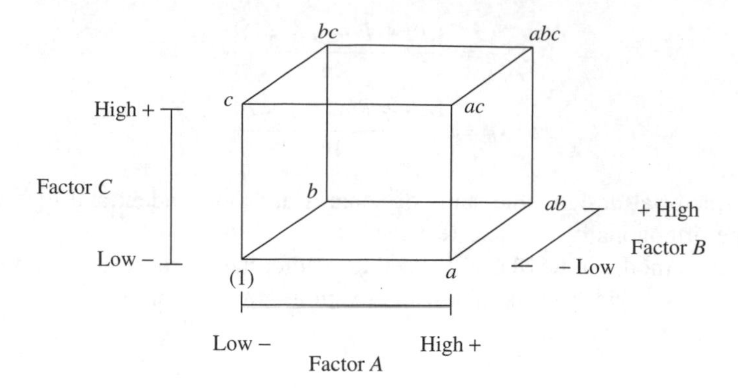

Figure 7.5 Geometric layout of 2^3 design space.

various approaches that we can take in terms of the analysis. One approach is to appeal to the *sparsity-of-effects principle* which proposes that most processes are dominated by main effects and lower-order interactions. Most high-order interactions are assumed to be negligible. Assuming negligible higher-order interactions, we can ignore these effects and use an error mean square that is obtained from pooling these higher-order interactions. Occasionally, however, higher-order interactions do occur and pooling is inappropriate. Daniel (1959) suggests the use of a normal probability plot for determining important effects. Negligible effects are normally distributed with mean zero and common variance σ^2 and will thus tend to fall along a straight line, whereas the important effects will have non-zero means and thus will not lie on a straight line.

7.3 Centre runs

The problem with unreplicated 2^k designs is that they offer no degrees of freedom with which to estimate experimental error variance. A method of replication that not only provides an independent estimate of error but also protects against curvature in the response surface is the use of *centre points*. Assuming the '+'/'−' coding of factor levels, centre points consist of n_c replicates run at the design point $x_i = 0$ $(i = 1, 2, \ldots, k)$. An advantage of adding replicate runs at the centre of the design space is that the centre points do not impact the usual effect estimates in the 2^k design.

For illustration, consider a 2^2 design with one observation at each of the four design points [(−1, −1), (−1, 1), (1, −1) and (1, 1)] and n_c observations at the centre point, (0, 0). The design and corresponding observations are in Figure 7.6. Experimental error can be estimated with $n_c - 1$ degrees of freedom and curvature can be determined by a single-degree-of-freedom contrast, which compares the average response at the four factorial points ($\overline{y}_F$) to the average response at the centre point ($\overline{y}_C$).

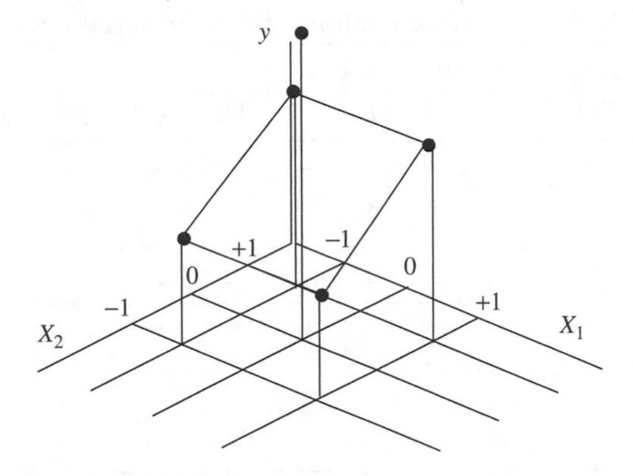

Figure 7.6 2^2 *design with centre runs.*

7.4 Blocking in 2^k factorial designs

As mentioned earlier, as k increases, the number of treatment combinations needed for the full factorial design gets large and it may be impossible to run all combinations under homogeneous experimental conditions. When this is a problem, *blocking* can help. Blocks are homogeneous groups of experimental units. For instance, blocks may be raw materials, plants, operators, or time. Consider a chemical process in which the engineer is interested in optimizing batch assay. Assay is thought to be a function of extent of reaction (A), wash flow rate (B), and wash contact time (C). Suppose that only eight batches can be run from a single hopper of raw material. If we are to replicate a 2^3 design then the logical thing to do would be to run each replicate of the design with a separate hopper of raw material. The runs within each block (hopper) would be made in random order.

It is often the case that the number of treatment combinations required for a single replicate of the full factorial is greater than the block size. For instance, consider the previous example and suppose that instead of the hopper producing eight batches, only four batches can be produced. *Confounding* is the design technique of arranging a complete factorial design in blocks where the block size is smaller than the number of total treatment combinations. The technique purposefully causes information about certain effects (usually higher-order interactions) to be indistinguishable from, or *confounded* with, blocks.

To illustrate, first consider the contrasts for a 2^3 design as given in Table 7.2. If only four batches can be run within each hopper of raw material, we must decide which treatment combinations should be run with each block. Figure 7.7 shows one possibility of assignment, where all runs with ABC at the high level (+) are in one block and all runs at the low level (−) are in the second block. Recall that an effect is estimated by taking the average response at the high level of the factor and subtracting the average response at the low level. Thus, to estimate an effect within a block, we need the effect to have both plus and minus treatment combinations

Table 7.2 Table of contrasts for 2^3 factorial design.

Treatment combination	Factorial effect							
	I	A	B	C	AB	AC	BC	ABC
a	+	+	−	−	−	−	+	+
b	+	−	+	−	−	+	−	+
c	+	−	−	+	+	−	−	+
abc	+	+	+	+	+	+	+	+
ab	+	+	+	−	+	−	−	−
ac	+	+	−	+	−	+	−	−
bc	+	−	+	+	−	−	+	−
(1)	+	−	−	−	+	+	+	−

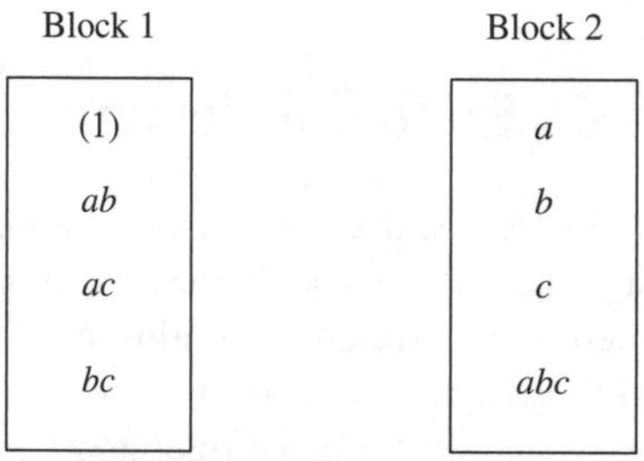

Block 1	Block 2
(1)	a
ab	b
ac	c
bc	abc

Figure 7.7 A 2^3 design in two blocks of size 4 with ABC confounded with blocks.

within the given block. Specifically, notice that in block 1, A has two treatment combinations at the high level (ab and ac) and two combinations at the low level [(1) and bc]. This is true about all effects except for ABC, where all high values of ABC are in block 1 and all low values of ABC are in block 2. Thus, we say that ABC is confounded with blocks since ABC cannot be estimated separately from a block effect.

7.5 Fractional factorial designs

To appreciate the importance of fractional factorial designs, recall the sparsity-of-effects principle that states that most processes are driven by main effects and two-factor interactions. Now consider a 2^6 design that requires 64 treatment combinations to be run. Of the 63 degrees of freedom, only six are used to estimate main effects and 15 are used to estimate two-factor interactions. The remaining 42 degrees of freedom are used to estimate three-factor and higher-order interactions. As k gets larger, more and more of these interactions may not be important. If the experimenter is willing to assume that certain higher-order interactions are negligible, then information about main effects and two-factor interactions can be obtained by running only a fraction of the full factorial experiment. These *fractional factorial* designs are widely used in industry. Fractional factorials are particularly useful in screening experiments when the purpose is to take a large number of factors and identify those factors with the largest effects. Once these important factors are identified, more detailed experiments can be done in the important factors.

7.5.1 The one-half fraction of the 2^k design

Recall the table of contrasts given for the 2^3 design in Table 7.2. Suppose that resources prohibit experimenters from running all eight treatment combinations and instead a design with only four runs is decided upon. Since only one-half of the treatment combinations are to be run, we refer to the design as a one-half fraction of the 2^3 design (denoted 2^{3-1}). Under the sparsity-of-effects principle, the one effect that is least likely to be important is the ABC interaction. Similar to the ideas discussed in blocking, the experimenter may decide to sacrifice his/her ability to estimate the ABC interaction and thus choose to run only those treatment combinations that have a plus sign in the ABC column (treatment combinations a, b, c, and abc). ABC is called the *generator* of this particular fraction. Often, the generator is referred to as a *word*. Since the identity column, I, is also always plus, we call $I = ABC$ the *defining relation* for the design. We say that ABC is *aliased* with the intercept since, by choosing ABC to generate the design, we have lost our ability to estimate it. In general, the defining relation is always the set of columns which are equal to the identity column I.

Since we are running only four of the eight possible treatment combinations, we have only three degrees of freedom with which to estimate main effects and two-factor interactions. Referring to Table 7.2, note that the contrast which is used to estimate the main effect A is given by

$$l_A = \frac{1}{2}(a - b - c + abc).$$

It can be shown that the contrast to estimate the BC interaction is also given by

$$l_{BC} = \frac{1}{2}(a - b - c + abc).$$

Thus, we have that $l_A = l_{BC}$ and we say that A is aliased with BC since it is impossible to distinguish between effects A and BC. It is easy to show that, for this design, $l_B = l_{AC}$ and $l_C = l_{AB}$. When estimating A, B, and C we are really estimating $A + BC$, $B + AC$, and $C + AB$, respectively. This is indicated by the notation $l_A = A + BC$, $l_B = B + AC$, and $l_C = C + AB$. To determine the aliases for a given effect we need only multiply the effect by the defining relation modulo 2. For example, in the present design, the aliases of A are found by

$$A * I = A * ABC = A^2BC = BC,$$

and thus $A = BC$. Software packages often summarise the alias structure in tabular form as:

$I = ABC$
$A = BC$
$B = AC$
$C = AB$

The alias structure in this form tells us the largest model we can estimate from the experiment. In this case, we can estimate a term associated with the intercept (I), the main effect for A, the main effect for B and the main effect for C.

The one-half fraction chosen with $I = +ABC$ is generally referred to as the *principal* fraction. If we selected the other one-half fraction, the fraction consisting of the treatment combinations with minus signs in the ABC column, we would have the *alternative* fraction, denoted $I = -ABC$. Here the alias structure would be found the same way, namely

$$A * I = A(-ABC) = -A^2BC = -BC$$

and, when estimating A, we are really estimating $A - BC$. In practice it does not matter which fraction is chosen (plus or minus), since both fractions together form a full 2^3 design. The implication is that if we at first decide to run the positive fraction and then subsequently run the negative fraction, we will obtain de-aliased estimates of all of the effects except the ABC interaction, which will be confounded with blocks.

7.5.2 Design resolution and constructing fractional factorials

One-half fractions readily extend to cases in which we have more than three factors. To determine the fractional factorial with the highest *resolution*, the highest-order interaction is used for the defining relation. For instance, if we have a 2^5 design with factors A, B, C, D, and E, then the defining relation is $I = ABCDE$ and we determine the alias structure through modulo 2 arithmetic. In this case we have that all main effects are aliased with four-factor interactions and all two-factor interactions are aliased with three-factor interactions. This 2^{5-1} design is referred to as a resolution V design. Resolution V designs are designs in which no main effect or two-factor interaction is aliased with any other main effect or two-factor interactions, but two-factor interactions are aliased with three-factor interactions. The 2^{5-1} resolution V design is denoted 2_{V}^{5-1}. Resolution IV designs are those designs in which no main effect is aliased with any other main effect or two-factor interactions but two-factor interactions are aliased with other two-factor interactions. Resolution III designs are those designs in which no main effect is aliased with another main effect but main effects are aliased with two-factor interactions. As we will see later, resolution III designs are useful in screening experiments. The higher the resolution, the less restrictive one needs to be regarding which interactions are negligible. Resolution is defined as the number of letters in the smallest word in the defining relation.

When there are five or more factors, it is often useful to construct smaller fractions of the 2^k design than the one-half fraction. For instance, if we have five factors, we may have enough resources to run an experiment with only eight runs. Thus, we would be interested in a 2^{5-2}, or a quarter fraction of the 2^5 factorial design. The quarter fraction, or 2^{k-2} design, is constructed by first writing down the runs for a full 2^{k-2} design and then associating the two additional columns with generators that are appropriately chosen interactions involving the first $k-2$ factors. Thus, a 2^{k-2} design involves two generators, $I = P$ and $I = Q$, which are referred to as the *generating relations*. The signs of P and Q (+ or −) determine which one of the quarter fractions is produced. All four fractions of the generators $\pm P$ and $\pm Q$ are members of the same family and the fraction involving all treatment combinations where P and Q are '+' is known as the *principal fraction*. Just as in the one-half fraction case, the *complete defining relation* for the design consists of all columns that are equal to the identity column. These columns will consist of P, Q, and the interaction PQ. For example, for the 2^{5-2} design if we select $ABDE$ and $ABCD$ as the *principal fractions*, then $ABDE(ABCD)$ mod $2 = A^2B^2CD^2E = CE$ is the generalised interaction. The *complete defining relation* is given by $I = ABDE = ABCD = CE$.

In determining the principal fractions, the goal is to do so in such a way as to have the highest possible design resolution. In the present case, our smallest word is CE and thus our design only has resolution II, implying that some main effects are aliased with other main effects. The optimal resolution for a 2^{5-2} design is resolution III and could be found for instance by using the generating relation $I = ABD$ and $I = ACE$ with generalised interaction $BCDE$. To find the aliases of any effect (such as A), multiply the effect (mod 2) by each word in the defining relation. For A this produces

$$A = BD = CE = BDE.$$

It is easy to show that no main effects are aliased with each other and that all two-factor interactions are aliased with other two-factor interactions. Thus, the defining relation $I = ABD = ACE = ABDE$ produces a resolution III design. Myers and Montgomery (2002: 179–180) provide a chart that lists generating relations for optimal resolution for designs involving up to 11 factors.

7.5.3 Resolution III designs

The sequential experimentation process often begins with an initial experiment that is a resolution III design in which the goal is to screen for important variables. These designs are cheap (require few runs) and often enable the practitioner to investigate up to $k = N - 1$ factors in only N runs, where N is a multiple of 4. Designs in which N is a power of 2 can be constructed by the methods previously discussed. Of particular importance are those designs requiring 4 runs for up to 3 factors (2^{3-1}_{III}), 8 runs for up to 7 factors (2^{7-4}_{III}), and 16 runs for up to 15 factors

(2_{III}^{15-11}). Designs in which the number of factors, k, is equal to the number of runs minus one ($N - 1$) are said to be *saturated*. Saturated designs are those in which there are no degrees of freedom with which to estimate experimental error. As an example, consider an eight-run design for studying seven factors, the 2_{III}^{7-4} design. The design is constructed by writing down the full design in A, B, and C and then associating the levels of the four additional factors with the interactions of the original three as follows: $D = AB$, $E = AC$, $F = BC$, and $G = ABC$. The design is given in Table 7.3. Notice that the seven degrees of freedom in this design may be used to estimate the seven main effects. Each of these effects will have 15 aliases and, assuming that the three-factor and higher interactions are negligible, the simplified alias structure is given in Table 7.5.

Table 7.3 The saturated 2_{III}^{7-4} design with generators $I = ABD$, $I = ACE$, $I = BCF$, and $I = ABCG$.

Treatment combination	Effects						
	A	B	C	$D = AB$	$E = AC$	$F = BC$	$G = ABC$
def	−	−	−	+	+	+	−
afg	+	−	−	−	−	+	+
beg	−	+	−	−	+	−	+
abd	+	+	−	+	−	−	−
cdg	−	−	+	+	−	−	+
ace	+	−	+	−	+	−	−
bcf	−	+	+	−	−	+	−
abcdefg	+	+	+	+	+	+	+

It is often of interest to estimate main effects free from any two-factor interactions. For instance, suppose that the design in Table 7.3 is run and, based on a normal probability plot, factors A, B, and D appear significant. However, since D is confounded with the AB interaction, it would also be correct to conclude that A, B, and AB are the important effects. To resolve this issue, the practitioner can systematically isolate certain effects by combining fractional factorial designs in which certain signs are switched. This procedure is known as *fold-over* and it is used in resolution III designs to break the links between main effects and two-factor interactions. In a *full fold-over*, we add to a resolution III fractional a second fraction in which the signs for all the factors are reversed. Table 7.4 is a fold-over of the design in Table 7.3. The combined design (16 runs) can then be used to estimate all main effects free of any two-factor interactions. Notice that when we combine the effect estimates from this second fraction with the effect estimates from the original fraction we find that the main effects have indeed been isolated from each other. For a more complete discussion of the fold-over strategy, see Montgomery (2005: 314–318).

We have just discussed resolution III designs for studying up to $k = N - 1$ variables in N runs, where N is a power of 2. Plackett and Burman (1946) propose

Table 7.4 Fold-over of the 2^{7-4}_{III} design in Table 7.3.

Treatment combination	Effects						
	A	B	C	$D = -AB$	$E = -AC$	$F = -BC$	$G = ABC$
def	+	+	+	−	−	−	+
afg	−	+	+	+	+	−	−
beg	+	−	+	+	−	+	−
abd	−	−	+	−	+	+	+
cdg	+	+	−	−	+	+	−
ace	−	+	−	+	−	+	+
bcf	+	−	−	+	+	−	+
abcdefg	−	−	−	−	−	−	−

Table 7.5 Alias structures for the designs in Tables 7.3 and 7.4.

Design from Table 7.3	Design from Table 7.4
$l_A \rightarrow A + BD + CE + FG$	$l'_A \rightarrow A - BD - CE - FG$
$l_B \rightarrow B + AD + CF + EG$	$l'_B \rightarrow B - AD - CF - EG$
$l_C \rightarrow C + AE + BF + DG$	$l'_C \rightarrow C - AE - BF - DG$
$l_D \rightarrow D + AB + CG + EF$	$l'_D \rightarrow D - AB - CG - EF$
$l_E \rightarrow E + AC + BG + DF$	$l'_E \rightarrow E - AC - BG - DF$
$l_F \rightarrow F + BC + AG + DE$	$l'_F \rightarrow F - BC - AG - DE$
$l_G \rightarrow G + CD + BE + AF$	$l'_G \rightarrow G - CD - BE - AF$

a family of resolution III fractional factorial designs where N is a multiple of 4. When N is a power of 2, these designs are identical to those discussed above. Plackett-Burman designs are of particular interest for $N = 12$, 20, 24, 28, and 36. For more information on the construction of Plackett–Burman designs along with an example, see Myers and Montgomery (2002: 190–192).

7.6 Process improvement with steepest ascent

After the initial screening experiment, interest focuses on estimating a model in the important factors. This model can then be used to construct a *path of steepest ascent* (or *descent*, depending on whether the goal is maximization or minimization). Series of experimental runs are conducted along this path until the response begins to deteriorate. At this point, another two-level fractional factorial design (generally of resolution IV) is conducted and another first-order model is fitted. A test for lack of fit is done and if the lack of fit is not significant, a second path, based on the new model, is computed. Experimental runs are conducted along this new path until process improvement begins to deteriorate again. At this point, we are likely to be quite close to the point of optimal operating conditions and a more detailed experiment is conducted. It should be noted that it is quite possible that only one path of ascent will be used as it is often the case that there are significant interactions or quadratic lack-of-fit contributions at the second stage. The path of steepest ascent is based on a first-order model and thus should not be used when curvature begins to play an important role. For a more detailed treatment of steepest ascent/descent, see Myers and Montgomery (2002: 203–228).

7.6.1 Second-order models and optimisation

The response surface will generally display curvature in the neighbourhood of process optimal settings. Although the true functional relationship between the response and the factors is unknown, a good approximation to this relationship when curvature is present is a second-order Taylor series approximation of the form

$$\begin{aligned} y = \beta_0 + \beta_1 x_1 + \beta_2 x_2 + \cdots + \beta_k x_k + \beta_{11} x_1^2 + \cdots + \beta_{kk} x_k^2 \\ + \beta_{12} x_1 x_2 + \beta_{13} x_1 x_3 + \cdots + \beta_{k-1,k} x_{k-1} x_k + \varepsilon. \end{aligned}$$

The second-order model is useful in determining optimal operating conditions and is easily accommodated via the use of a wide variety of experimental designs. To estimate a second-order model, the design must have at least as many distinct treatment combinations as terms to estimate in the model and each factor must have at least three levels.

The most commonly used design to estimate a second-order model is the *central composite design* (CCD). The two-factor CCD is given in Table 7.6. The CCD is comprised of three components: a full 2^k factorial, a one-factor-at-a-time array of *axial* points, and a set of centre runs. The axial points lie on the axes defined by the design variables, and the specific choice of α depends on the researcher but

Table 7.6 Central composite design for two factors.

X_1	X_2
−1	−1
−1	1
1	−1
1	1
$-\alpha$	0
α	0
0	$-\alpha$
0	α
0	0

is generally taken to be one of three possible choices: $\alpha = 1$ for a face-centred cuboidal design regions, $\alpha = \sqrt{k}$ for spherical design regions, or $\alpha = n_f^{0.25}$ for a rotatable CCD where n_f denotes the number of factorial runs used in the design. A face-centred cube has all of the treatment combinations except the centre runs on the surface of a cube. Likewise, a spherical CCD has all treatment combinations except the centre runs on the surface of a sphere. The rotatable CCD produces prediction variances that are functions only of the experimental error, σ^2, and the distance from the centre of the design. The reader may find a more detailed discussion of the various central composite designs and rotatability in Myers and Montgomery (2002: 321–342).

Another popular second-order design is the Box–Behnken design (BBD) proposed by Box and Behnken (1960). BBDs represent a family of three-level designs that are highly efficient for estimating second-order response surfaces. The formulation of these designs is accomplished by combining two-level factorial designs with incomplete block designs in a specific manner. Although many BBDs can be constructed by combining factorial designs with balanced incomplete block designs, sometimes the construction must involve a partially balanced incomplete block design instead. In the construction this implies that not every factor will occur in a two-level factorial structure the same number of times with every other factor. The CCD and BBD are also quite comparable in size for larger numbers of factors. Borkowski (1995a, 1995b) develops prediction variance properties of the CCD and BBD and then uses variance dispersion plots (see Giovannitti-Jensen and Myers, 1989) to compare the two designs in terms of minimum, maximum and average spherical prediction variances. Borkowski notes that the two designs possess good prediction variance properties, especially when a sufficient number of centre runs are used. For a more complete discussion on BBDs, see Robinson (2007).

7.7 Optimal designs and computer-generated designs

Standard industrial designs such as the CCD and BBD, along with their variations, are widely used in practice because of the flexibility afforded by them and their appealing statistical properties. Generally, if the design region is spherical or cuboidal, standard designs are applicable. There are situations, however, when a problem calls for the use of a design which is non-standard. Instead of submitting the experimental conditions to a standard design format, one is better off using a design that is tailored to the experimental situation. Tailor-made designs are often referred to as *computer-generated designs*. Montgomery (2005: 439–441) outlines specific situations which call for the use of a computer-generated experiment:

1. *An irregular experimental region*. If the region of interest is not spherical or cuboidal due to constraints on the ranges of the factors, the region may be *irregular*. For instance, certain combinations of the experimental factors may be prohibitive to run.

2. *A non-standard model.* Generally the experimenter will use a first- or second-order model in the experimental factors. However, if there is a highly non-linear relationship between the response and factors, some sort of higher-order polynomial or even non-linear regression model may fit better than a lower-order polynomial. In these situations, there are no standard designs.

3. *Unusual sample size requirements.* In some situations, the user may wish to use a design such as the CCD but resources are limited and not all of the runs required by a CCD can be used. As such, a computer-generated design would be called for. Popular software packages for generating designs include SAS JMP, Design-Expert, MINITAB, and the SAS procedure PROC OPTEX.

The genesis of computer-generated designs can be attributed to work by Kiefer (1959, 1961) and Kiefer and Wolfowitz (1959). Computer-generated designs rely on the user providing an objective function which reflects the interests of the experiment, the design region of interest, a reasonable model which relates the response to the set of design factors, the number of experimental runs, and information as to whether or not blocking is needed. Using this information, the computer uses a search algorithm and a candidate set of design points (generally a fine grid of the design region) to select an appropriate design. It is important to note that the quality of the design selected is heavily dependent upon the quality of the user's model. Heredia-Langner *et al.* (2003) consider the use of genetic algorithms for selecting optimal designs when the user is interested in protecting against model misspecification. Good references on the basics of using genetic algorithms as a means of computer-generated designs include Borkowski (2003) and Heredia-Langner *et al.* (2004).

Regarding objective functions that summarise the statistical property of interest for a given experimental design, there are several which are widely used and available in software packages. These optimality criteria are referred to as *alphabetic optimality criteria*, and they are best motivated by considering the regression model which is often specified for describing phenomena in industrial experiments. Recall that most industrial experiments assume the following regression model for relating the response to the set of design factors:

$$\mathbf{y} = \mathbf{X}\boldsymbol{\beta} + \boldsymbol{\varepsilon},$$

where $\mathbf{X}$ is the model matrix, $\boldsymbol{\beta}$ is the vector of coefficients expanded to model form, and $\boldsymbol{\varepsilon}$ denotes the $N \times 1$ set of model errors assumed to be NID($0,\sigma^2$). The vector of estimated regression coefficients is obtained via least squares and is given by

$$\hat{\boldsymbol{\beta}}_{\text{ols}} = (\mathbf{X}'\mathbf{X})^{-1}\mathbf{X}'\mathbf{y}.$$

The covariance matrix of the estimated model coefficients is given by $\text{Var}(\hat{\boldsymbol{\beta}}_{\text{ols}}) = (\mathbf{X}'\mathbf{X})^{-1}\sigma^2$. The predicted value of the mean response at any location is given by

$$\hat{y}_0 = \mathbf{x}_0'\hat{\boldsymbol{\beta}}_{\text{ols}} = \mathbf{x}_0'(\mathbf{X}'\mathbf{X})^{-1}\mathbf{X}'\mathbf{y},$$

where $\mathbf{x}_0$ is the point of interest in the design space expanded to model form. The prediction variance at $\mathbf{x}_0$ is then given by

$$\mathrm{Var}(\hat{\mathbf{y}}_0) = \mathbf{x}_0'(\mathbf{X}'\mathbf{X})^{-1}\mathbf{x}_0\sigma^2.$$

Objective functions for determining the optimal design involve working with the determinant of the variance–covariance matrix of the parameter estimates (D-optimality), the sum of the variances of the estimated regression coefficients (A-optimality), the integrated prediction variance throughout the design region (IV-optimality), and maximum prediction variance throughout the design region (G-optimality).

The D-criterion has traditionally been the objective function of choice for selecting designs that yield precise estimates of model parameters. The D-criterion is defined in terms of the scaled moment matrix. For completely randomised designs, the moment matrix is $\mathbf{X}'\mathbf{X}/\sigma^2$, and scaling by σ^2/N (observation error divided by the design size) yields the scaled moment matrix, $\mathbf{M} = \mathbf{X}'\mathbf{X}/N$. When the user specifies the D-criterion as an objective function, the computer search algorithm finds the set of design points that maximises the determinant of the scaled moment matrix.

If interest is in finding a design with precise estimates of the predicted mean throughout the design space, G- and IV-values are popular choices for objective functions. Similar to the scaling done in determining the D-criterion, the prediction variance is scaled by the observation error divided by the design size, yielding the scaled prediction variance (SPV)

$$SPV = \frac{N}{\sigma^2}\mathbf{x}_0'(\mathbf{X}'\mathbf{X})^{-1}\mathbf{x}_0.$$

The IV-criterion then involves choosing the design whose integrated SPV across the design region is smallest, whereas the G-criterion involves choosing the design whose maximum SPV is at a minimum. See Montgomery (2005: 442–444) for an illustration of the use of the D-criterion in selecting an optimal experimental design.

7.8 Graphical techniques for comparing experimental designs

Alphabetical criteria, such as D-, IV-, or G-efficiencies, are often employed when evaluating designs. However, the decision of a *best* design is typically more complicated than can be summarised by a single number. For instance, the practitioner may be interested in predicting and thus the desire is to find a design that has an appealing distribution of prediction variance throughout the region. In these situations, graphical tools are more informative for comparing competing designs and allow us to better understand the prediction performance of these designs. Giovannitti-Jensen and Myers (1989) proposed the variance dispersion graph (VDG) to evaluate prediction properties of response surface designs. These two-dimensional plots show the SPV patterns for a given model throughout the design space. They have the SPV on the vertical axis and a shrunken measure of the design space on the horizontal axis. By using shrinkage factors, that is, multipliers of the original design space and ranging from 0 to 1, where 0 indicates centre and 1 represents the original outline of the given space (see Khuri *et al.*, 1999), the VDGs can be extended to designs with cuboidal and irregular constrained design regions. For an illustration of the use of VDGs, see Myers and Montgomery (2002: 403–405). Examples of VDGs can be found in Myers *et al.* (1992), Borkowski (1995a, 1995b), and Trinca and Gilmour (1998). Goldfarb *et al.* (2003) developed three-dimensional VDGs to evaluate prediction performances of mixture-process experiments. They evaluated mixture and process components of the design space separately to allow better understanding of the relative performance of designs in different regions of the design space. Liang *et al.* (2006a) developed three-dimensional VDGs for evaluating the performance of industrial split-plot designs.

Another popular graphical tool for comparing competing designs is the fraction of design space (FDS) plot, developed by Zahran *et al.* (2003). In the FDS plot, SPV values are plotted against the fraction of the design space that has SPV at or below the given value. The theoretical solution of the FDS, which ranges from 0 to 1, is calculated by the equation $FDS = \frac{1}{\psi}\int \overset{\cdots}{_A} \int dx_k \ldots dx_1$ for a given design with k variables, where ψ is the volume of the entire design region and $A = \{(x_1, \ldots, x_k) : \nu(\mathbf{x}) < \upsilon\}$ is the region with SPV values less than or equal to the given cut-off value, υ; $\nu(\mathbf{x})$ is the expression to calculate SPV values. Goldfarb *et al.* (2004) developed the FDS plots for mixture and mixture-process experiments with irregular regions. Liang *et al.* (2006b) developed the FDS plots for industrial split-plot experiments.

7.9 Industrial split-plot designs

Many industrial experiments involve two types of factors, some with levels hard or costly to change and others with levels relatively easy to change. Typical examples of hard-to-change factors include humidity, pressure and temperature. When hard-to-change factors exist, it is in the practitioner's best interest to minimise the number of times the levels of these factors are changed. A common strategy is to run all combinations of the easy-to-change factors for a given setting of the hard-to-change factors. Such a strategy results in a split-plot design and is often more economically realistic than complete randomisation of factor level combinations.

In split-plot experiments, the levels of hard-to-change factors are randomly applied to the whole plot units. The separate randomisations in split-plot experiments leads to correlated observations within the same whole plot, and this correlation must be accounted for not only when conducting inferences but also when determining an optimal design. Many efficient designs have been developed for completely randomised designs, such as factorials and fractional factorials for first-order models and the CCD and BBDs for second-order models. The determination of optimal split-plot designs is a topic that has received considerable attention in the recent literature. Bingham and Sitter (1999, 2001), and Kulahci *et al.* (2006) consider the use of minimum aberration criteria for determining two-level fractional factorial screening experiments. Parker *et al.* (2007) develop a class of split-plot designs in which the ordinary least squares estimates of model parameters are equivalent to the generalised least squares estimates. Goos and Vandebroek (2001, 2003, 2004) propose algorithms for determining D-optimal split-plot designs. Trinca and Gilmour (2001) propose designs for restricted randomisation cases building by strata. Letsinger *et al.* (1996) compare the performance of several second-order designs (CCD, BBD) within the split-plot setting. McLeod and Brewster (2006) consider blocked fractional factorial split-plot experiments in the robust parameter design setting. Ganju and Lucas (1999) point out that split-plot designs chosen by random run order are not as appealing as those chosen by a good design strategy. Liang *et al.* (2006a, 2006b) considered graphical techniques for assessing competing split-plot designs.

7.10 Supersaturated designs

In many practical situations, the resources are such that one is forced to study $k = N - 1$ factors, where N denotes the number of unique design runs. These types of experimental designs are known as *saturated* designs since the number of parameters to estimate in the regression model is equal to the number of experimental runs. A topic which has received attention in the literature during recent years is the use of *supersaturated designs* in which the number of design factors is $k > N - 1$. When the number of factors in the experiment exceeds the number of runs, the design matrix cannot be orthogonal and factor effects are not independent. Supersaturated designs are created so as to minimise the non-orthogonality between factors. This notion was first introduced by Satterthwaite (1959) and later resurrected by Lin (1993). Supersaturated designs are generally constructed using computer search algorithms that seek to optimise a well-chosen objective function. A survey on the construction of supersaturated designs can be found in Lin (2000). A popular construction method for supersaturated designs involves the use of Hadamard matrices where one enumerates the two-factor interactions of certain Hadamard matrices (see Deng and Tang, 2002). The analysis of supersaturated designs is often characterised by high Type I and Type II error rates and thus the philosophy in analysing these designs should be to eliminate inactive factors and not to clearly identify a few important factors (for more details, see Holcomb *et al.*, 2003).

7.11 Experimental designs for quality improvement

Statistical experimental design is widely used for manufacturing process development and optimisation. The objective of a process robustness study is to provide information to the experimenters that will enable them to determine the settings for the control factors (factors whose levels remain fixed in the experiment and process) that will substantially reduce the variability transmitted to the response variable(s) of interest by the noise factors (factors whose levels are fixed in an experiment but fluctuate at random in the process). Consequently, the study of interactions between control and noise factors is of considerable importance.

Taguchi (1986, 1987) introduced a system of experimental designs for process robustness studies. Crossed array designs were created by forming separate designs for the control and noise factors (often referred to as inner and outer arrays, respectively) and then forming the Cartesian product of these two designs. Crossed array designs have at least two weaknesses. First, they often require a large number of runs. In some industrial settings (such as process industries, semiconductor manufacturing) the number of runs would often be prohibitive. Second, the inner array is usually a three-level (or mixed-level) fraction of resolution III, and the outer array is a two-level design. Accordingly, the experimenters may estimate the linear and quadratic effects of the control factors but not the two-factor interactions between the control factors.

Many authors have suggested placing both the control factors and the noise factors in a single design, often called a combined array. For examples, see Sacks *et al.* (1989) and Borkowski and Lucas (1997). Using the combined array design, a single model containing both the control factors and the noise factors can be fitted to the response of interest.

Borkowski and Lucas (1997) propose the use of *mixed resolution* designs for use in robust parameter design settings. A mixed resolution design is any 2^{k-p} fractional factorial design which satisfies the following three properties: (1) among the control factors, the design is at least resolution V; (2) among the noise factors, the design is at least resolution III; (3) none of the two-factor control-by-noise interactions are aliased with any main effect or any two-factor control-by-control interactions. When second-order terms are introduced for the control factors, Borkowski and Lucas (1997) propose *composite designs* which are similar in structure to the central composite designs of Box and Wilson (1951). Borror *et al.* (2002) compared the performance of several composite designs to CCDs, computer-generated designs, and high D-efficiency designs via variance dispersion plots. In general, they found that the composite designs performed as well as or better than the competitors considered. For an overview of experimental design in process robustness studies, see Robinson *et al.* (2004).

7.12 The future

In this chapter, I have briefly discussed the planning and running of industrial experiments and the associated analyses. The methods are those that are most commonly used today. More will come. Many researchers continue to develop the methods and invent new ones according to the needs of industrial practitioners. On-line control of multivariate processes is an area in which we may expect some valuable developments.

References

Bingham, D. and Sitter, R.R. (1999) Minimum-aberration two-level fractional factorial split-plot designs. *Technometrics*, **41**, 62–70.

Bingham, D. and Sitter, R.R. (2001) Design issues for fractional factorial experiments. *Journal of Quality Technology*, **33**, 2–15.

Borkowski, J.J. (1995a) Minimum, maximum, and average spherical prediction variances for central composite and Box–Behnken designs. *Communications in Statistics: Theory and Methods*, **24**, 2581–2600.

Borkowski, J.J. (1995b) Spherical prediction-variance properties of central composite and Box–Behnken designs. *Technometrics*, **37**, 399–410.

Borkowski, J.J. (2003) Using a genetic algorithm to generate small exact response surface designs. *Journal of Probability and Statistical Science*, **1**, 65–88.

Borkowski, J.J. and Lucas, J.M. (1997) Designs of mixed resolution for process robustness studies. *Technometrics*, **39**, 63–70.

Borror, C.M., Montgomery, D.C., and Myers, R.H. (2002), Evaluation of statistical designs for experiments involving noise variables. *Journal of Quality Technology*, **34**, 54–70.

Box G.E.P. and Behnken D.W. (1960) Some new three-level designs for the study of quantitative variables. *Technometrics*, **2**, 455–475.

Box, G.E.P. and Wilson, K.B. (1951) On the experimental attainment of optimum conditions. *Journal of the Royal Statistical Society, Series B*, **13**, 1–45.

Daniel, C. (1959), Use of half-normal plots in interpreting factorial two-level experiments. *Technometrics*, **1**, 311–342.

Deng, L.-Y. and Tang, B. (2002) Design selection and classification for Hadamard matrices using generalised minimum aberration criteria. *Technometrics*, **44**, 173–184.

Ganju, J. and Lucas, J.M. (1999) Detecting randomisation restrictions caused by factors. *Journal of Statistical Planning and Inference*, **81**, 129–140.

Giovannitti-Jensen, A., and Myers, R.H. (1989) Graphical assessment of the prediction capacity of response surface designs. *Technometrics*, **31**, 159–171.

Goldfarb, H.B., Borror, C.M., Montgomery, D.C., and Anderson-Cook, C.M. (2003) Three-dimensional variance dispersion graphs for mixture-process experiments. *Journal of Quality Technology*, **36**, 109–124.

Goldfarb, H.B., Anderson-Cook, C.M., Borror, C.M., and Montgomery, D.C. (2004) Fraction of design space to assess the prediction capability of mixture and mixture-process designs. *Journal of Quality Technology*, **36**, 169–179.

Goos, P. and Vandebroek, M. (2001) Optimal split-plot designs. *Journal of Quality Technology*, **33**, 436–450.

Goos, P. and Vandebroek, M. (2003) D-optimal split-plot designs with given numbers and sizes of whole plots. *Technometrics*, **45**, 235–245.

Goos, P. and Vandebroek, M. (2004) Outperforming completely randomised designs. *Journal of Quality Technology*, **36**, 12–26.

Heredia-Langner, A., Carlyle W.M., Montgomery C.C., Borror C.M., and Runger G.C. (2003) Genetic algorithms for the construction of D-optimal designs. *Journal of Quality Technology*, **35**, 28–46.

Heredia-Langner, A., Montgomery C.C., Carlyle W.M., and Borror C.M. (2004) Model robust designs: A genetic algorithm approach. *Journal of Quality Technology*, **36**, 263–279.

Holcomb, D.R., Montgomery, D.C., and Carlyle, W.M. (2003) Analysis of supersaturated designs. *Journal of Quality Technology*, **35**, 13–27.

Khuri, A.I., Harrison, J.M. and Cornell, J.A. (1999) Using quantile plots of the prediction variance for comparing designs for a constrained mixture region: An application involving a fertilizer experiment. *Applied Statistics*, **48**, 521–532.

Kiefer, J. (1959) Optimum experimental designs. *Journal of the Royal Statistical Society, Series B*, **21**, 272–304.

Kiefer, J. (1961) Optimum designs in regression problems. *Annals of Mathematical Statistics*, **32**, 298–325.

Kiefer, J. and Wolfowitz, J. (1959) Optimum designs in regression problems. *Annals of Mathematical Statistics*, **30**, 271–294.

Kuehl, R.O. (2000) *Design of Experiments: Statistical Principles of Research Design and Analysis*, 2nd edition. Duxbury Press, New York.

Kulahci, M., Ramirez, J.G., and Tobias, R. (2006) Split-plot fractional designs: Is minimum aberration enough? *Journal of Quality Technology*, **38**, 56–64.

Letsinger, J.D., Myers, R.H., and Lentner, M. (1996) Response surface methods for bi-randomisation structure. *Journal of Quality Technology*, **28**, 381–397.

Liang, L., Anderson-Cook, C.M., Robinson, T.J., and Myers, R.H. (2006) Three-dimensional variance dispersion graphs for split-plot designs. *Journal of Computational and Graphical Statistics*, **15**, 757–778.

Liang, L., Anderson-Cook, C.M., and Robinson, T.J. (2006) Fraction of design space plots for split-plot designs. *Quality and Reliability Engineering International*, **22**, 1–15.

Lin, D.K.J. (1993) A new class of supersaturated designs. *Technometrics*, **35**, 28–31.

Lin, D.K.J. (2000) Recent developments in supersaturated designs. In H. Park and G.G. Vining (eds), *Statistical Process Monitoring and Optimisation*, pp. 305–319. Marcel Dekker, New York.

McLeod, R.G. and Brewster, J.F. (2006) Blocked fractional factorial split-plot experiments for robust parameter design. *Journal of Quality Technology*, **38**, 267–279.

Montgomery, D.C. (2005) *Design and Analysis of Experiments*, 6th edition. John Wiley & Sons, Inc., Hoboken, NJ.

Myers, R.H. and Montgomery, D.C. (2002) *Response Surface Methodology: Process and Product Optimisation Using Designed Experiments*, 2nd edition. John Wiley & Sons, Inc., New York.

Myers, R.H., Vining, G., Giovannitti-Jensen, A., and Myers, S.L. (1992) Variance dispersion properties of second-order response surface designs. *Journal of Quality Technology*, **24**, 1–11.

Parker, P.A., Kowalski, S.M., and Vining G.G. (2007) Construction of balanced equivalent estimation second-order split-plot designs. *Technometrics*, **49**, 56–65.

Plackett, R.L., and Burman, J.P. (1946), The design of optimum multifactorial experiments. *Biometrika*, **33**, 305–325.

Rao, G., and Saxena, S.C. (1993) Prediction of flue gas composition of an incinerator based on a nonequilibrium-reaction approach. *Journal of Waste Management Association*, **43**, 745–752.

Robinson, T.J. (2007) Box–Behnken designs. In *Encyclopedia of Statistics in Quality and Reliability,* Wiley: London (in press).

Robinson, T.J., Borror, C.M., and Myers, R.H. (2004) Robust parameter design: A review. *Quality and Reliability Engineering International*, **20**, 81–101.

Ryan, T.P. (2007) *Modern Experimental Design*. Hoboken, NJ: Wiley-Interscience.

Sacks, J., Welch, W.J., Mitchell, T.J., and Wynn, H.P. (1989) Design and analysis of computer experiments. *Statistical Science*, **4**, 409–435.

Sattherthwaite, F.E. (1959) Random balance experimentation (with discussion). *Technometrics*, **1**, 111–137.

Taguchi, G. (1986) *Introduction to Quality Engineering*. UNIPUB/Kraus International, White Plains, NY.

Taguchi, G. (1987) *System of Experimental Design: Engineering Methods to Optimise Quality and Minimise Cost.* UNIPUB/Kraus International, White Plains, NY.

Trinca, L.A. and Gilmour, S.G. (1998) Variance dispersion graphs for comparing blocked response surface designs. *Journal of Quality Technology*, **30**, 314–327.

Trinca, L.A. and Gilmour, S.G. (2001) Multi-stratum response surface designs. *Technometrics*, **43**, 25–33.

Zahran, A.R., Anderson-Cook, C.M., and Myers, R.H. (2003) Fraction of design space to assess prediction capability of response surface designs. *Journal of Quality Technology*, **35**, 377–386.

8

Data mining for business and industry

Paola Cerchiello, Silvia Figini and Paolo Giudici

8.1 What is data mining?

Data mining is the process of selection, exploration, and modelling of large quantities of data to discover regularities or relations that are at first unknown with the aim of obtaining clear and useful results for the owner of the database.

The aim of data mining is to obtain results that can be measured in terms of their relevance for the owner of the database (business advantage). Data mining is defined as the process of selection, exploration, and modelling of large quantities of data to discover regularities or relations that are initially unknown, in order to obtain clear and useful results for the owner of the database.

In the business context, the utility of the result becomes a business result in itself. Therefore what distinguishes data mining from statistical analysis is not so much the amount of data analysed or the methods used but that what we know about the database, the means of analysis, and the business knowledge are integrated. To apply a data mining methodology means:

- following an integrated methodological process that involves translating the business needs into a problem which has to be analysed;
- retrieving the database needed for the analysis; and

Statistical Practice in Business and Industry Edited by S.Y. Coleman, T. Greenfield, D.J. Stewardson and D.C. Montgomery

- applying a statistical technique implemented in a computer algorithm with the final aim of achieving important results useful for taking a strategic decision.

The strategic decision, in turn, will require new measurements and consequently new business objectives, thus starting 'the virtuous circle of knowledge' induced by data mining (Berry and Linoff, 1997).

There are at least three other aspects that distinguish the statistical analysis of data from data mining. First, all data mining is concerned with analysing great masses of data. This implies new considerations for statistical analysis. For example, for many applications it may be computationally impossible to analyse or even access the whole database. Therefore, it becomes necessary to have a sample of the data from the database being examined. This sampling must done with the data mining aims in mind and, therefore, the sample data cannot be analysed with the traditional statistical sampling theory tools.

Second, as we shall see, many databases do not lead to the classical forms of statistical data organisation. This is true, for example, of data that comes from the internet. This creates the need for appropriate analytical methods to be developed, which are not available in the field of statistics.

One last but very important difference that we have already mentioned is that data mining results must be of some consequence. This means that constant attention must be given to business results achieved with the data analysis models.

From an operational point of view, data mining is a process of data analysis that consists of a series of activities that go from the definition of the objectives of the analysis, to the analysis of the data and the interpretation and evaluation of the results.

The various phases of the process are as follows:

1. Definition of the objectives for analysis.

2. Selection, organisation and pre-treatment of the data.

3. Exploratory analysis of the data and their subsequent transformation.

4. Specification of the statistical methods to be used in the analysis phase.

5. Analysis of the data based on the methods chosen.

6. Evaluation and comparison of the methods used and the choice of the final model for analysis.

7. Interpretation of the model chosen and its subsequent use in decision processes.

For a more detailed description of data mining and the data mining process, see Giudici (2003).

8.2 Data mining methods

The data mining process is guided by applications. For this reason the methods used can be classified according to the aim of the analysis. With this criterion, we can distinguish three main classes of methods that can be exclusive or correspond to distinct phases of data analysis.

8.2.1 Descriptive methods

The aim of these methods (also called symmetrical, unsupervised or indirect) is to describe groups of data more briefly. This may involve both the synthesis of the observations, which are classified into groups not known beforehand (cluster analysis, Kohonen maps), as well as the synthesis of the variables that are connected among themselves according to links unknown beforehand (association methods, log-linear models, graphical models). In this way, all the variables available are treated at the same level and there are no hypotheses of causality.

8.2.2 Predictive methods

With these methods (also called asymmetrical, supervised or direct) the aim is to describe one or more of the variables in relation to all the others. This is done by looking for rules of classification or prediction based on the data. These rules help us to predict or classify the future result of one or more response or target variables in relation to what happens to the explanatory or input variables. The main methods of this type are those developed in the field of machine learning such as neural networks (multilayer perceptrons) and decision trees, but also classical statistical models such as linear and logistic regression models.

8.2.3 Local methods

Whereas descriptive and predictive methods are concerned with describing the characteristics of the database as a whole (global analysis), local methods are used to identify particular characteristics related to subset interests of the database (local analysis). Examples of this type of analysis include association rules for analysing transactional data, or the identification of anomalous observations (outliers).

We think that this classification is exhaustive, especially from a functional point of view. However, there are further distinctions discussed in the literature on data mining that the reader can refer to for further information; see, for example, Hand *et al.* (2001), Hastie *et al.* (2001) and Han and Kamber (2001).

8.3 Data mining applications

There are many applications of data mining. We now describe those that are most frequently used in the business field.

8.3.1 Market basket analysis

Market basket analysis is used to examine sales figures so as to understand which products were bought together. This information makes it possible to increase sales of products by improving offers to customers and promoting sales of other products associated with them.

8.3.2 Web usage mining

Web usage mining is the analysis of data traffic generated by the internet. Three main areas can be distinguished:

- web content mining, whose aim is to describe (often by means of text mining tools) the content of websites;
- web structure mining, where the aim is to look at the link structure of the internet; and
- web usage mining, where the aim is to show how information related to the order in which the pages of a website are visited can be used to predict the visiting behaviour on the site.

Web usage mining is the most important from a business viewpoint. For example, the data analysed from an e-commerce site may make it possible to establish which pages influence electronic shopping of what kind of products.

8.3.3 Profiling

Profiling leads from data to classification of individual units (clients or visitors) based on their behavioural information. With this information, it is possible to get a behavioural segmentation of the users that can be used when making marketing decisions.

8.3.4 Customer relationship management

Statistical methods are used in customer relationship management to identify groups of homogeneous customers in terms of buying behaviour and socio-demographic characteristics. The identification of the different types of customers makes it possible to draw up a personalised marketing campaign and to assess the effects of this, as well as looking at how offers can be changed.

8.3.5 Credit scoring

Credit scoring is a procedure that gives each statistical unit (customer, debtor, business) a score. The aim of credit scoring is to associate each debtor with a numeric value that represents his/her creditworthiness. In this way, it is possible to decide whether or not to give someone credit based on his/her score.

In this chapter we focus on web mining, which is a very recent research area with considerable potential for application in the fields of e-commerce and web categorisation.

8.4 A case study in web usage mining

In the last few years there has been an enormous increase in the number of people using the internet. Companies promote and sell their products on the Web, institutions provide information about their service, and individuals exploit personal Web pages to introduce themselves to the whole internet community. We will show how information about the order in which the pages of a website are visited can be profitably used, by means of statistical graphical models, to predict visitor behaviour on the site.

Every time a user visits a website, the server keeps track of all the actions in a *log file*. What is captured is the 'click flow' (clickstream) of the mouse and the keys used by the user while navigating inside the site. Usually every mouse click corresponds to the viewing of a web page. Therefore, we can define the clickstream as the sequence of web pages requested. The sequence of pages seen by a single user during his navigation around the Web identifies a user session. Typically, the analysis concentrates only on the part of each user session concerning access at a specific site. The set of pages seen, inside a user session and within a specific site, is known as the *server session*.

All this information can be profitably used to efficiently design a website. A web page is well designed if it is able to attract users and direct them easily to other pages within the site. This is the main problem faced by Web mining techniques. An important application of data mining techniques is to discover usage patterns from Web data, to design a website optimally and to better satisfy the needs of different visitors.

The purpose of our analysis is to use Web clickstream data to understand the most likely paths of navigation within a website, with the aim of predicting, possibly on-line, which pages will be seen, having seen a specific path of other pages before. Such analysis can be useful to understand, for instance, the probability of seeing a page of interest (such as the purchasing page in an e-commerce site) when approached from another page, or the probability of entering (or exiting) the website from any particular page.

In the description of the case study, we shall follow the steps of the data mining process.

8.4.1 Exploratory data analysis

The data set that we consider for the analysis is derived from a log file concerning an e-commerce site. For a more detailed description of the data, see Giudici (2003). Hits on the website were registered in a log file for about two years, from 30 September 1997 to 30 June 1999. The log file was then processed to produce a data set which contains the user id (c_value), a variable with the date and

Table 8.1 Extract from the example data set. Paolo Giudici (2003) Applied Data Mining: Statistical Methods for Business and Industry. Reproduced with permission from John Wiley and Sons, Ltd.

c_value	c_time	c_caller
70ee683a6df...	14OCT97:11:09:01	home
70ee683a6df...	14OCT97:11:09:08	catalog
70ee683a6df...	14OCT97:11:09:14	program
70ee683a6df...	14OCT97:11:09:23	product
70ee683a6df...	14OCT97:11:09:24	program

Source: Giudici (2003: 230, Table 8.1).

the instant the visitor linked to a specific page (c_time) and the web page seen (c_caller). Table 8.1 is a small extract from the available data set, corresponding to one visit. The table shows that the visitor, identified as 70ee683a6df, entered the site on 14 October 1997 at 11:09:01, visited, in sequence, the pages home, catalog, program, product, program, and left the website at 11:09:24.

The data set in its entirety contains 250 711 observations, each corresponding to a click, that describe the navigation paths of 22 527 visitors among the 36 pages which compose the site of the webshop. The visitors are taken as unique; that is, no visitor appears with more than one session. On the other hand, we note that a page can occur more than once in the same session.

We also note that other variables can be obtained from the log file. For instance, the total time of a clickstream session (*length*), the total number of clicks made in a session (*clicks*), and the time at which the session starts (*start*, where 0 represents midnight of the preceding day). Another variable (*purchase*) indicates whether the session led to at least one commercial transaction.

In this case-study, we are mainly concerned with understanding the navigation patterns and, therefore, we will consider 36 binary variables that describe whether each web page is visited at least once (when they take the value 1) or not (0).

To give an idea of the types of pages, we now list some of the most frequent ones.

Home: the homepage of the website.

Login: a user enters his name and other personal information, during the initial registration, to access certain services and products reserved for customers.

Logpost: prompts a message that informs whether the login has been successful or has failed.

Logout: the user can leave his personal details on the login page.

Register: so as to be recognised later, the visitor has to provide a user ID and a password.

Regpost: shows partial results of the registration, requesting missing information.

Results: once registration is accomplished, this page summarises the information given.

Regform1: the visitor inserts data that enables him/her to buy a product, such as a personal identification number.

Regform2: the visitor contracts to accept the conditions for on-line commerce help: it also deals with questions that may arise during the navigation through the website.

Fdback: allows return to the page previously visited.

Fdpost: allows return to a page previously seen, in specified areas of the site.

News: presents the latest available offerings.

Shelf: contains the list of programs that can be downloaded from the website.

Program: gives detailed information on the characteristics of the software programs that can be bought.

Promo: gives an example (demo) of the peculiarities of a certain program.

Download: allows download of software of interest.

Catalog: contains a complete list of the products on sale in the website.

Product: shows detailed information on each product that can be purchased.

P_Info: gives detailed information on terms of payment.

Addcart: where the virtual basket can be filled with items to be purchased.

Cart: shows what items are currently in the basket.

Mdfycart: allows changes to the current content of the basket, such as removing items.

Charge : indicates the cost of items in the basket.

Pay_Req: shows the amount finally due for the products in the basket.

Pay_Res: the visitor agrees to pay, and payment data are inserted (such as the credit card number).

Freeze: where the requested payment can be suspended, for example to add new products to the basket.

Our main aim is to discover the most frequent sequence rules among the 36 binary variables describing whether any single page has been visited. To reach valid conclusions, the data must be homogeneous with respect to the behaviour described by each row (corresponding to a visitor). To assess whether the available data set is homogeneous we have run an outlier detection and a cluster analysis on the visitors. We refer the reader to Chapter 8 of Giudici (2003) for details.

On removing the outliers, the initial 22 527 visitors in the data set were reduced to 22 152. The cluster analysis indicated that four clusters would be optimal. The cluster analysis confirmed the heterogeneity of behaviours. In order to find navigation patterns, we decided to analyse only one cluster: the third. This choice is subjective but, nevertheless, has two important peculiarities. First, the visitors in this cluster stay connected for a long time and visit many pages. This allows us to explore the navigation sequences between the web pages. Second, this cluster has a high probability of purchase; it seems important to consider the typical navigation pattern of a group of high purchasers.

Therefore, in the following sections, we shall consider a reduced data set, corresponding to the third cluster, with 1240 sessions and 21 889 clicks.

8.4.2 Model building

We now introduce our proposed methodology, based on graphical models. We begin by recalling association and sequence rules, typically used in the context of web usage mining.

An *association rule* is a statement between two sets of binary variables (item sets) A and B, that can be written in the form $A \rightarrow B$, to be interpreted as a logical statement: if A, then B. If the rule is ordered in time we have a sequence rule and, in this case, A precedes B.

In web clickstream analysis, a sequence rule is typically *indirect:* namely, between visiting page A and visiting page B other pages can be seen. On the other hand, in a *direct* sequence rule A and B are seen consecutively.

A *sequence rule* is an algorithm that searches for the most interesting rules in a database. To find a set of rules, statistical measures of 'interestingness' have to be specified. The measures more commonly used in web mining to evaluate the importance of a sequence rule are the indices of support and confidence.

In this case, we shall mainly consider the confidence index. The confidence for the rule $A \rightarrow B$ is obtained by dividing the number of server sessions which satisfy the rule by the number of sessions containing the page A. Therefore, the confidence approximates the conditional probability that, in a server session in which page A has been seen, page B is subsequently requested.

For more details the reader can consult a recent text on data mining, such as Han and Kamber (2001) or, from a more statistical viewpoint, Hand *et al.* (2001), Hastie *et al.* (2001) and Giudici (2003).

The main limit of these indices, which are extremely flexible and informative, is that, as descriptive indices, they allow valid conclusions to be drawn for the observed data set only. In other words, they do not reliably predict the behaviour of new users. A more general model is needed. Different solutions have been presented. For a review, see Giudici (2003). The solution we present here is based on graphical models.

A graphical model (such as Lauritzen, 1996) is a family of probability distributions incorporating the conditional independence assumptions represented by a graph. It is specified by a graph that depicts the local relations among the variables (these are represented by nodes). In general, direct influence of a variable on another is indicated by a directed edge (an arrow), while a symmetric association is represented by an undirected edge (a line). We can use a graph to identify conditional independence between the variables examined. For more details, including the conditional independence rules, see Lauritzen (1996).

We can distinguish four different types of graph and correspondingly different types of graphical model:

1. *Undirected graphs* contain only undirected edges to model symmetric relations among the variables. They give rise to Markov networks.

2. *Acyclic directed graphs* contain only directed edges. They are used to model asymmetric relations among the variables. They give rise to acyclic directed graphical models, also called Bayesian networks.

3. *Graphs with multi-directional edges* are directed graphs that may contain multi-directional arrows. They give rise to dependency networks.

4. *Chain graphs* contain both undirected and directed edges, and, therefore, can model both symmetric and asymmetric relationships. They give rise to chain models.

Directed graphical models (such as Bayesian networks and dependency networks) are powerful tools for predictive data mining because of their fundamental assumption of causal dependency between variables. However, symmetric graphical models may be more useful in the preliminary phase of analysis, because they can show the main relevant associations, with a limited number of assumptions.

In our case study, we are mainly concerned with symmetric graphical models and we compare them with other data mining methodologies. Other types of graphical models have been fitted to web usage mining data. See Giudici (2003) on Bayesian networks and Heckerman *et al.* (2000) on dependency networks.

The main idea now is to introduce dependence between time-specific variables. In each session, for each time point i, here corresponding to the ith click, there is a corresponding discrete random variable, with as many possible values as the number of pages (these are *named states of the chain*). The observed ith page in the session is the observed realisation of the Markov chain, at time i, for that session. Time can go from $i = 1$ to $i = T$, and T can be any finite number. Note that a session can stop well before T: in this case the last page seen is called an absorbing state (end_session for our data).

A Markov chain model establishes a probabilistic dependence between what was seen before time i and what will be seen at time i. In particular, a first-order Markov chain, which is the model we consider here, establishes that what is seen at time i depends only on what was seen at time $i - 1$. This short-memory dependence can be assessed by a transition matrix that establishes the probability of going from any one page to any other in one step only. For 36 pages there are $36 \times 36 = 1296$ probabilities of this kind.

The conditional probabilities in the transition matrix can be estimated from the available conditional frequencies. If we add the assumption that the transition matrix is constant in time (homogeneity of the Markov chain), as we shall do, we can use the frequencies of any two adjacent pairs of time-ordered clicks to estimate the conditional probabilities.

Note the analogy of Markov chains with direct sequences. Conditional probabilities in a first-order Markov model correspond to the confidence of second-order direct sequence rules and, therefore, a first-order Markov chain is a model for direct sequences of order 2. Furthermore, it can be shown that a second-order Markov model is a model for direct sequences of order 3, and so on. The difference is that the Markov chain model is based on all data, that is, it is a global and not a local model. This is, for example, reflected in the fact that Markov chains consider all pages and not only those with a high support. Furthermore, the Markov model is a probabilistic model and, as such, allows inferential results to be obtained.

To build a Markov chain model (first-order and homogeneous) we reorganised the data set, obtaining a new one in which there are two variables: *page_now* and *page_tomorrow*. The variable *page_now* indicates what is visualised by the visitor at a certain click and *page_tomorrow* what is visualised immediately afterwards. More details on such data sets are given in Giudici (2003).

We illustrate our methodology with data extracted from a log file concerning an e-commerce site. It corresponds to a cluster of 1240 visitors, nearly homogeneous as regards time of visits, number and length of clicks and propensity to make e-commerce purchases.

To model the data, we considered four main classes of statistical models: sequence rules, decision trees, symmetric graphical models and Markov chains. We used MIM software for the construction of the symmetric graphical models.[1] We referred to SAS Enterprise Miner for association rules and decision trees. We used SAS, with some additional instructions written in the SAS IML programming language, for Markov chains.

We first consider the comparison of graphical models with more conventional data mining models, such as sequence rules and classification trees. Later on we shall dwell upon the application of Markov chains and their comparison with sequence rules.

For the first comparison we considered two contexts of analysis. In the first, to obtain a useful comparison between graphical model and decision trees, we introduced a target variable, *purchase*, to describe whether or not a client buys the products sold. Because of the limitations of the version of MIM we had available, we took a sample of 500 observations from the 1240 available. A forward selection procedure, using a significance level of 5 %, gave us the structure in Figure 8.1. Notice that the variable *purchase* has only one link (direct association), with the variable *addcart*. In Figure 8.2 we compare this result with those from a classification tree model. This gives a similar result; in fact *addcart* is the most important variable in the first split of the tree. Therefore we can conclude that the results from graphical models and decision trees give consistent results.

In a second context of analysis we excluded the target variable *purchase* to compare graphical models and sequence rule models. We used the same sample of data as in the first context. To summarise the main results, in Figure 8.3 we present the selected undirected graphical model (using a forward procedure and 5 % significance level). We also attach to each edge the higher of the two confidence indices and conditional probabilities associated with it. In Figure 8.3 we show such statistics for only the two strongest associations, *catalog–program* and *program–product*. While confidence indices are based on the results from the

---

[1]MIM is a Windows program for graphical modelling. It is the only available software supporting graphical modelling with both discrete and continuous variables. It can be downloaded at http://www.hypergraph.dk/.

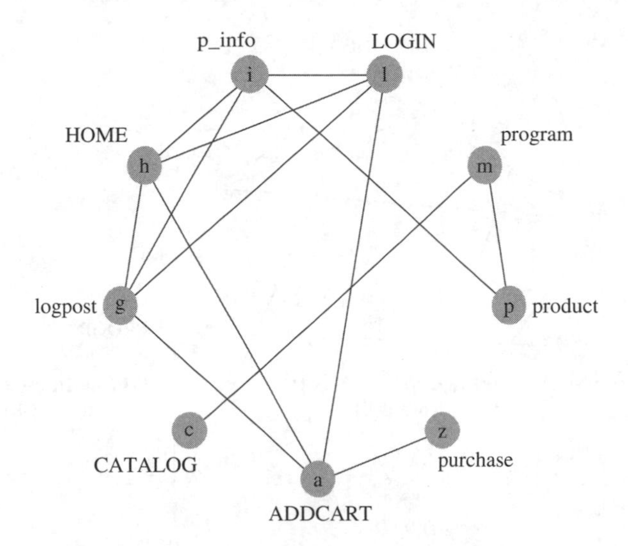

Figure 8.1 Graphical model with the variable purchase.

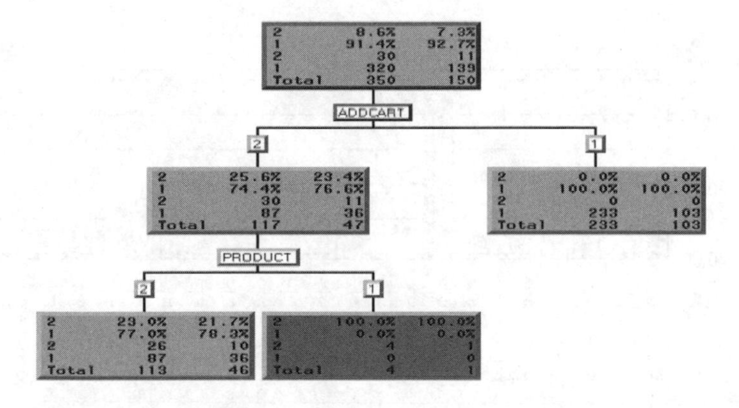

Figure 8.2 Decision tree with the variable purchase.

a priori algorithm of SAS Enterprise Miner (such as Giudici, 2003), conditional probabilities can be derived from the fitted contingency tables produced by MIM.

When we compare confidence indices with conditional probabilities, we see that the results from graphical models are similar to those from sequence rules. This means that, for data of this kind, sequence rules, based on marginal associations, can satisfactorily detect association structures. They are also much easier to calculate and interpret.

We now consider the application of Markov chains. We first calculated the transition matrix that we could show as a table with 37 rows (corresponding to the 36 web pages and a deterministic variable, describing *start_session*) and 37 columns

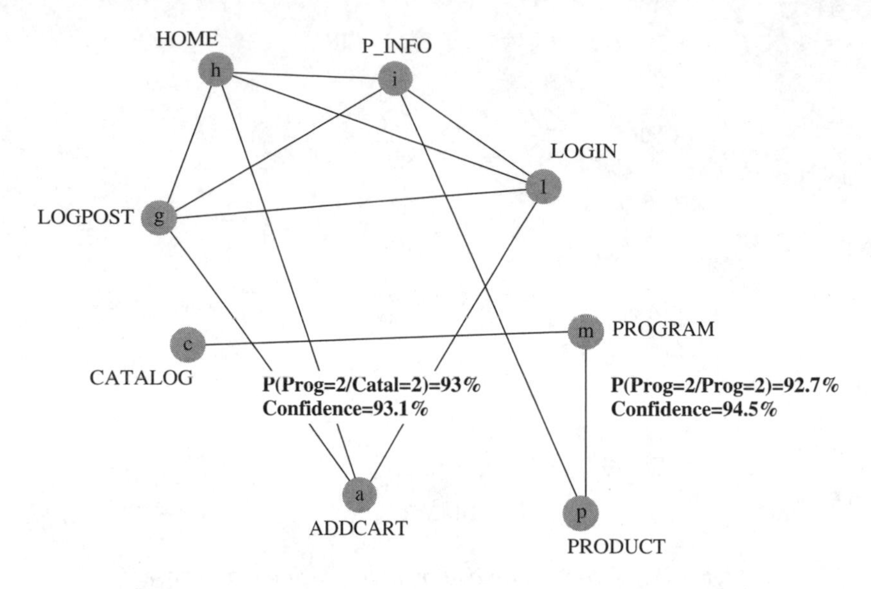

Figure 8.3 Graphical model with confidence indices and conditional probabilities.

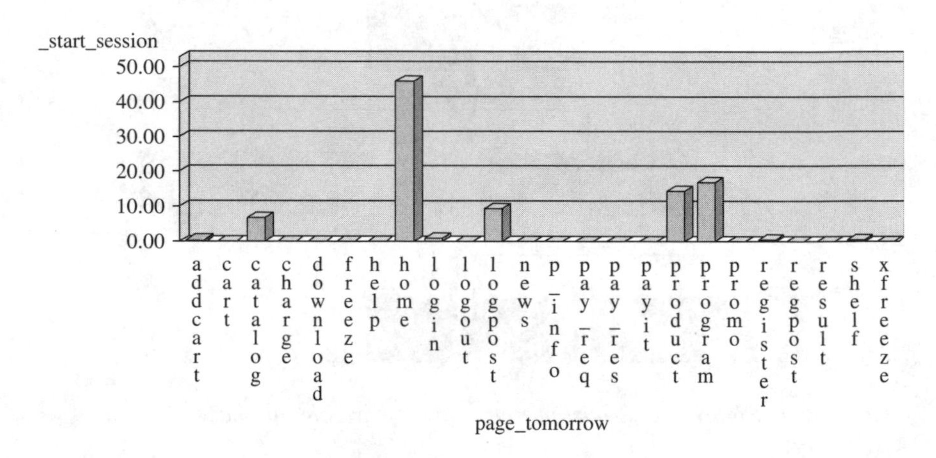

Figure 8.4 Entrance probabilities estimated by the Markov chain model (Giudici, 2003: 248, Figure 8.6).

(the 36 pages and a deterministic variable, describing *end_session*). Therefore, the transition matrix contains a total of $37 \times 37 = 1369$ estimated probabilities.

For simplicity, we do not show the whole transition matrix, but rather consider some conclusions that can be drawn from it. First of all, we can estimate where the visitor will be most likely to enter the site. For this, we must consider the transition probabilities of the *start_session* row. We show these probabilities graphically in Figure 8.4 (omitting those estimated to have zero probability). Notice that the most frequent entrance page is *home* (48.81 %), followed by *program* (17.02 %), *product* (14.52 %), *logpost* (9.19 %) and *catalog* (6.77 %). This is consistent with the nature of the visitors belonging to the third cluster. We remark that, differently from what happens with association rules, the transition probabilities sum to 1, by row and column.

We can then consider the most likely exit pages. To do this we must estimate the transition probabilities of the *end_session* column. Figure 8.5 shows these probabilities graphically (excluding those estimated to have zero probability). Notice that the most likely exit page is *product* (20.81 %), followed by *logpost* (12.10 %), *download* (10.56 %), *home* (9.19 %) and *p_info* (6.94 %). Compare Figures 8.4 and 8.5 to see that, while there are a few most likely entrance points, the distribution of exit points is more sparse.

Finally, from the transition matrix, we can establish a path that connects nodes through the most likely transitions. One possibility is to proceed forward from *start_session*. From *start_session*, we can connect it to the page with which *start_session* has the highest probability of transition: this is the *home* page. Next follows *program*, then *product* and then *P_info*. From *P_info* the most likely transition is back to *product* and, therefore, the path ends.

We can compare what the previous path would be, using the confidence indices from the a priori sequence rule algorithm. Initially, from *start_session*, the highest confidence is reached for *home* (45.81 %, as occurs for Markov chains), followed by *program* (20.39, more than with Markov chains), *product* (78.09 %, as against 14.52 %). The next most probable link is now with *addcart* (28.79 %) rather than *link*. The differences with Markov chains are because direct sequences are considered for only those pages that pass a certain criterion of interestingness.

For more details and comparison of Markov chain models with sequence rules, see Chapter 8 of Giudici (2003).

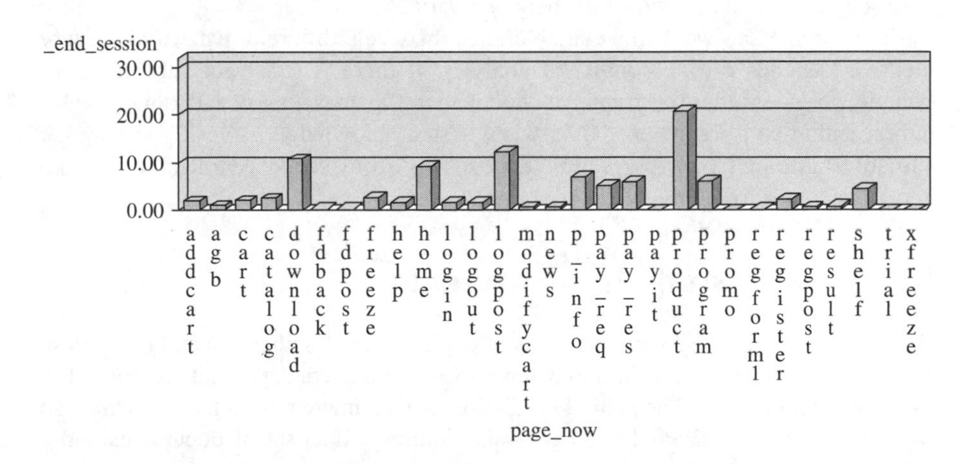

Figure 8.5 Exit probabilities estimated by the Markov chain model (Giudici, 2003: 248, Figure 8.7).

8.4.3 Results and conclusions

In our analysis we have considered different models for pattern discovery in Web mining. Our main objective is to infer, on the basis of the observed data, sequences of web pages that are often seen by a visitor to a website. This information can then be used either to devise specific direct marketing campaigns or simply to redesign the website, with the aim of attracting a wider range of visitors.

The models we have compared belong to two main families: non-inferential and inferential. We have used descriptive (non-inferential) statistical measures, aimed at capturing efficiently the most relevant marginal associations between groups of variables (item sets). The measures we used are association rules and classification trees.

The main advantage of descriptive measures is that they are relatively simple to compute and interpret. However, they have the disadvantage of being based on the sample at hand. A more serious drawback, which holds only for association rules (which indeed is a computational advantage), is that they are local, that is, based on subsets of the data. Correlation between subsets is not fully taken into account. Furthermore, it is rather difficult to develop measures of overall quality of an output set of association rules and, therefore, model assessment becomes rather subjective.

In this chapter we propose inferential models for web mining pattern discovery, based on graphical models and on Markov chain models. We have chosen graphical modelling methods as they advantageously combine global modelling with local computations (see Lauritzen, 1996). The advantage of using such an approach is that it gives coherent predictions, whose uncertainty can be evaluated easily, for instance, by means of a confidence interval. Furthermore, model diagnosis and comparison can proceed in a rigorous and consistent manner. The possible disadvantage of our approach is the need to come up with a structure of some kind, to describe constrained relationships between variables.

To summarise, we believe that choice between different pattern discovery methods depends on the context of analysis. If there is sufficient subject-matter knowledge to entertain a graphical model then we may use it and obtain better interpretation and predictions. Otherwise, if such knowledge is weak, we may get valuable (although perhaps preliminary) results from sequence rules, which are generally easier to obtain.

8.5 A case study in text mining

Text mining is a process with two areas of application: classification and prediction. The aim of classification is to find batches of terms and concepts that are present in a document collection. The goal of prediction is to estimate new behaviours through supervised techniques. Prediction analysis requires a data set of documents and a target variable.

In a well-known context, where communication with the customer is direct and definite, non-structured data analysis can be done through the classical methods of information retrieval and semantic mapping of the text. Information retrieval is a field that deals with the problem of storage, representation and retrieval of documents. It is not limited to textual document processing. Every information retrieval system compares demand, expressed in a specific language, with document representations, either with the documents themselves (direct representation) or with their surrogates (indirect representation).

Today, business data processing systems are based on structured data and do not allow the correct analysis of customer information, which arrives in textual form through different channels (such as e-mails, phone calls, interviews, notes), ignoring in this way significant data for the comprehension of customers' profiles.

It is not correct to compare text mining with information retrieval. The aim of information retrieval is to find those documents that contain the key words of the query made to retrieve documents. We can compare information retrieval with an on-line analytical processing technique, which answers the user's questions and returns existing information. On the other hand, text mining emphasises the search for many rules in documents to reveal new knowledge.

In a text analysis, we can usually find the following problems:

- Because of its intrinsic ambiguity, it is difficult to reorganise the textual language to make it comprehensible to the analysis software.

- The information contained in the document is hidden and there are different ways to express the same concept.

- It is difficult to represent abstract concepts.

Recently there has been great attention to the development of text mining technologies because:

- The quantity of textual data exceeds the human capability to analyse it.

- There are now software tools that can extract information from textual data and reorganise it into data patterns that can be analysed through classical data mining algorithms.

The aim of this work is to develop and test a predicting text mining process in a customer relationship learning system.

8.5.1 Exploratory data analysis

The text mining process has two phases: pre-processing of the document followed by analysis.

Pre-processing starts with data cleaning and is followed by set data reduction and analysis. Data are first read and then tokenised[2] to separate text strings and

[2]In computing, a token is a categorized block of text, usually consisting of indivisible characters known as lexemes. A lexical analyser initially reads in lexemes and categorizes them according to function, giving them meaning. This assignment of meaning is known as tokenization.

find single words. A tokenised document contains few very common words and words without information (stop words), some words with medium frequency and a lot of words that are not very frequent.

When the document's characteristics are identified, we can choose a list of keys to retrieve information from the document. This list can be indexed and refined, so that the retrieval of a specific document can be faster.

When the frequency pattern for single words is ready, we can go on to data reduction. For this purpose, there is a stop list that contains a series of terms removable from the analysis, for example conjunctions, punctuation and some verbs. For each problem it is possible to personalise the stop list to reduce noise in the document.

The result of this process is a matrix **A**: each column is a document vector whose elements represent the frequency of each term. An example is given in Table 8.2.

Table 8.2 Data after text parsing.

	Document 1	Document 2	Document 3	Document 4
Term 1	1	0	3	4
Term 2	2	0	4	6
Term 3	0	0	9	2
Term 4	2	7	12	8

The matrix **A** contains many null frequencies, so it is very difficult to analyse it. To continue with the analysis, the data must be reduced to a matrix with few null elements. Using singular value decomposition (SVD), the matrix can be decomposed into three matrices,

$$\underset{n\times p}{\mathbf{A}} = \underset{n\times p}{\mathbf{U}}\ \underset{p\times p}{\mathbf{\Sigma}}\ \underset{p\times p}{\mathbf{V}},$$

where **U** and **V** are orthonormal (each column is orthogonal and the Euclidean norm of each column is equal to 1) and $\mathbf{\Sigma}$ is a non-negative diagonal matrix. Thus, by selecting the document vectors corresponding to the K largest eigenvalues, it is possible to reduce the noise and get only the important information. The previous decomposition is based on all available documents, say, p (for more details, see Seber, 1984).

A can be approximated by a lower-dimensional projection, for example in $k < p$ principal components:

$$\underset{n\times p}{\mathbf{B}} = \underset{n\times k}{\mathbf{T}}\ \underset{k\times k}{\mathbf{\Gamma}}\ \underset{k\times p}{\mathbf{Z}^T}.$$

To illustrate this methodology, we consider a real data example that concerns the classification of some products. The data are in English and come from an American database about a financial product.

We have data for about 450 new items and we have to classify them into three different product categories, represented by three binary variables which take the value 1 if the item belongs to the product category, and 0 if not. The text variable represents the character string which has to be analysed through the parsing phase and the SVD. The target variable represents the product categories. We use a frequency table to describe these. Table 8.3 shows that the product category that contains most of the items is that of *liquidity*, with a target frequency equal to 42.92 %.

Table 8.3 Frequency tables of the three product categories.

Balance of payments	Frequency	Percent	Cumulative percent
0	329	77.11	77.11
1	109	24.89	100.00
Agricultural goods	Frequency	Percent	Cumulative percent
0	294	67.12	67.12
1	144	32.88	100.00
Liquidity	Frequency	Percent	Cumulative percent
0	250	57.08	57.08
1	188	42.92	100.00

After descriptive analysis, we sample the data to split the data set into training, validation and test parts. The data can then be analysed. After the data partition, we proceed with the *parsing* phase, using the research and personalization of a stop list (a data set containing those words which must then be removed). The parsing phase is before the SVD phase. The SVD phase allows a considerable reduction of data. We obtained a reduction of the number of documents from 450 to 50 principal components.

8.5.2 Model building

Hitherto, we have considered the data setting phase. We now proceed with the predicting phase, using a data mining technique based on a memory-based reasoning (MBR) algorithm.

The first step in memory-based reasoning is to identify similar cases from experience, and then to apply the information from these cases to the problem at hand. The method uses two operations: the *distance function*, which assigns a distance between any two records, and the *combination function*, which combines neighbouring records. We need a good definition of distance. We define it in a metric space in which, for any two points of our database (such as A and B), the following properties hold:

- non-negativity, $D(A,B) \geq 0$;
- nullity, $D(A,A) = 0$;

- symmetry, $D(A,B) = D(B,A)$;
- triangular inequality, $D(A,B) \leq D(A,C) + D(C,B)$.

In particular, the properties of symmetry and the triangular inequality guarantee that the MBR algorithm will find well-defined local K-neighbourhoods. The cardinality K represents the complexity of the nearest-neighbour model; the higher the value of K, the less adaptive the model. In the limit, when K is equal to the number of observations, the nearest-neighbour fitted value coincides with the sample mean. In our application, the best MBR algorithm was implemented with the Euclidean distance and eight neighbourhoods.

We choose the type of distance function according to the type of data we have to analyse. For categorical data the correct choice is uniform metrics, while for continuous data it is best to consider the absolute value, or the Euclidean metrics.

In our application the available variables are divided into the explanatory variables (text = X) and the target variable (binary variable = Y). We collect a sample of observations in the form (text, target) to form the training data set. For this training data, we introduce a distance function between the X values of the observations. With this, we can define, for each observation, a neighbourhood formed by the observations that are closest to it, in terms of the distance between the X values.

We used MBR for predictive classification. In particular, for each observation, Y, we determine its neighbourhood as before. The fitted probabilities of each category are calculated as relative frequencies in the neighbourhood.

Another point in favour of MBR, in this application to textual data, is that the algorithm does not require a probability distribution.

Our MBR model, called MBR_1, uses as input variables the columns given by the SVD analysis and as target variable the *balance of payments*. To assess model results we calculated loss function measures using validation data for the target variable.

The first evaluation tool we consider is the confusion matrix. This allows model diagnostics by comparison between the predicted values and the observed ones. More generally, a confusion matrix can be calculated for many cut-off values. In Table 8.4 we show it for a 50 % cut-off threshold. This table indicates that, generally speaking, the model classifies well the events and non-events. In particular, considering the 88 items in the validation data set, the model can correctly classify 61 non-events and 21 events, while it makes mistakes in predicting events that were observed as non-events and vice versa.

We can calculate further statistical indices based on the confusion matrix. Suppose have the confusion matrix in Table 8.5, where $A + B$ are the predicted elements equal to 1 and $C + D$ those equal to 0. We define two relative indices which can evaluate the quality of the algorithm: the *precision*, given by $A/(A + B)$, which is the number of correctly classified events equal to 1 compared with the sum of all the predicted elements equal to 1; and the *recall*, $A/(A + C)$, which is the number of correctly classified events equal to 1 compared with the sum of all the observed elements equal to 1. These indices, constant under trade-off, are the

Table 8.4 Confusion matrix with a 50 % cut-off.

```
----------------- thresh=50 -------------

        The FREQ Procedure

      Table of actual by predict

actual     predict

Frequency|
Percent  |
Row Pct  |
Col Pct  |0        |1        |  Total
---------+---------+---------+
0        |      61 |       5 |     66
         |   69.32 |    5.68 |  75.00
         |   92.42 |    7.58 |
         |   98.39 |   19.23 |
---------+---------+---------+
1        |       1 |      21 |     22
         |    1.14 |   23.86 |  25.00
         |    4.55 |   95.45 |
         |    1.61 |   80.77 |
---------+---------+---------+
Total          62        26       88
            70.45     29.55   100.00
```

Table 8.5 Theoretical confusion matrix.

	1	0
1	*A*	*C*
0	*B*	*D*

coordinates used to define a receiver operating characteristic (ROC) curve, which helps us to choose the best cut-off and increase the predictive ability of our model. The ROC curve for our MBR_1 model in Figure 8.6 shows that, using precision and recall values, our model performs well.

We used the same process to predict the classification of new items in the *agricultural goods* and *liquidity* categories. In Table 8.6 we list the precision and recall results of data set training, validation and test. The high precision and recall values show that the best predictions are with the settings for *liquidity*, using the MBR algorithm. The settings for *balance of payments* are weaker.

To overcome this weakness, we tried other predictive data mining algorithms, such as neural networks and tree models. The neural network that we used had one hidden layer with *balance of payments* as target variable and the SVD columns as input. The classification tree used the same target and input variable and *entropy* as the splitting criterion.

We now wish to compare the implemented models with suitable measures of goodness of fit for the *balance of payments* target. The predictive comparison among the models can be made up of precision and recall as before. The results

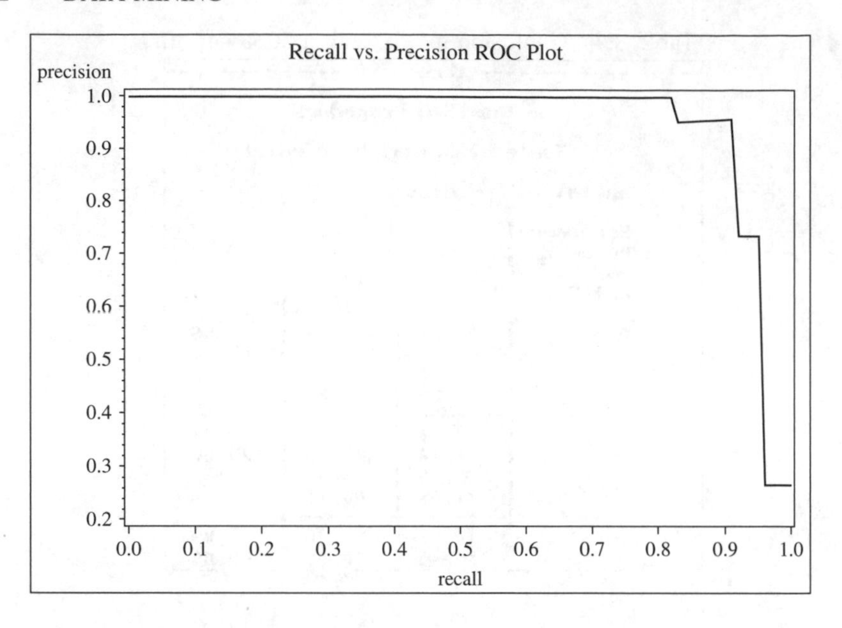

Figure 8.6 ROC plot for model MBR_1. Paolo Giudici (2003) Applied Data Mining: Statistical Methods for Business and Industry. Reproduced with permission from John Wiley and Sons, Ltd.

Table 8.6 Precision and recall for each target variable.

Balance of payments	Training	Precision = 0.93	Recall = 0.89
	Validation	Precision = 0.81	Recall = 0.95
	Test	Precision = 0.96	Recall = 0.96
Agricultural goods	Training	Precision = 0.99	Recall = 0.95
	Validation	Precision = 1	Recall = 0.93
	Test	Precision = 0.97	Recall = 0.97
Liquidity	Training	Precision = 0.99	Recall = 0.96
	Validation	Precision = 0.97	Recall = 0.93
	Test	Precision = 1	Recall = 0.97

are in Table 8.7, which shows that the best model is the MBR. We also used other models, such as logistic regression, but the results were much worse.

We now compare models with the lift chart. The lift chart puts the observations in the validation data set into increasing or decreasing order on the basis of their score, which is the probability of the response event (success), as estimated from the training data set. It subdivides these scores into deciles and then calculates and graphs the observed probability of success for each of the decile classes in the validation data set. A model is valid if the observed success probabilities follow the same order as the estimated probability. To improve interpretation, a model's lift

Table 8.7 Comparison models.

MODEL	PRECISION	RECALL
MBR	0.95652	0.95652
NEURAL	0.95	0.82609
TREE	0.88	0.95652

chart is usually compared with a baseline curve, for which the probability estimates are known in the absence of a model, that is, by taking the mean of the observed success probabilities (for more details, see Giudici, 2003). Figure 8.7 shows the results of model comparison for our problem. It is evident that, in the first four deciles, neural networks are better than the trees and the MBR algorithm. The

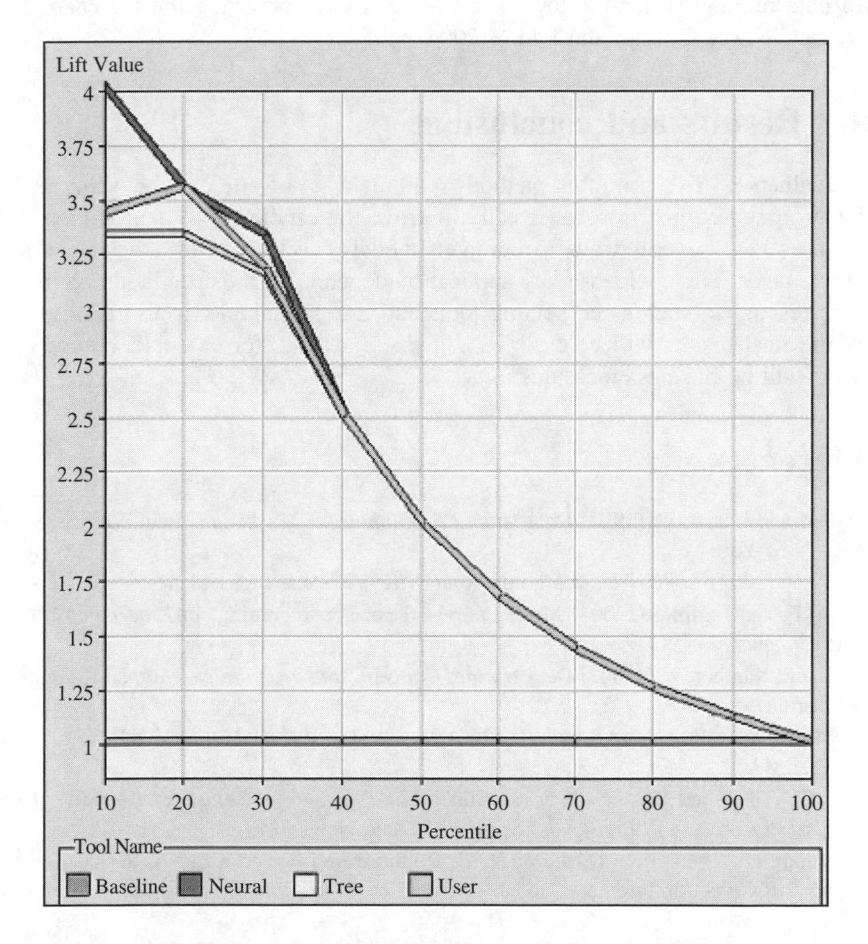

Figure 8.7 Lift chart. Paolo Giudici (2003) Applied Data Mining: Statistical Methods for Business and Industry. Reproduced with permission from John Wiley and Sons, Ltd.

lift value reached by the networks is above 4.00 This means that, by using this algorithm instead of MBR, we can improve our prediction by about 0.5 points.

In Figure 8.7, considering our observations, classified in percentiles, we measure on the ordinate the percentage of target events we can capture. Here we can see that the number of captured target events seems to be similar and the neural networks are perhaps better, but only after considering the first 30 % of observations. We can thus say that to improve the prediction based on the MBR algorithm, it is possible to use neural networks.

This work shows that it is possible to use predictive techniques on textual data. Starting from SVD results, we showed a series of important analyses to characterise and solve real problems on textual data. We proposed an MBR algorithm and, by using precision and recall concepts, we compared alternative predictive models with data mining evaluation tools such as the ROC curve and the lift chart. For more details, see Giudici and Figini (2004).

8.6 Results and conclusions

The evaluation of data mining methods requires a lot of attention. A valid model evaluation and comparison can greatly improve the efficiency of data mining. We have presented several ways to compare models, each with its advantages and disadvantages. The choice for any application depends on the specific problem and on the resources (such as computing tools and time) available. It also depends on how the final results will be used. In a business setting, for example, comparison criteria will be business measures.

References

Berry, M and Linoff, G. (1997) *Data Mining Techniques for Marketing*. John Wiley & Sons, Inc., New York.

Giudici, P. (2003) *Applied Data Mining*. John Wiley & Sons, Ltd, Chichester.

Giudici, P. and Figini S. (2004) Metodi previsivi per il text mining. In *Proceedings of the SAS Campus*, pp. 59–66.

Han, J. and Kamber, M. (2001) *Data Mining: Concepts and Techniques*. Morgan Kaufmann, San Francisco.

Hand, D.J., Mannila, H. and Smyth, P. (2001) *Principles of Data Mining*. MIT Press, Cambridge, MA.

Hastie, T., Tibshirani, R., Friedman, J. (2001) *The Elements of Statistical Learning: Data Mining, Inference and Prediction*. Springer-Verlag, New York.

Heckerman, D., Chickering, D.M., Meek, C., Rounthwaite, R. and Kadie, C. (2000) Dependency networks for inference, collaborative filtering and data visualization. *Journal of Machine Learning Research,* **1**, 49–75.

Lauritzen, S.L. (1996) *Graphical Models*. Oxford University Press, Oxford.

Seber, G.A.F. (1984) *Multivariate Observations*. John Wiley & Sons, Inc., New York.

9

Using statistical process control for continual improvement

Donald J Wheeler and Øystein Evandt

Process behavior charts, also known as control charts, are more than mere process monitors. They have a central role in learning how to operate any process up to its full potential, they complement and complete the work done by research and development, and by combining both the process performance and the process potential in one simple graph they are the locomotive of continual improvement.

This chapter begins with the specification approach to the problem of variation. Section 9.2 outlines Shewhart's alternate approach, while Sections 9.3 and 9.4 illustrate the two basic process behavior charts. Section 9.5 shows both how to learn from the charts and how the charts complement experimental studies. Section 9.6 combines the specification approach with Shewhart's approach to obtain four possibilities for any process. Section 9.7 outlines why and how Shewhart's simple approach will work with all kinds of data, while Section 9.8 lists some common misunderstandings about statistical process control (SPC).

9.1 The problem of variation

For the past two hundred years manufacturing has been focused on the mass production of interchangeable parts. This approach places a premium on each part fitting in with other parts so that assemblies will function properly. However, the fact that no two things are exactly alike prevents us from making completely uniform parts,

Statistical Practice in Business and Industry Edited by S.Y. Coleman, T. Greenfield, D.J. Stewardson and D.C. Montgomery

so we have had to settle for making similar parts. Having said this, the question becomes: how similar is similar? One of the earliest attempts to answer this question was the use of specifications. Those parts that deviated from the target value by at most a given amount were said to be acceptable, while those that deviated by more were said to be unacceptable.

While the specification approach to variation will let us characterise outcomes as acceptable and unacceptable, it does nothing to help us produce good outcomes. It merely lets us separate the good stuff from the bad stuff. So we ship the good stuff and then we try to figure out what we can do with the bad stuff. If we cannot make enough good stuff, we often speed up the line and then find out that we have a higher percentage of bad stuff than before. All of which results in what the legendary W. Edwards Deming (1900–1993) referred to as 'the Western approach to quality: burn the toast and scrape it'.

By focusing on the outcomes rather than on the process that produced those outcomes the specification approach seeks to deal with variation after the fact rather than at the source. When quality is thought of as something that is achieved by scraping burnt toast then it will, of necessity, be thought of as being contrary to productivity. The flow diagram in Figure 9.1 shows an operation where the quality audit figures were improved by increasing the size of the rework department. While the company continued to ship conforming product to their customers, and while they had good quality audit values, they went out of business using this approach. So, while specifications may be necessary to separate the good stuff from the bad stuff, when it comes to dealing with the problem of reducing variation around the target value in an economic manner the specification approach is not sufficient.

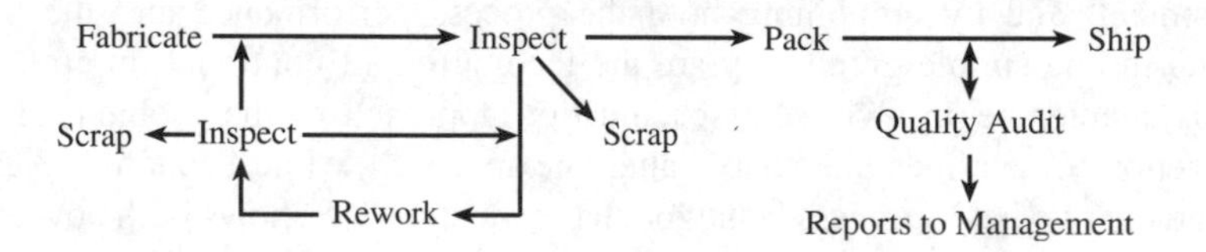

Figure 9.1 A typical Western manufacturing process.

The specification approach is a cut-your-losses approach: it seeks to identify those items that are so far off-target that it will be cheaper to scrap them than to use them. It is a compromise that was created to deal with the fact that we could not produce a completely uniform product stream, which was the original idea behind interchangeable parts.

9.2 Shewhart's approach

Walter Shewhart (1891–1967) came to understand the shortcomings of the specification approach when working at Western Electric and Bell Labs in the 1920s. In his work Shewhart realised that all variation could be classified as being either routine or exceptional, and this distinction provided the key to working with process variation in a rational way.

Every process will always display some amount of routine variation. Routine variation can be considered as inherent background variation that is part of all processes in this world. For Shewhart this variation was the inevitable result of a large number of cause-and-effect relationships where no one cause-and-effect relationship is dominant. As a consequence, routine variation will always be homogeneous and predictable within limits.

Exceptional variation, on the other hand, was thought of as being due to one or more dominant cause-and-effect relationships which are not being controlled by the manufacturer. Shewhart called such dominant, uncontrolled factors *assignable causes*, and argued that the *erratic behavior* they introduce into a process can be easily detected. Whenever erratic process behavior is detected, efforts to identify the underlying assignable causes, and to bring these causes under the control of the manufacturer, will generally prove to be economical. This economic benefit comes from the dominant, uncontrolled nature of assignable causes. When such dominant cause-and-effect relationships are brought under the control of the manufacturer, major sources of variation are removed from the production process. This allows more economical process operation, with less nonconforming product, and greater product uniformity. Reductions in the process variation of 50–80 % are not unusual.

- A process that displays nothing but routine variation is consistent over time and is therefore predictable within limits. Because of this consistency we may use the past as a guide to the future. Such a process may be said to be *statistically stable*.

- A process that displays exceptional variation will change from time to time, and may be said to be *statistically unstable*. Because of these changes we cannot depend upon the past to be a reliable guide to the future. However, the detection of exceptional variation does present an opportunity for process improvement.

Shewhart developed several different techniques for separating the routine variation that is inherent in all processes from any exceptional variation that may be present. Among these techniques was one which placed limits around running records – the idea being that since exceptional variation is dominant, it will be found outside limits that bracket the routine variation, as illustrated in Figure 9.2. Thus the main problem that remained was how to define these limits.

If the limits in Figure 9.2 are too close together we will get *false alarms* (or false signals) when routine variation causes a point to fall outside the lines by chance. We can avoid this mistake entirely by computing limits that are far apart. But if we have the limits too far apart we will *miss some signals* of exceptional variation.

Thus, Shewhart sought to strike a balance between the economic consequences of these two mistakes. In his ground-breaking book, *Economic Control of Quality of Manufactured Product* (1931), he came up with a way of computing limits to reliably bracket virtually all of the routine variation, but little more. This technique

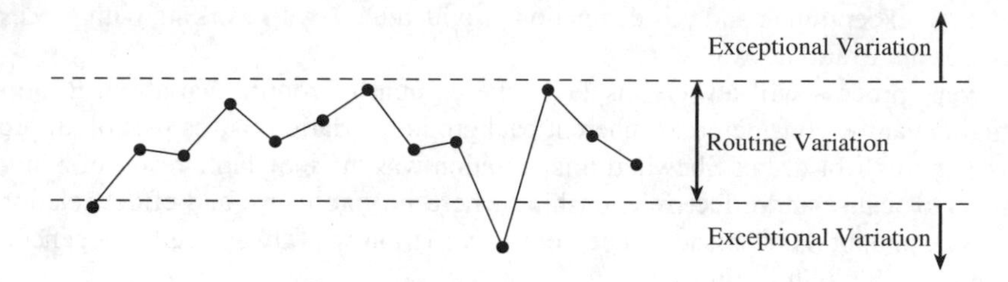

Figure 9.2 Separating routine variation from exceptional variation.

of computing limits allows one to separate routine variation from exceptional variation, thereby identifying potential signals that deserve to be investigated.

Once Shewhart found out how to filter out virtually all of the routine variation, he knew that any remaining values could be reasonably interpreted as signals of exceptional variation. The twin foundations of Shewhart's approach were (1) the use of symmetric, three-sigma limits based upon (2) the within-subgroup variation. Why symmetric, three-sigma limits work is explained in Section 9.7. In using the within-subgroup variation to filter out the routine variation Shewhart's approach is built on the same foundation as the analysis of variance and other commonly used techniques of statistical analysis.

The running record with Shewhart's limits attached is usually known as a control chart. However the present authors prefer the term *process behavior chart,* partly because this term better describes what the chart really depicts, and partly because of the many misleading connotations attached to the word 'control' in situations where such charts are useful.

To explain the motivation for this terminology we return to Shewhart's original definition: 'A phenomenon will be said to be controlled when, through the use of past experience, we can predict, at least within limits, how the phenomenon may be expected to vary in the future.' Any way you read this statement, the essence of Shewhart's concept of control is predictability. He was not concerned with choosing a probability model, or with statistical inferences, but simply with the idea of whether or not the past can be used as a guide to the future. We could easily paraphrase Shewhart as follows: A *process* will be said to be *predictable* when, through the use of past experience, we can *describe*, at least within limits, how the *process* will *behave* in the future. Hence the 'control chart' has always been a device for characterising the behavior of a process and so the authors prefer to call it what it is, a process behavior chart.

9.3 Learning from Process Data

Shewhart decided to examine the time-order sequence of the data for clues to the existence of dominant causes of exceptional variation. Rather than collecting the data into one big histogram and looking at the shape, he decided to arrange the data in a series of small subgroups, where each subgroup would consist of a few values that were logically collected under the same conditions, and then to use average variation within these subgroups to estimate the amount of routine variation present in these data. If the average within-subgroup variation could completely explain the variation in the data set as a whole, then the underlying cause system of variation was likely to be a homogeneous set of common causes of routine variation. However, if the variation for the data set as a whole exceeds that which can be explained by the variation within the subgroups, then the cause system is likely to contain assignable causes in addition to the common causes.

Thus, Shewhart's solution to the problem was to look at the data *locally* rather than *globally*. Based on how the data were obtained, he would organise them into small, logically consistent, local subgroups, and then compute local, within-subgroup summaries of variation for each subgroup. These local summaries would capture the routine variation due to the many common causes that are always present, but they would also be relatively uncontaminated by any less frequent, but dominant, assignable causes of exceptional variation. Therefore, by characterising the routine variation of the common causes, Shewhart's local analysis provided a way to detect the exceptional variation that would indicate the presence of dominant assignable causes of exceptional variation.

This process is illustrated in Figure 9.3. On the left-hand side periodic, local subgroups of four values each are obtained from a predictable process. These subgroups are summarised by their averages and ranges, which are plotted in the charts at the bottom. Since the process on the left is predictable, the averages and ranges stay within their limits.

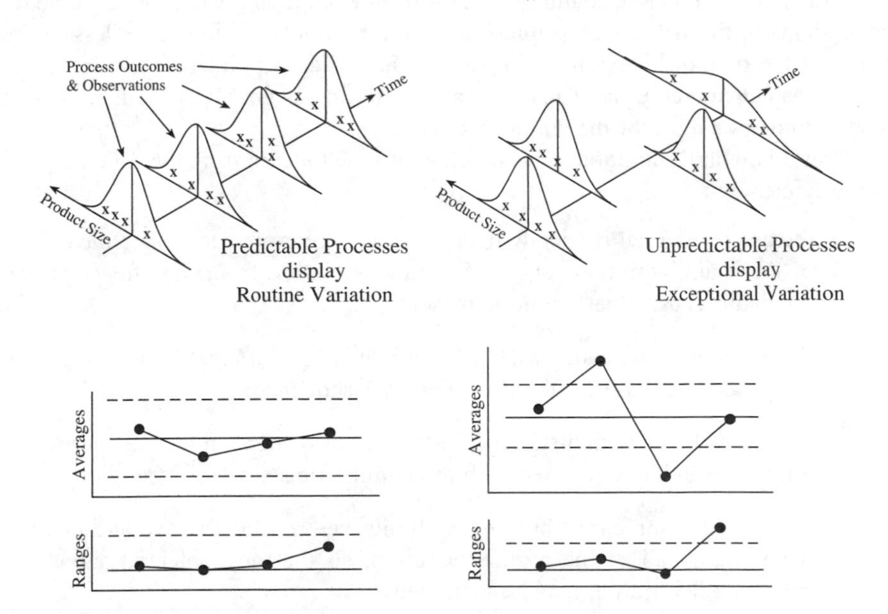

Figure 9.3 Using subgroups to characterise predictable and unpredictable processes.

On the right-hand side periodic, local subgroups of four values each are obtained from an unpredictable process. The subgroup averages and subgroup ranges do not stay within their limits on the charts, thereby signaling the unpredictable nature of the process on the right. Changes in location show up on the average chart, while changes in dispersion show up on the range chart. Sections 9.3 and 9.4 discuss how to calculate the limits.

The first premise of Shewhart's solution to the problem of learning from the data is therefore known as *rational subgrouping*. We have to organise the data into logical subgroups, and this will always involve an element of judgment. The idea of using local subgroups is to capture the routine, background variation within those subgroups. Once we have done this we can characterise that routine variation and use it as a filter to separate out any exceptional variation that is present in the data set as a whole. Therefore, in defining our subgroups, we will need to take into account the way the data were obtained, the way the process is operated, the types of cause-and-effect relationships that might be present, and any other relevant contextual information. Moreover, we should be able to explain the basis for our subgrouping. Rational subgroups are created on the basis of judgment, using process knowledge and experience, in order to capture the routine background variation that is due to the common causes.

Having partitioned the data into rational subgroups, and having used the variation within those subgroups to characterise the routine variation, there remains the problem of how to separate the routine variation from any exceptional variation that may be present. To do this Shewhart chose to use symmetric, three-sigma limits as his filter to separate routine variation from exceptional variation. Symmetric three-sigma limits will cover virtually all of the routine variation regardless of the shape of the overall histogram. This means that when a point falls outside these limits it is considerably more likely to be due to an assignable cause than it is to be due to one or more of the common causes.

Thus, Shewhart's solution to the problem of how to learn from the data involved four key elements:

- the use of rational (logically homogeneous) subgroups to isolate the effects of the many common causes of routine variation from the effects of any dominant causes that might be present;
- the plotting of the data, or the subgroup averages, along with the ranges, in a logical order (most often the time-order sequence);
- the use of local, within-subgroup estimates of dispersion to characterise the routine variation associated with the common causes;
- the use of symmetric, three-sigma limits centred on the average to filter out virtually all of the routine variation, so that any potential signals of exceptional variation may be identified.

These four elements are the hallmarks of Shewhart's solution. We focus on them in this chapter because they are often overlooked in discussions of SPC.

9.4 The average and range chart

Shewhart used his solution to develop five different techniques for examining data for clues to the nature of the underlying cause system. His Criterion I was an average and root mean square deviation chart. Because of the complexity of computing the root mean square deviation in the days before electronic calculators, this chart was modified in practice to become an average and range chart. Furthermore, the average and range chart was also easier to explain to nontechnical personnel, which is still true today.

The data in Table 9.1 consist of 44 determinations of the resistivity of an insulating material measured in megohms. To place these data on an average and range chart we will need to partition them into logically homogeneous subgroups. As Shewhart expressed it, when we place two values in the same subgroup we are making a judgment that the two values were collected under essentially the same conditions. This requirement of homogeneity within the subgroups will favor small subgroup sizes. At the same time, increasing the subgroup size will increase the sensitivity of the average chart. With these contradictory pressures, it turns out that sample sizes of 4 and 5 are quite common. In this case Shewhart arranged these 44 values in time order and then divided them into 11 subgroups of size 4. The averages and ranges for each subgroup are shown in Table 9.1.

Table 9.1 Forty-four determinations of resistivity arranged into 11 subgroups.

5045	4290	3980	3300	5100	4635	4410	4725	4790	4110	4790
4350	4430	3925	3685	4635	4720	4065	4640	4845	4410	4340
4350	4485	3645	3463	5100	4810	4565	4640	4700	4180	4895
3975	4285	3760	5200	5450	4565	5190	4895	4600	4790	5750
4430.0	***4372.5***	***3827.5***	***3912.0***	***5071.25***	***4682.5***	***4557.5***	***4725.0***	***4733.75***	***4372.5***	***4943.75***
1070	***200***	***335***	***1900***	***815***	***245***	***1125***	***255***	***245***	***680***	***1410***

The average and range chart (Figure 9.4) starts with two running records: one for the subgroup averages and one for the subgroup ranges. The average values for each running record are commonly used as the central lines. For Table 9.1 the grand average is 4512, while the average range is 753. The three-sigma limits for each portion of the chart are placed symmetrically on either side of the central lines. The formulas for these limits have the following form:

$$\text{Grand Average} \pm 3\ \text{Sigma}(\overline{X}),$$

$$\text{Average Range} \pm 3\ \text{Sigma}(R).$$

Special scaling factors were developed during World War II to facilitate the conversion of the average range into the appropriate measurements of dispersion seen in the formulas above. For an average and range chart these scaling factors are known as A_2, D_3, and D_4. Tables of these and other scaling factors are commonly

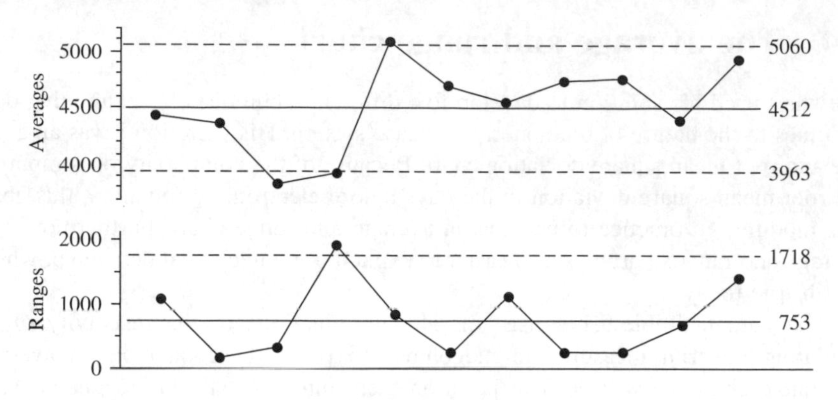

Figure 9.4 Average and range chart for the subgroups of Table 9.1.

found in SPC textbooks. Some simple algebra gives the following formulas:

$$\text{Grand Average} \pm 3\ \text{Sigma}(\overline{X}) = \text{Grand Average} \pm A_2\ \text{Average Range},$$

$$\text{Average Range} + 3\ \text{Sigma}(R) = D_4\ \text{Average Range},$$

$$\text{Average Range} - 3\ \text{Sigma}(R) = D_3\ \text{Average Range}.$$

From Table 9.2, for $n = 4$, these scaling factors are $A_2 = 0.729$ and $D_4 = 2.282$. (Since ranges cannot be negative the range chart will not have a lower limit until $n > 6$.) When combined with the grand average and the average range of data from Table 9.1, these formulas result in the limits shown in Figure 9.4.

The limits in Figure 9.4 represent the routine variation. Points outside the limits of the average chart indicate changes in location. Points above the upper range

Table 9.2 Scaling factors

For average and range charts using the average range				For average and standard deviation charts using the average standard deviation			
n	A_2	D_3	D_4	n	A_3	B_3	B_4
2	1.880	–	3.268	2	2.659	–	3.267
3	1.023	–	2.574	3	1.954	–	2.568
4	0.729	–	2.282	4	1.628	–	2.266
5	0.577	–	2.114	5	1.427	–	2.089
6	0.483	–	2.004	6	1.287	0.030	1.970
7	0.419	0.076	1.924	7	1.182	0.118	1.882
8	0.373	0.136	1.864	8	1.099	0.185	1.815
9	0.337	0.184	1.816	9	1.032	0.239	1.761
10	0.308	0.223	1.777	10	0.975	0.284	1.716

limit indicate excessive variation within the subgroups. In this case the four points outside the limits of Figure 9.4 were interpreted as evidence that this process was subject to one or more assignable causes. These assignable causes were identified and when the levels of these assignable causes were controlled the variation in the resistivity readings was cut in half. Thus Figure 9.4 shows a process where the changes were large enough that it was economical to spend the time and effort to identify the assignable causes and to remove their effects.

Moreover, since finding and controlling assignable causes of exceptional variation involves nothing more than operating a process up to its full potential, these improvements do not generally require any substantial capital outlay. In most cases any expenditures can be easily covered out of the operating budget. This is why points outside the limits on a process behavior chart are opportunities for improvement that should always be investigated.

Since the scaling factors are specific to the way the within-subgroup variation is summarised, there are several different sets of scaling factors. Two of the more commonly used sets are given in Table 9.2.

The use of local, within-subgroup variation is one of the hallmarks of Shewhart's solution, therefore the reader is cautioned against the use of global measures of dispersion to obtain limits. (Global measures of dispersion will be severely inflated by any exceptional variation that is present, resulting in limits that will be too wide to work properly.) The correct ways of computing limits for average charts will *always* use the average or median of k within-subgroup measures of dispersion. Since virtually all of the software available today contains options which compute the limits incorrectly, it is imperative that you verify that you are getting the correct limits from your software. (To do this, use a data set that contains points outside the limits on the average chart, compute the limits by hand, and then compare these with the limits produced by the various options in your software. Any software option that does not result in limits that are reasonably close to your limits is wrong.)

9.5 The chart for individual values and moving ranges

Shewhart's approach was later extended to create charts for individual values. Given a sequence of values that are logically comparable, we can characterise the short-term variation by using the absolute differences between successive values, which are commonly known as the *two-point moving ranges*. The rationale behind this choice is that when the successive values are logically comparable, most of these moving ranges will tend to represent the routine variation, and their average will provide a reasonable estimate thereof. Therefore, the average moving range, with an effective subgroup size of $n = 2$, is used to establish limits for both the individual values and the moving ranges. The resulting chart for individual values and moving ranges is commonly known as an *XmR* chart.

To illustrate the *XmR* chart we shall use the percentage of incoming shipments for one electronics plant that were shipped by premium freight. These values are shown in Table 9.3 along with their moving ranges. Had they been tracking these data on an *XmR* chart they might have computed limits at the end of Year One.

Table 9.3 Premium freight percentages and moving ranges

Year One			Year Two			Year Three		
Month	*X*	*mR*	Month	*X*	*mR*	Month	*X*	*mR*
			Jan	6.7	0.1	Jan	8.3	0.6
			Feb	4.5	2.2	Feb	8.7	0.4
			Mar	6.1	1.6	May	9.4	0.7
			Apr	5.7	0.4	Apr	8.0	1.4
May	6.1		May	5.9	0.2	May	11.8	3.8
Jun	6.0	0.1	Jun	6.8	0.9	Jun	13.2	1.4
Jul	5.8	0.2	Jul	6.1	0.7	Jul	11.8	1.4
Aug	4.8	1.0	Aug	6.6	0.5			
Sep	3.6	1.2	Sep	7.0	0.4			
Oct	4.7	1.1	Oct	9.2	2.2			
Nov	5.9	1.2	Nov	9.7	0.5			
Dec	6.8	0.9	Dec	8.9	0.8			

At the end of Year One the average is 5.46 and the average moving range is 0.81. Three-sigma limits for the *X* chart may be found using:

$$\text{Average} \pm 3\ \text{Sigma}(X) = \text{Average} \pm 2.660\ \text{Average Moving Range}$$
$$= 5.46 \pm 2.660 \times 0.81 = 3.31 \text{ to } 7.61.$$

(The value of 2.660 is yet another of the many scaling factors associated with process behavior charts. This one is for using the average moving range to obtain limits for individual values.) These limits for individual values are commonly known as *natural process limits*. The *X* chart will consist of the running record for *X* with the average and the natural process limits added.

Since the two-point moving ranges define a series of subgroups of size 2, the *mR* chart will look like a range chart for $n = 2$. It will consist of the running record of the moving ranges with the average moving range and the upper range limit added. In this case,

$$URL = D_4\ \text{Average Moving Range} = 3.268\ \text{Average Moving Range}$$
$$= 3.268 \times 0.81 = 2.65.$$

Based on the last three values in Table 9.3, the General Manager asked for an explanation for the increase in premium freight. The *X* chart in Figure 9.5 shows

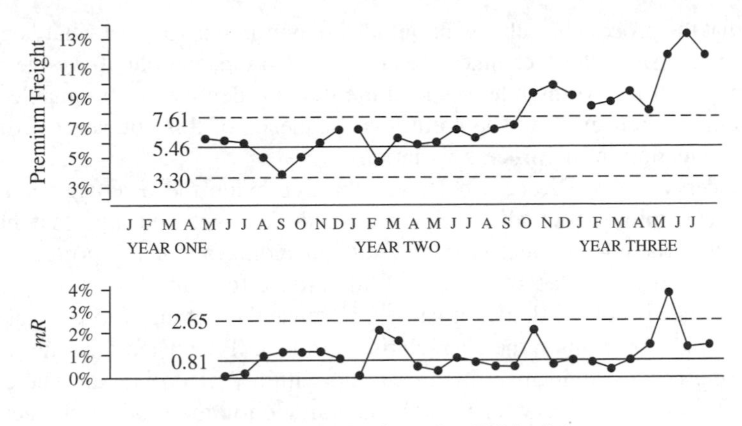

Figure 9.5 XmR chart for Table 9.3.

that he should have been asking this question in October of Year Two! The moving range above the upper range limit identifies the jump in May of Year Three as being too large to be due to routine variation. This process was not only gradually drifting upward, but was also subject to sudden and dramatic changes in level.

This example also illustrates what Shewhart observed: one need not wait until large amounts of data have been obtained – useful limits may be found using as few as eight data. This is possible because, rather than being used to estimate some parameter, the limits are merely being used to characterise the process behavior. To achieve this objective, high precision is not required.

9.6 So what can we learn from a process behavior chart?

The computations described in the previous sections have a very special property – they allow us to obtain reasonable limits for the routine variation even when the data come from a process that displays exceptional variation. As a result, the limits approximate what a process can do when it is operated with maximum consistency. They characterise the ideal of what you are likely to see when the process is operated with minimum variation. They effectively define the process potential.

At the same time, the running records display the actual process performance. If points fall outside the limits then the process is not being operated up to its full potential. By displaying both the process potential and the process performance on the same graph, a process behavior chart provides a way to judge to what extent a process is falling short of achieving its full potential.

Moreover, when a process is not operating up to its full potential, those points that fall outside the limits are signals of exceptional variation that deserve to be investigated. By pinpointing those occasions where the effects of assignable causes are most likely to be found the process behavior chart provides a methodology for process improvement. According to Shewhart, points outside the limits are not only signals that the process has changed, but also opportunities to learn what is causing these changes. Since these changes are large enough to stand out above the routine variation, it will be worthwhile to spend the time to identify the assignable causes and to remove their effects. The natural consequence of this course of action is a dramatic reduction in the process variation.

To understand how process behavior charts provide the leverage for process improvement, think about all of the cause-and-effect relationships that have an impact upon a single product characteristic (including all of the primary factors and their interaction effects as well). With little effort this list will commonly contain dozens, if not hundreds, of causes. Fortunately, not all of these causes will have the same impact upon the characteristic. In fact, if we listed them in the order of their impact we would most likely end up with a Pareto diagram where a few of the causes would account for most of the variation in the product characteristic. Since it costs money to control any cause-and-effect relationship, we will want to control the critical few and ignore the trivial many.

To this end R&D will perform experiments to identify the critical few. However, R&D will have to do this with limited time and limited resources. As a result, there will always be some cause-and-effect relationships that do not get studied. In Figure 9.6 R&D studied what appeared to be the top ten cause-and-effect relationships and found factors 5, 1, and 7 to have the dominant impacts upon the product characteristic. So they told Manufacturing to control factors 5, 1, and 7. Manufacturing went into production while holding factors 5, 1, and 7 at the recommended levels and found that the product characteristic varied too much to meet the specifications. Next they spent the money to control factor 4, but they still had too much variation in the product.

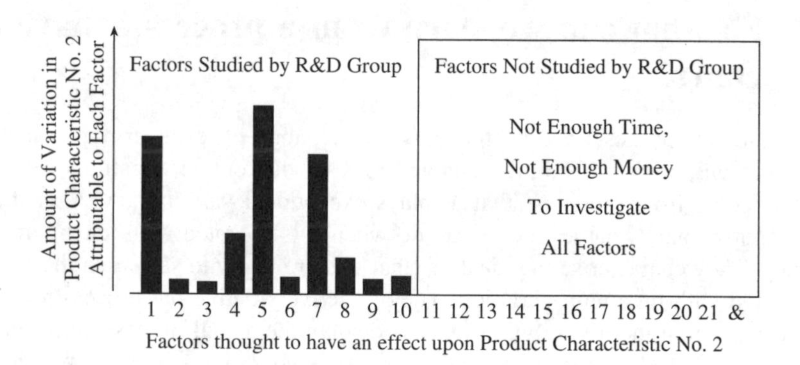

Figure 9.6 Identifying which factors to control in production.

Why could Manufacturing not make good stuff? The answer is seen in Figure 9.7. When factor 14 goes south, the process goes south. When factor 16 goes on walkabout, the process also goes on walkabout. While factors 14 and 16 were not studied by R&D, and while you may not know their names, they will still have their impact upon production. It is precisely these unknown, dominant

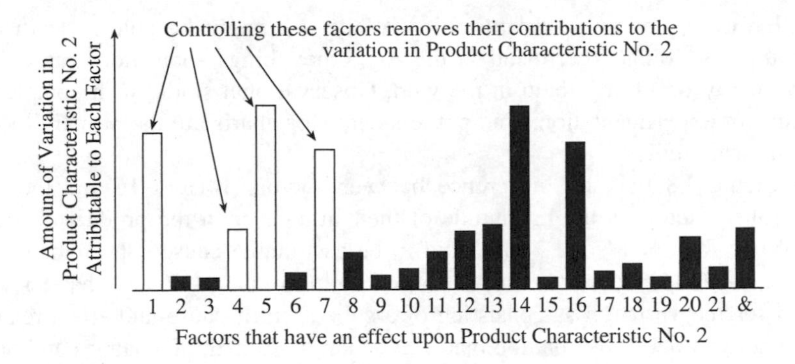

Figure 9.7 The rest of the story.

cause-and-effect relationships that can be found and identified by using process behavior charts.

Once we identify a dominant factor and control the level of that factor, it will no longer contribute to the variation in the product stream. Virtually all of the variation in the product stream will come from the 'uncontrolled factors'. As seen in Figure 9.8, as long as there are dominant cause-and-effect relationships in the set of uncontrolled factors, you will have high payback opportunities for process improvement.

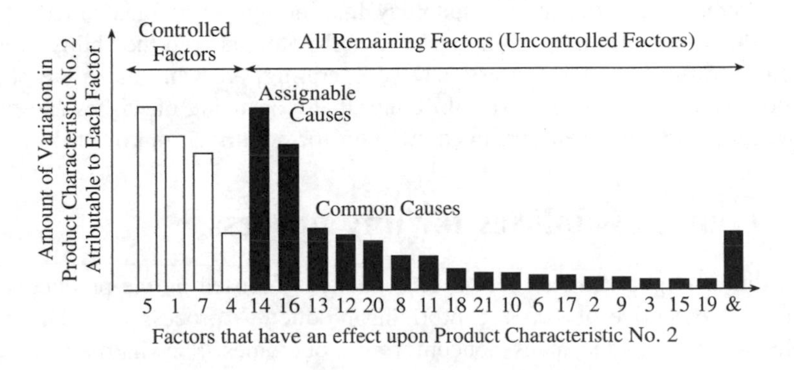

Figure 9.8 Controlled and uncontrolled factors.

By identifying the dominant cause-and-effect relationships that remain in the set of uncontrolled factors and then moving them over into the set of controlled factors, you can remove substantial amounts of variation from the product stream. This is how process behavior charts and observational studies complement a program of research and experimentation. By their very nature, experiments cannot study all of the factors – some factors will always be held constant or ignored. On the other hand, observational studies listen to the process while it is under the influence of all of the cause-and-effect relationships. While we might be able to learn a lot about lions by doing experiments at the zoo, other things about lions can only be discovered by observing them in the wild. Observational studies can complement programs of experimentation. And process behavior charts are the primary tool for observational studies.

In Figure 9.8 the only difference between factors 14 and 16 and the other uncontrolled factors is the magnitude of their impact upon the product characteristic. While factors 14 and 16 are said to be assignable causes, the remainder of the uncontrolled factors can be said to be common causes. Recall that Shewhart defined routine variation as consisting of a collection of cause-and-effect relationships where no one cause predominates. As soon as something changes and one of these common causes raises its head and begins to dominate the others it becomes an assignable cause.

This means that as you remove assignable causes from the set of uncontrolled factors you will reduce the level of variation in the process. When you lower the water level you will often find more rocks. As the process variation is reduced, that which originally looked like routine variation may turn out to contain some signals of exceptional variation. As these new sources of exceptional variation are investigated more assignable causes may be identified, resulting in new opportunities for process improvement. This peeling of successive layers of the onion is the way that many have learned to use the process behavior charts as the locomotive of continual improvement.

These three elements of Shewhart's approach, the definition of the ideal of what the process can do, the methodology for finding the assignable causes that prevent the process from operating up to its full potential, and the ability to judge how close to the ideal the process may be operating, are what make the process behavior chart approach so powerful. Rather than consisting of wishes and hopes, it provides a complete, self-contained program for continual improvement.

9.7 Four possibilities for any process

The traditional approach based on specifications is focused on the product stream and has the objective of 100 % conforming product. A process is said to be 'in trouble' when it has too many nonconforming outcomes. Shewhart's approach is focused on the process behavior and has the objective of operating a process up to its full potential. Throughout his books, Shewhart used 'trouble' as a synonym for an unpredictable process. When these two definitions of trouble are combined we end up with four possibilities for any process:

- conforming product and predictable process (no trouble);
- nonconforming product and predictable process (product trouble);
- conforming product and unpredictable process (process trouble);
- nonconforming product and unpredictable process (double trouble).

These four possibilities can and should be used to characterise every process. Moreover, any attempt to characterise a process using only one definition of trouble will be incomplete and unsatisfactory. As we will see, the four possibilities of Figure 9.9 provide a simple way to determine what type of improvement effort is appropriate. In our discussion of these four states we shall begin with the no trouble state.

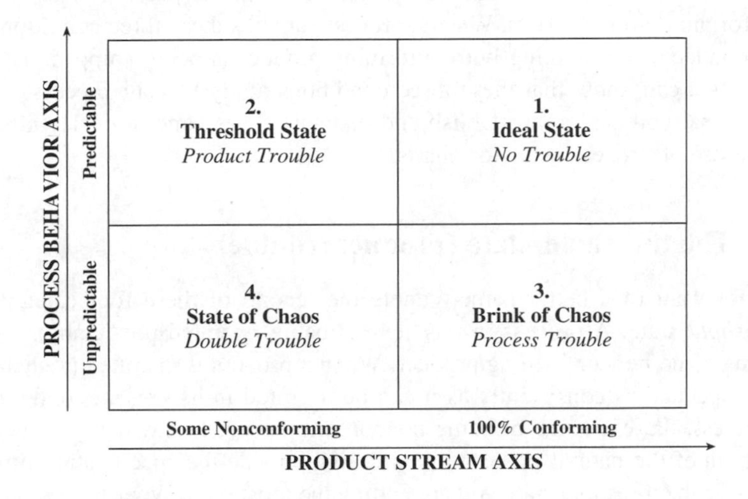

Figure 9.9 Combining the two approaches to variation.

9.7.1 The ideal state (no trouble)

For want of a better name, denote the first of these four categories as the *ideal* state. A process in this state is being operated predictably and is producing 100 % conforming product. Operating a process predictably requires constancy of purpose and the effective use of process behavior charts. The conformity of the product will be the result of having natural process limits (limits for individual values) that fall inside the specification limits.

When your process is in the ideal state, you and your customer can expect the conformity of the product to continue as long as the process is operated predictably. Since the product stream for a predictable process can be thought of as being homogeneous, the measurements taken to maintain the process behavior chart will also serve to characterise the product produced by the predictable process.

How does a process arrive at this ideal state? Only by satisfying three conditions:

- The management must establish and maintain an environment where it is possible to operate the process in a predictable and consistent manner. The operating conditions or target values cannot be selected or changed arbitrarily.

- The process average must be maintained reasonably close to the appropriate target value.

- For those characteristics having specifications, the natural process limits must fall inside the specification limits.

Whenever one of these conditions is not satisfied, the possibility of shipping nonconforming product exists. When a process satisfies these three conditions, you can be confident that nothing but conforming product is being shipped. The only way that you can know that these three conditions apply to your process, and the only way that you can both establish and maintain these conditions day after day, is by the use of process behavior charts.

9.7.2 The threshold state (product trouble)

Again, for want of a better name, denote the second of these four categories as the *threshold* state. A process in this state is being operated predictably, but it is producing some nonconforming product. When a process is operated predictably it is being operated as consistently as it can be operated in its present configuration. Nevertheless, the existence of some nonconforming product will be the result of one or both of the natural process limits falling outside the specification limits.

Thus, in the threshold state you are getting the most out of your process, yet that is still not good enough to keep you out of product trouble. You cannot simply wait for things to spontaneously improve, so you will have to intervene to either change the process or change the specifications before you can operate in the ideal state.

If the nonconforming product occurs because the process average is not where it needs to be, then you will need to find some way of adjusting and maintaining the process aim. Once again, the same process behavior chart that facilitates the predictable operation can facilitate adjusting the aim.

If the nonconforming product occurs because the process shows too much variation, then you will need to try to reduce the process variation. Since a process that is operated predictably is one that is already being operated with minimum variation (given its current configuration), the only way to reduce the process variation will be to change the process itself in some major way. As you experiment with major process changes the process behavior chart will allow you to evaluate the effects of your changes. Thus, process behavior charts will help you not only achieve a predictable process, but also in moving the process from the threshold state to the ideal state. It should also be mentioned that you are more likely to get useful results when experimenting with a process after it has been brought into a predictable state by means of a process behavior chart, than if you are experimenting with the process while it is unpredictable. This is so because any process changes that are observed while one is experimenting with an unpredictable process can be due to changes in the unknown assignable causes, rather than being the result of the experimental changes.

Since making a major change in the process will often require a lot of work and expense you may, instead, try to get the specifications relaxed. With the process behavior chart to demonstrate and define the consistent voice of the process, you will at least have a chance to get the customer to agree to a change in the specifications. Without the process behavior chart to demonstrate the predictability of your process you will have no basis for asking for a change in the specifications.

As always, a short-term solution to the existence of nonconforming product is to use 100 % inspection. However, as has been proven over and over again, 100 % screening of product is imperfect and expensive. The only way to guarantee that you will not ship any nonconforming product is to avoid making any in the first place. Sorting should be nothing more than a stopgap measure – not a way of life.

Thus, process behavior charts are not only essential in getting any process into the threshold state, but they are also critical in any attempt to move from the threshold state to the ideal state.

9.7.3 The brink of chaos (process trouble)

The third state could be labeled the *brink of chaos*. Processes in this state are being operated unpredictably even though the product stream currently contains 100 % conforming product. With the traditional view the fact that you have 100 % conforming product is considered to be evidence that the process is 'operating OK'. Unfortunately, this view inevitably leads to benign neglect, which, in conjunction with the unpredictable operation of the process, results in a process whose conformity can disappear at any time. (Have you noticed that trouble tends to appear *suddenly*?)

When a process is operated unpredictably it is subject to the effects of assignable causes. These effects can best be thought of as changes in the process that seem to occur at random times. So while the conformity to specifications may lull the producer into thinking all is well, the assignable causes will continue to change the process until it will eventually produce some nonconforming product. The producer will suddenly discover that he is in product trouble, yet he will have no idea of how he got there, nor any idea of how to get out of trouble. (Of course the sense of being in trouble will cause 'solutions' to be tried until, by some means, the conformity improves. However, if the issue of unpredictable operation is not addressed the process will be doomed to recurring bouts of 'trouble'.) The change from 100 % conforming product to some nonconforming product can come at any time, without the slightest warning. When this change occurs the process will be in the state of chaos.

Thus, there is no way to predict what a process in the brink of chaos will produce tomorrow, or next week, or even in the next hour. Since the unpredictability of such a process is due to assignable causes, and since assignable causes are dominant cause-and-effect relationships that are not being controlled by the manufacturer, the only way to move from the brink of chaos to the ideal state is first to eliminate the effects of the assignable causes. This requires that they be identified, and the operational definition of an assignable cause is the occurrence of signals on the process behavior chart.

9.7.4 The state of chaos (double trouble)

The *state of chaos* exists when an unpredictable process is producing some nonconforming product. The nonconforming product will alert the producer to the fact that there is a problem. The fact that the process is being operated unpredictably means that some of the dominant causes of variation that govern the process have not yet been identified. Thus, a manufacturer whose process is in the state of chaos knows that there is a problem, but he usually does not know what to do to correct it. Moreover, efforts to correct the problem will ultimately be frustrated by the random changes in the process which result from the presence of the assignable causes. When a needed modification to the process is made, its effect may well be short-lived because unknown assignable causes continue to change the process. When an unnecessary modification is made, a fortuitous shift by the assignable causes may mislead everyone. No matter what is tried, nothing works for long because the process is always changing. As a result, people finally despair of ever operating the process rationally, and they begin to talk in terms of 'magic' and 'art'.

As may be seen in Figure 9.10, the only way to make any progress in moving a process out of the state of chaos or the brink of chaos is first to eliminate the assignable causes. This will require the disciplined and effective use of process behavior charts. As long as assignable causes are present, you will find your improvement efforts to be like walking in quicksand. The harder you try, the more deeply mired you become.

9.7.5 The effect of entropy (the cause of much trouble)

All processes for which specification limits are defined belong to one of the four states of Figure 9.10. But processes do not always remain in one state. It is possible for a process to move from one state to another. In fact there is a universal force acting on every process that will cause it to move in a certain direction. That force is *entropy* and it continually acts upon all processes to cause deterioration and decay, wear and tear, breakdowns, and failures.

Entropy is relentless. Because of it every process will naturally and inevitably migrate toward the state of chaos. The only way this migration can be overcome is by continually repairing the effects of entropy. Of course, this means that the assignable causes of effects for a given process must be known before they can be removed. With such knowledge, the repairs are generally fairly easy to make. On the other hand, it is very difficult to repair something when you are unaware of the cause of the deficiency. Yet if the effects of entropy are not repaired, they will come to dominate the process and force it inexorably toward the state of chaos.

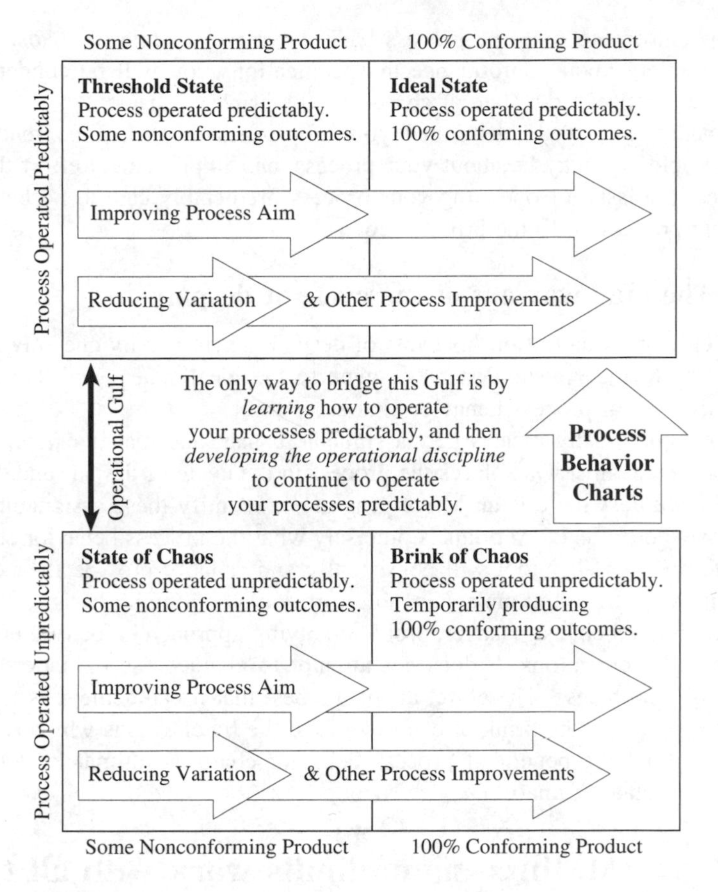

Figure 9.10 Improvement efforts and process behavior charts.

9.7.6 The cycle of despair (the result of the old definition of trouble)

Since everybody knows that they are in trouble when their processes are in the state of chaos, it is inevitable that problem-solvers will be appointed to drag the process out of the state of chaos. With luck, these problem-solvers can get the process back to the brink of chaos – a state which is erroneously considered to be 'out of trouble' in most operations.

Once they get the process back to the brink of chaos the problem-solvers are sent off to work on other problems. As soon as their backs are turned, the process begins to move back down the entropy slide toward the state of chaos.

New technologies, process upgrades, and all the other 'magic bullets' which may be tried can never overcome this cycle of despair. You may change technologies – often a case of jumping out of the frying pan and into the fire – but the benign neglect which inevitably occurs when the process is on the brink of chaos will allow entropy to drag the process back down to the state of chaos. Thus, if you focus solely upon conformance to specifications you will be condemned to forever cycle between the state of chaos and the brink of chaos.

No matter how well intentioned your improvement efforts, no matter how knowledgeable you may be about your process, any improvement effort that does not address the issue of operating your process predictably can do no better than to get your process up to the brink of chaos.

9.7.7 The only way out of the cycle of despair

There is only one way out of this cycle of despair. There is only one way to move a process up to the threshold state, or even to the ideal state – and that requires the effective use of process behavior charts.

Entropy continually creates new assignable causes and makes existing causes grow worse over time. This places the process in the cycle of despair and dooms it to stay there unless there is an active program to identify these assignable causes and remove their effects. And this is precisely what the process behavior chart was created to do. No other tool will consistently and reliably provide the necessary information in a clear and understandable form on a continuing basis.

The traditional chaos-manager, problem-solving approach is focused upon conformance to specifications. It does not attempt to characterise or understand the behavior of the process. Therefore, about the best that it can achieve is to get the process to operate in the brink of chaos some of the time. This is why any process operated without the benefit of process behavior charts is ultimately doomed to operate in the state of chaos.

9.8 How can three-sigma limits work with all types of data?

All of the limits on the various process behavior charts are computed according to the same principle; they are all symmetric, three-sigma limits. But how can a single, unified approach work with different variables and with different types of data? To discover the answer to this question it is instructive to compare Shewhart's approach to the problem of separating routine variation from exceptional variation with the traditional approach of statistical inference. To see the difference in these two approaches consider the equation:

$$\int_A^B f(x)\,dx = P.$$

The statistical approach to this equation can be summarised as follows:

- *Fix* the value of P to be some value close to 1.0 (such as 99 %).
- Then, for a *specific* probability model $f(x)$.

- find the corresponding critical values A and B.
- Values outside the interval of A to B are considered to be exceptional values for this specific probability model.

Shewhart looked at this structure, which was already well defined by the 1920s, and noted that any probability model must be interpreted as a limiting characteristic of an infinite sequence of data, and therefore cannot be said to characterise any finite portion of that sequence. Moreover, since histograms will always have finite tails, we will never have enough data to ever fully specify a probability model $f(x)$ to use in the argument above.

However, Shewhart realised that an *approximate* solution could be obtained by reversing the whole process: Instead of fixing the value for P, he chose to fix the critical values A and B. Thus we have Shewhart's approach to the equation above:

- *Fix* the values of A and B such that,
- for *any* probability model $f(x)$,
- the corresponding value of P will be reasonably close to 1.00.
- Then values that fall outside the interval of A to B may be considered to be exceptional values regardless of what probability model might characterise the routine variation.

Thus, Shewhart's approach to the equation above is exactly the reverse of the traditional statistical approach. Given Shewhart's approach, the only thing that remained to do was to choose A and B, and for this Shewhart simply used empirical evidence. Based on data sets from existing processes at Western Electric and throughout the Bell System, Shewhart quickly determined that symmetric, three-sigma limits would filter out virtually all of the routine variation, but very little more, and thereby allow us to detect any exceptional variation that might be present.

In support of this empirical choice, consider how symmetric, three-sigma limits work with the six standardised probability models in Figure 9.11. These models range from the uniform distribution to the exponential distribution. The proportion of the total area covered by the symmetric, three-sigma limits is shown for each distribution.

Empirical evidence suggests that placing A and B at a distance of three estimated standard deviations on either side of the average will result in values of P that are reasonably close to 1.00. Extensive studies using hundreds of probability models have verified this. With mound-shaped probability models symmetric, three-sigma limits will have values of P that exceed 0.98. With all but the most extreme J-shaped probability models the use of symmetric, three-sigma limits will result in values of P that exceed 0.975. And with bimodal models that are not heavily skewed, three-sigma limits will have a value of P that is 1.00. It is only with the most extreme J-shaped distributions or with heavily skewed bimodal distributions that three-sigma limits result in values of P that are less than 0.975.

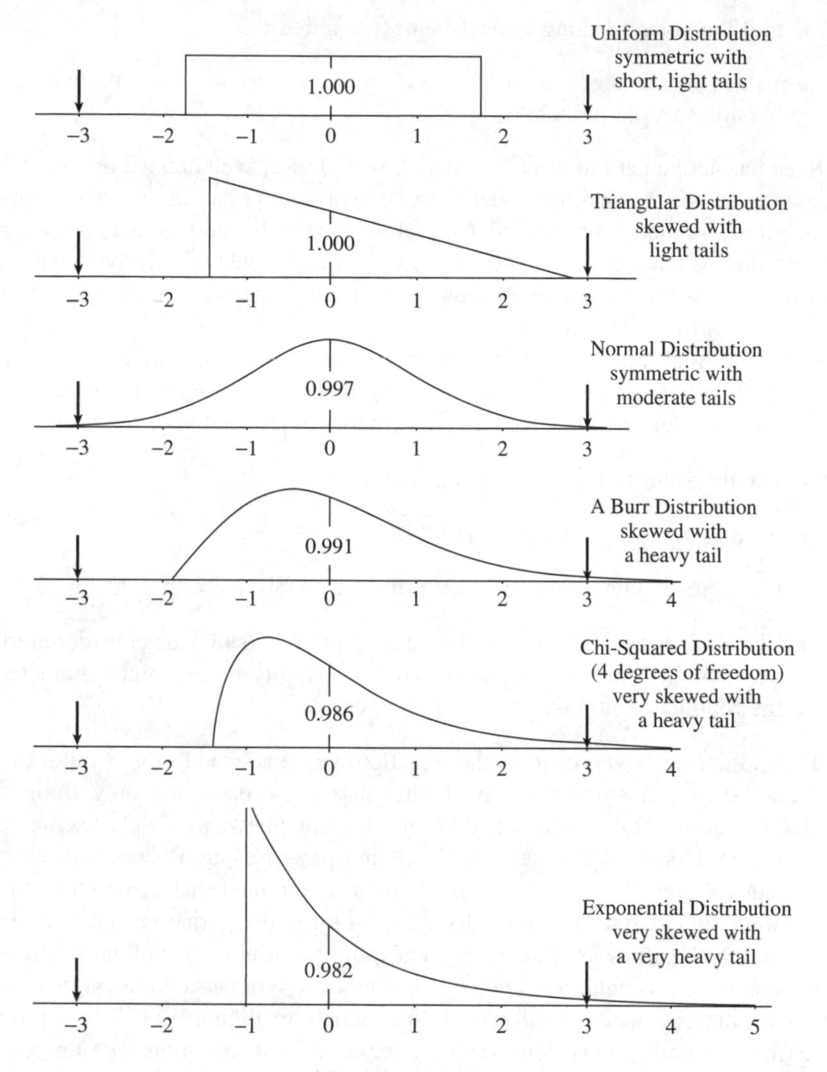

Figure 9.11 The coverage provided by symmetric, three-sigma limits.

However, since these last two classes of models are seldom reasonable models for a *homogeneous* data set, this is not a problem in practice.

This means that the great advantage of Shewhart's approach is that one does not need to make any assumption about the type of probability distribution involved. Further, one may use the standard scaling factors for computing the chart limits because they are robust to nonnormality. (When Irving Burr derived the scaling factors for 27 nonnormal distributions he found that they varied by less than 2.3 % across a very wide range of probability models (Wheeler, 2004: 124).)

Thus, the use of symmetric, three-sigma limits based on local, within-subgroup estimates of dispersion provides what is, in effect, a universal filter for separating the routine variation from any exceptional variation that is present. This approach is not dependent on distributional assumptions in any critical way. Once the four key elements listed at the end of Section 9.2 are satisfied you have a powerful technique for separating exceptional variation from routine variation.

Finally, since the first of these four key elements is rational subgrouping, this is a technique that does not automate well. While software to create the charts is easy to obtain, the organisation of the data in such a way that the charts will be useful is where thought and specific process knowledge are required. While this will involve some effort, it is effort that has a very large potential payback.

9.9 Some misunderstandings about SPC

While some say that the data have to come from a normal distribution in order to be placed on a process behavior chart, Shewhart wrote: 'We are not concerned with the functional form of the universe, but merely with the assumption that a universe exists.'

While some say that the process behavior chart works because of the central limit theorem, this statement ignores the robustness of three-sigma limits. When you have already filtered out approximately 99 to 100 % of the routine variation by brute force using three-sigma limits, you do not need to appeal to any distributional properties to know that the points outside the limits are potential signals of exceptional variation. While the central limit theorem is true, it is not the actual foundation of the process behavior chart.

While some say that you have to have independent observations, and you cannot place autocorrelated data on a process behavior chart, Shewhart did. In fact the data in Table 9.1 are part of an autocorrelated data set given in Shewhart's first book. There he showed how such data could be used to reduce the process variation by over 50 %.

While some say that your process has to be predictable before you can put your data on a process behavior chart, many examples are given in Shewhart's books where this is not the case. This misunderstanding most often comes from a failure to understand either the use of within-subgroup variation or the use of rational, homogeneous subgroups. Perhaps the most common mistake here is the use of a global standard deviation statistic to compute limits. Using the global standard deviation when it should not be used is one of the oldest mistakes in statistical analysis. It dates back to 1840 and is known as the *Quetelet fallacy*, after the Belgian statistician Lambert A.C. Quetelet (1796–1874).

While some say that the standard deviation statistic will provide a better estimate of dispersion than the range, they are overlooking the fact that we will be working with small subgroups. As can be seen in Figure 9.12, the normal-theory probability models for bias-adjusted ranges and bias-adjusted standard deviations are virtually identical for small subgroup sizes.

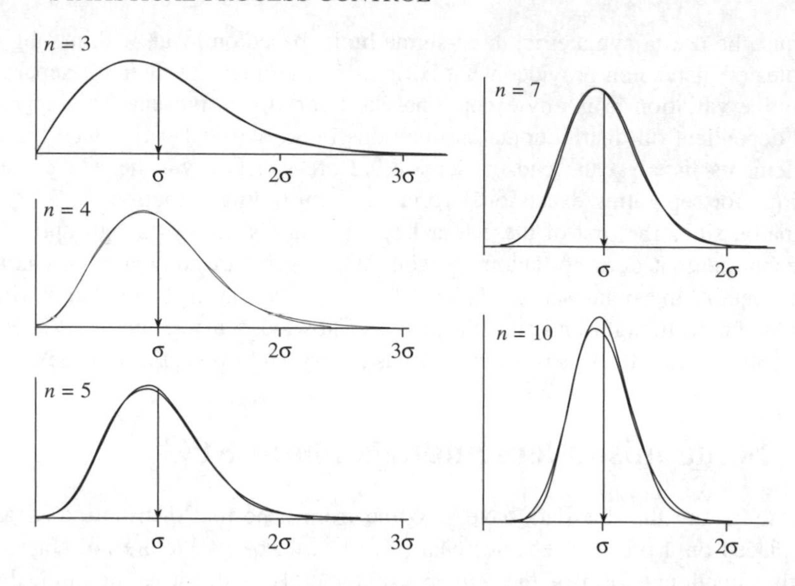

Figure 9.12 Bias-adjusted distributions of ranges and standard deviation statistics.

9.10 Shewhart's approach to learning from existing data

SPC was designed as a technique for analyzing existing data. It works when used with historical data, and it works the same way when used on the line in real time. Each time you add a point to a process behavior chart you are performing an analysis. You are determining if the current value is likely to have come from the same universe as the baseline values, or if the current value is evidence that the underlying process has changed. In real time such a signal of a change can be used both to adjust the process and to justify looking for the assignable cause of the process change. (It is the second of these two actions that makes the process behavior chart into the locomotive of continual improvement.) Thus, the basic question asked by SPC is not whether or not a particular probability model is appropriate for the universe, but whether or not a single universe exists.

If the data are reasonably homogeneous the techniques of statistical inference may be used to characterise the underlying process. But if the data are not reasonably homogeneous then you need to identify those assignable causes that are changing the process outcomes without your knowledge. No statistical inferences are needed in order to do this. In fact, in the absence of a reasonably homogeneous data set, no statistical inferences are even meaningful. Thus, the first question in any data analysis must be the question of homogeneity for the data, and this is the question asked by SPC. While process behavior charts are simple, the questions they address are profound. This is why any analysis of data from an existing process that does not begin with a process behavior chart is fundamentally flawed.

Both the average and range chart and the *XmR* chart provide the ability to examine a set of data for clues to the existence of dominant causes of variation. This ability is based upon the rational organisation of the data. Rational subgroups are those that allow local measures of dispersion to be used to compute limits that will fully characterise the routine variation. When this approach is applied in a careful and thoughtful manner, it will allow you to unlock secrets that would otherwise remain hidden.

The techniques of SPC turn out to be less dependent upon distributional assumptions and the like than are techniques which seek to develop a mathematical model based upon the data. Rather than seeking to fit a model to your data, SPC is concerned with getting your current process to operate up to its full potential. This can be done without fitting a model to the data. Process behavior charts provide the feedback needed in order to know if your process is being operated on target and with as little variation as possible, and they help you to identify those obstacles that prevent you from achieving this. They provide the operational definition of how to get the most out of your process.

References

Shewhart, W. (1931) *Economic Control of Quality of Manufactured Product*. Van Nostrand, New York. Reissued by the American Society for Quality, Milwaukee, WI, 1980.

Wheeler, D.J. (2004) *Advanced Topics in Statistical Process Control*, 2nd edition. SPC Press, Knoxville, TN.

Further reading

Grant, E.L. and Leavenworth, R.S. (1996) *Statistical Quality Control*, 7th edition. McGraw-Hill, New York.

Montgomery, D.C. (2005) *Introduction to Statistical Quality Control*, 5th edition. John Wiley and Sons, Hoboken, NJ.

Shewhart, W. (1939) *Statistical Method from the Viewpoint of Quality Control*. The Graduate School, Department of Agriculture, Washington. Reissued by Dover Publications, New York, 1986.

Wheeler, D.J. (2000) *Normality and the Process Behavior Chart*. SPC Press, Knoxville, TN.

Wheeler, D.J. and Chambers, D.S. (1992) *Understanding Statistical Process Control*, 2nd edition. SPC Press, Knoxville, TN.

10

Advanced statistical process control

Murat Kulahci and Connie Borror

10.1 Introduction

Statistical process control consists of statistical methods that are useful in monitoring product or process quality. Control charts are the most common tools for monitoring quality and were first introduced by Walter A. Shewhart in the 1920s and 1930s. Several univariate control charts are referred to as Shewhart control charts.

Variability will always be present in a production process, regardless of how well maintained the process may be. *Inherent variability* is the natural variability in the process and can be thought of as occurring by chance. If a production process is operating with only *chance causes of variability* present, the process is said to be operating in a state of statistical control. When variability is the result of some unusual occurrence or disturbance in the process, the process is considered out of statistical control. In these situations, there may be an *assignable cause of variability* present in the process. Examples of assignable causes could include machine wear-out in a manufacturing process, defective materials, transactional errors in a business process, or recording errors.

In this chapter, several important topics in advanced statistical process control will be presented. The topics will include the phases of control charting efforts, performance measures for control charts, two univariate control charts, two multivariate control charts, a comparison of engineering process control and statistical process control, and statistical process control with autocorrelated data. Examples and recommendations are provided for the methods discussed. For complete development of these topics, see Montgomery (2005).

Statistical Practice in Business and Industry Edited by S.Y. Coleman, T. Greenfield, D.J. Stewardson and D.C. Montgomery

10.2 Phases of process monitoring and control

Statistical process monitoring and control efforts are often summarised in the form of a control chart. A control chart as given in Figure 10.1 typically consists of a centre line, and upper and lower control limits. The observations are then plotted on the control chart. In the most basic case, the process is deemed 'out of control' when an observation falls outside the control limits. Hence it can be argued that, with each observation, we are making a yes/no decision to label the process in or out of control. At first glance, this seems to be somewhat similar to hypothesis testing – a controversial statement within the statistical process control community. We will discuss this issue further in the next section. It can nevertheless be seen that the construction of a control chart would require the knowledge about the statistical properties of the 'in control' process. So that we can tell what is bad, we have to first define what is good. Consequently, the implementation of most statistical process monitoring and control schemes usually has two very distinct phases.

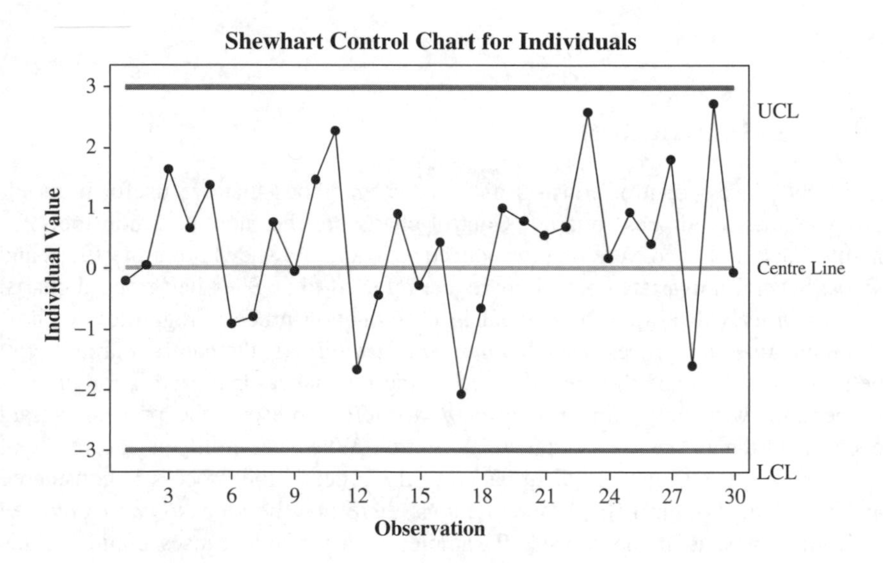

Figure 10.1 Typical statistical process control chart.

10.2.1 Phase I

Phase I in statistical process monitoring and control is about defining the characteristics of an in-control process to form the basis for the on-line (or real-time) monitoring in Phase II. This is achieved (usually off-line) by determining the statistical distributional properties of the data coming from an in-control process. We can at this stage use already available historical data or collect new data. For the latter, we should also consider issues such as sampling frequency and subgroup size, as they will be of concern in the next phase as well. It should be noted that this phase is about exploration rather than monitoring. It is in this retrospective phase, through careful analysis of the available data, that we calculate the parameters such as the control limits of our control chart. The procedure to construct the control chart can be greatly simplified if the data indeed come from an in-control process. This, however, is typically not the case in practice. Therefore there are several issues that have to be addressed at this stage, depending on the nature of the available data. If new data are to be collected, we are often limited to a relatively low number of subgroups, say, 20–25 (see Montgomery, 2005). This will certainly not allow for a sensible determination of statistical control of the process. If, on the other hand, historical and relatively abundant data are available, more subgroups can be used to determine the state of the process. An in-control process will also signal every now and again. The average run length, as discussed in the next section, provides a good measure of how often an observation from an in-control process will fall outside the control limits. Hence with historical data we would expect some out-of-control signals even if the process is in control. There are usually three different approaches recommended in this situation. In the first approach, we investigate the out-of-control observations and if assignable cause is found, eliminate it and bring the process back in statistical control. Of course for historical data, identification of the assignable cause can be a problem. In the second approach, we simply ignore the out-of-control signals and establish the control limits based on the available data. Finally, we can omit the out-of-control observations from the data set and recalculate the control limits. It should be clear by now that assuming that the process is in control at this stage of Phase I can be a bit far fetched. That is why the limits established in this stage are often called the *trial control limits*. Consequently, many authors argue for the need for a second stage within Phase I during which the real-time monitoring of the process is started using the trial control limits (Alt, 1985; Palm, 2000). During this stage any out-of-control signal is carefully analyzed and assignable causes are eliminated to bring the process back in statistical control. This stage can take several cycles and iterations until enough evidence is gathered in terms of rare occurrence of out-of-control signals to point out that the process is indeed in statistical control and that the current control limits can be used in Phase II.

10.2.2 Phase II

It is in this phase that the actual process monitoring begins. The control limits established at the end of the previous phase are used to detect future assignable causes based on the out-of-control signals. As in stage 2 of Phase I, these assignable causes are eliminated and the process is brought back to statistical control. Depending on the prolonged state of the process, some changes in the sampling frequency and/or subgroup size can, with appropriate adjustments to the control limits, be made during this phase.

It should be cautiously noted that a statistical control chart should in general not be used as a diagnostic tool. Its function is often limited to detect the presence of an assignable cause and alert the user accordingly. It is then the user's job to identify the nature of the assignable cause and eliminate it. In many instances tinkering with the process based on the out-of-control signals only can be detrimental. Instead a group of experts, such as process engineers and operators, should be involved in finding out the root cause of the out-of-control signal and the ways to eliminate it. The out-of-control action plans (OCAPs) as discussed in Montgomery (2005) can be very useful in expediting this process, which may be performed off-line. These plans are often based on the analysis of known failure modes and provide the manual for the identification and the elimination of the assignable cause in an expert system-like fashion. It is very important to have an OCAP at the beginning of any statistical process monitoring and control efforts and to modify/expand it in time as new failure modes are discovered.

10.3 Performance measures of control charts

The performance of a control chart is often evaluated based on two issues: how frequently the control chart signals for an in-control process (*false alarm rate*); and how slow the control chart is in detecting a specific change in the process (*excessive delay in detection*). Unfortunately, the performance of a control chart in general cannot be improved vis-à-vis one of these issues without negatively affecting the other.

For example, in a Shewhart control chart, to avoid any false alarms, the control limits can be extended beyond the conventional three-sigma limits. But since the control limits are now made wider, it will take more time to detect a change in the process, hence causing excessive delay in detection. This concept is closely related to the Type I and Type II errors in hypothesis testing. Some authors in fact suggest that a control chart is nothing but 'a sequence of hypothesis tests'. This is somewhat true for Phase II where statistical properties such as the distribution of the in-control process are assumed to be known and new observations are tested to check whether they are coming from the hypothesised distribution (that is in control) or not (that is out of control). However, this relationship between the control charting and hypothesis testing gets blurry for Phase I where the whole process takes a more exploratory nature rather than testing any hypothesis. For further discussion, see Woodall (2000) and Hoerl and Palm (1992). Nevertheless since the performance of a control chart is often attributed and checked for Phase II during which on-line monitoring of the process is pursued, we will in the following discussion draw parallels from hypothesis testing when we talk about the false alarm rate and excessive delay in detection, and define two different performance measures: in-control average run length and out-of-control average run length.

The average run length (ARL) is in general defined as the expected number of samples selected before an out-of-control signal is observed. For an in-control process we would rather have the ARL as high as possible. We will subsequently determine values of the tuning parameters for the control charts of interest. For example, for the univariate EWMA chart discussed in the next section, the combination of λ and L values is determined so that they will result in the acceptable in-control ARL.

For illustration purposes, consider a local lottery where the probability of winning a small prize is 0.001. This means that we would expect on the average one out of 1000 people who bought a lottery ticket to win a prize. The way we calculate this number is by taking the inverse of the probability of a person winning a prize: $1/0.001 = 1000$. Similarly the average run length is calculated as

$$\text{ARL}_0 = \frac{1}{p},$$

where p is the probability of any one point plotting beyond one of the control limits. If the process is really in control but the control chart gives an out-of-control signal, then a *false alarm* has occurred. To illustrate, consider a standard $\overline{x}$ control chart with control limits set at three standard deviations. The probability that a single sample mean plots beyond the control limits when the process really is in control can be shown to be $p = 0.0027$. This can be shown from the fact that the sample average will approximate to a normal distribution according to the central limit theorem. As a result, the *in-control average run length* for the $\overline{x}$ control chart is

$$\text{ARL}_0 = \frac{1}{p} = \frac{1}{0.0027} \cong 370.$$

This indicates that one should expect to see a point plot beyond the control limits on the average every 370 samples, even if the process is in statistical control. That is, a false alarm would occur around every 370 samples or observations taken. A large value of the ARL_0 is desirable if the process is in control. From hypothesis testing, we can also show that the average run length of an in-control process is simply

$$\text{ARL}_0 = \frac{1}{\alpha},$$

where α is the probability that an observation from an in-control process will fall outside the control limits or the Type I error.

If the monitored process is out of control, there may be a shift in the process mean for example, then a small ARL value is desirable. That is, we would prefer our control chart to detect the change as soon as possible right after the change occurs. A small ARL indicates that the control chart will produce an out-of-control signal soon after the process has shifted. The out-of-control ARL is

$$\text{ARL}_1 = \frac{1}{1-\beta},$$

where β is the probability of not detecting a shift on the first sample after a shift has actually occurred, or $1 - \beta$ is the probability of detecting the shift on the first sample following the shift. In other words, ARL_1 is the expected number of samples observed before a shift is detected when the process is out of control. The goal is to have a small ARL_1. For more details on the development and use of ARLs for control charts, see Hawkins (1993), Lucas and Saccucci (1990) and Montgomery (2005).

It should be noted that for a control chart with given control limits ARL_0 is unique. That is, once the control limits are determined, ARL_0 can be calculated from the probability of a sample from an in-control process plotting outside these limits. Hence ARL_0 is a performance measure for an in-control process. ARL_1, however, is shift-specific. That is, many ARL_1 values can be obtained for various values of expected (or suspected) shifts in the process. In fact, it is often a good practice to obtain many ARL_1 values and plot them against the suspected shift. This is also called the operating characteristic curve of the control chart for a fixed sample size. These curves can be generated for various sample sizes and subsequently be used for selecting the 'right' sample size (Montgomery, 2005).

One of the criticisms that ARL as a performance measure has received over the years is the fact that its calculation is highly affected by the uncertainties in the parameter estimation (Wheeler, 2000; Quesenberry, 1993). ARL_0 values, for example, do show great variation depending on the degrees of freedom used to determine the control limits. The use of the probability of an alarm (or no alarm) on a single sample as a performance measure is therefore suggested. A more extensive graphical approach can be found in the 'waterfall' charts introduced by Box *et al.* (2003). Despite all its shortcomings, however, ARL still seems to be the preferred performance measure for statistical process control charts.

10.4 Univariate control charts

Univariate control charts are used to monitor a single variable or attribute of interest. Shewhart control charts for this situation are presented in Chapter 9. In this section, univariate control charts for variables data and attributes data are presented using two more advanced control charts. The control charts to be discussed are:

- exponentially weighted moving average (EWMA);
- cumulative sum (CUSUM).

Recall that variables data are data where an actual measurement has been taken on the item(s) or unit(s) being sampled. They are often referred to as measurement data. Attributes data arise in situations where the quality characteristic of interest is simply to classify the measurement or unit into a single category. To illustrate, a part may be measured, but only classified as defective/nondefective, conforming/nonconforming, or pass/fail based on the resulting measurement. Furthermore, if the measurement process involves counting the number of nonconformities on an item or unit, the resulting data are classified as attribute. EWMA and CUSUM control charts can be used to monitor processes with either variables data or attributes data.

10.4.1 Exponentially weighted moving average control charts

Shewhart control charts as presented in Chapter 9 discussed monitoring various univariate processes. These charts are most useful if the normality assumption is satisfied for the quality characteristic of interest or if shifts in the process mean to be detected are not small. Except when additional run rules such as Western Electric rules are introduced, Shewhart control charts have been shown to be poor in monitoring small shifts in the process mean because they rely exclusively on the current observation (see Gan, 1991; Hawkins, 1981, 1993; Lucas, 1976; Montgomery, 2005; Woodall and Adams, 1993).

The EWMA control chart is one alternative to the Shewhart control chart when small shifts in the mean are important to detect in a process. The EWMA control chart was first introduced by Roberts (1959). An EWMA statistic is defined by

$$Z_t = \lambda X_t + (1 - \lambda) Z_{t-1}, \tag{10.1}$$

where λ is a smoothing constant and tuning parameter ($0 < \lambda \leq 1$), X_t is the current observation, and Z_{t-1} is the previous EWMA statistic. The control chart is initialised with $Z_0 = \mu_0$, where μ_0 is the process or target mean. If the target mean is unknown, then $\overline{x}$ can be used as the estimate for μ_0. The EWMA statistic includes information from past observations in addition to the current observation, X_t. The EWMA statistic given in (10.1) is plotted on a control chart with control limits placed on the values of Z_t. The process is considered out of control if one or more Z_t values plot beyond the control limits.

For an in-control process, it can be shown that the expected value of the EWMA statistic is

$$E(Z_t) = \mu_0$$

with variance

$$\mathrm{Var}(Z_t) = \sigma_{z_t}^2 \left(\frac{\lambda}{2 - \lambda} \right) [1 - (1 - \lambda)^{2t}]. \tag{10.2}$$

For large values of t, the exact EWMA variance in (10.2) approaches its asymptotic value,

$$\mathrm{Var}(Z_t) = \sigma_{z_t}^2 \left(\frac{\lambda}{2 - \lambda} \right).$$

Asymptotic control limits are constant over time. Either the exact or asymptotic variance of the EWMA statistic may be used in control limit calculations. The control limits for the EWMA control chart using the exact variance are

$$\mathrm{UCL} = \mu_0 + L\sigma \sqrt{\left(\frac{\lambda}{2 - \lambda} \right) [1 - (1 - \lambda)^{2t}]},$$

$$\mathrm{LCL} = \mu_0 - L\sigma \sqrt{\left(\frac{\lambda}{2 - \lambda} \right) [1 - (1 - \lambda)^{2t}]}.$$

The control limits using the asymptotic variance are

$$\text{UCL} = \mu_0 + L\sigma\sqrt{\left(\frac{\lambda}{2-\lambda}\right)},$$

$$\text{LCL} = \mu_0 - L\sigma\sqrt{\left(\frac{\lambda}{2-\lambda}\right)},$$

where L is the width of the control limits. For a given smoothing constant λ, the value L is chosen to give the desired in-control ARL. Research has shown that small values of λ ($0.05 \leq \lambda \leq 0.25$) work well in practice. Corresponding values of L that will provide a desirable in-control ARL are often $2.6 \leq L \leq 3$. For complete details on the development of the EWMA control chart, see Crowder (1989), Lucas and Saccucci (1990), Montgomery (2005), and Testik *et al.* (2006).

Example 10.1 Chouinard (1997) presents a study on the failure rate of a search radar from the CP140 Aurora maritime patrol aircraft. The failure rate is an important performance measure that must be monitored and is given by

$$FR = \frac{\text{number of items failed}}{\text{total operating time}}.$$

The failure rate is recorded monthly and representative failure rates for 24 months are shown in Table 10.1. A baseline failure rate is known from historical data and is assumed to be 2.158, so $\mu_0 = 2.158$. From past experience it is believed that the standard deviation is $\sigma = 0.5$. If the average failure rate shifts from this target by one process standard deviation in either direction, then the engineers would like to detect this shift quickly. If the resulting shift is in the upward direction, this could indicate a significant increase in the failure rate. As a result, the process would be investigated quickly and the assignable causes found for increasing the failure rate. On the other hand, if the shift is in the downward direction, the engineer would also like to determine if this is significant and determine the cause of the decrease in failure rate. This determination could lead to process changes that will result in a permanent decreased failure rate. The engineer would like to construct an EWMA control chart for this data using $\lambda = 0.10$ and $L = 2.814$. These values are not arbitrary, but chosen based on design criteria for EWMA control charts (see Crowder, 1989; Lucas and Saccucci, 1990; Montgomery, 2005).

Solution

The target process mean is $\mu_0 = 2.158$ with $\sigma = 0.5$. With $L = 2.814$ and $\lambda = 0.10$, the EWMA statistic from equation (10.1) is

$$\begin{aligned} Z_t &= \lambda X_t + (1-\lambda)Z_{t-1} \\ &= 0.10X_t + (1-0.10)Z_{t-1} \\ &= 0.10X_t + 0.90Z_{t-1}. \end{aligned}$$

Table 10.1 Failure rate data for Example 10.1.

Observation	Measurement, X_t	Z_t
1	2.21	2.16320
2	2.31	2.17788
3	1.88	2.14809
4	2.11	2.14428
5	2.00	2.12985
6	1.75	2.09187
7	2.05	2.08768
8	1.88	2.06691
9	2.07	2.06722
10	2.15	2.07550
11	2.51	2.11895
12	2.43	2.15006
13	2.41	2.17605
14	2.18	2.17644
15	1.80	2.13880
16	1.80	2.10492
17	2.33	2.12743
18	2.70	2.18469
19	2.78	2.24422
20	3.00	2.31980
21	2.97	2.38482
22	3.30	2.47633
23	3.22	2.55070
24	3.11	2.60663

To illustrate, consider the first observation, $X_1 = 2.21$. The process is initialised with $Z_0 = \mu_0 = 2.158$. As a result,

$$\begin{aligned} Z_1 &= \lambda X_1 + (1 - \lambda) Z_0 \\ &= 0.10 \times 2.21 + 0.90 \times 2.158 \\ &= 2.163. \end{aligned}$$

The remaining EWMA statistics are calculated similarly and given in Table 10.1 with the original data. The control limits are:

$$\text{UCL} = \mu_0 + L\sigma\sqrt{\left(\frac{\lambda}{2-\lambda}\right)} = 2.158 + 2.814 \times 0.5\sqrt{\frac{0.10}{2-0.10}} = 2.481,$$

$$\text{Centerline} = 2.158,$$

$$\text{LCL} = \mu_0 - L\sigma\sqrt{\left(\frac{\lambda}{2-\lambda}\right)} = 2.158 + 2.814 \times 0.5\sqrt{\frac{0.10}{2-0.10}} = 1.832.$$

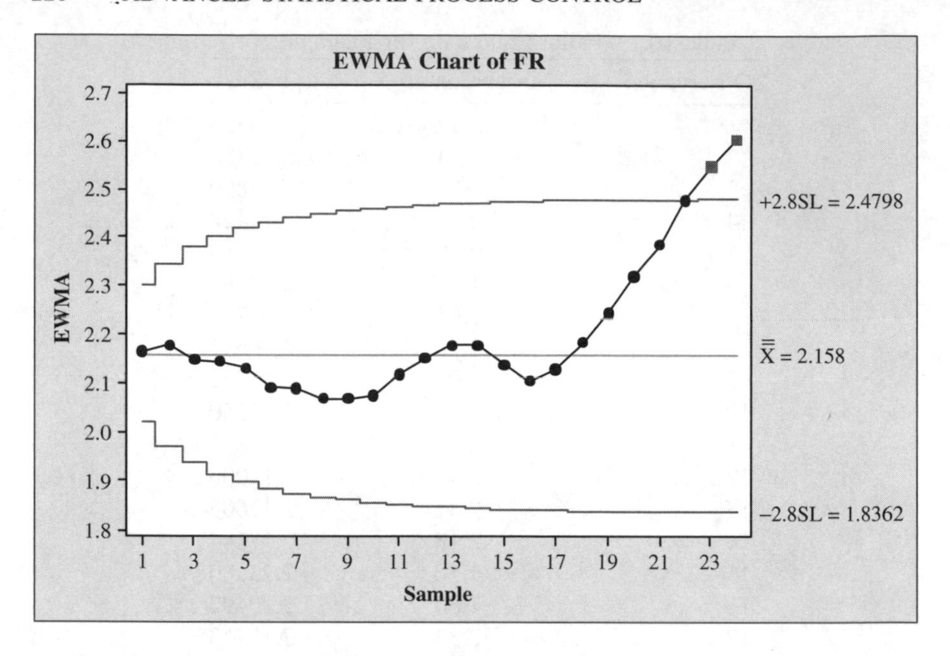

Figure 10.2 EWMA control chart for Example 10.1.

The control limits and the EWMA statistics, Z_t, are plotted on the EWMA control chart displayed in Figure 10.2. The process is not in control. The control chart has signaled at the 23rd month.

10.4.2 Cumulative sum control charts

A second alternative to Shewhart control charts is the CUSUM control chart. This was first introduced by Page (1954) and consists of information from successive sample observations. The *cumulative sums* of deviations of the observations from a target value are plotted on a control chart. Specifically, the quantity

$$C_i = \sum_{j=1}^{i} (\overline{x}_j - \mu_0)$$

is monitored and plotted on a chart against the subgroup i, where C_i is the sum of the deviations from target for all samples up to and including the ith sample, $\overline{x}_j$ the sample mean of the jth sample, and μ_0 the target value.

The cumulative sums C_i will be approximately zero as long as the process average remains at or around the target value μ_0. If the process has shifted away from its target value, then C_i increases in absolute value. Furthermore, if the shift away from the target is in the upward direction, C_i will become a large positive quantity. If the shift away from the target is in the downward direction, C_i will become a large negative value. Because the CUSUM incorporates information from past observations, it can detect small shifts in the process more rapidly than the standard Shewhart chart. The CUSUM control charts can be used for monitoring subgroup data, individuals data, and attribute data. The two-sided CUSUM control chart for subgroup data will be presented next. One-sided CUSUM charts can also be constructed if the practitioner is interesting in only one particular direction of shift. See Montgomery (2005) or the other references provided for a complete discussion of one-sided CUSUM.

The cumulative sum monitoring technique involves two statistics, C_i^+ and C_i^- (see Lucas, 1976; Montgomery, 2005). C_i^+ is the sum of deviations *above* the target mean and is referred to as the *one-sided upper CUSUM*. C_i^- is the sum of deviations *below* the target mean and is referred to as the *one-sided lower CUSUM*. The quantities C_i^+ and C_i^- are

$$C_i^+ = \max[0, x_i - (\mu_0 + K) + C_{i-1}^+],$$

$$C_i^- = \max[0, (\mu_0 - K) - x_i + C_{i-1}^-],$$

where C_i^+ and C_i^- are initial values of the CUSUM with $C_0^+ = C_0^- = 0$ and x_i is the ith observation. The constant, K, is referred to as the *reference value* and calculated as

$$K = \frac{|\mu_1 - \mu_0|}{2},$$

where μ_0 is the target mean and μ_1 is the out-of-control mean that we are interested in detecting.

The reference value is added to avoid false alarms. Consider, for example, C_i^+ without the reference value, that is,

$$C_i^+ = \max[0, x_i - \mu_0 + C_{i-1}^+].$$

Since we are using C_i^+, this means that we are aiming to detect an upward shift in the mean to μ_1. Now assume that most of the several consecutive observations coming from an in-control process happen to be above μ_0, causing C_i^+ without the reference value to go up and perhaps to cross the control limit to trigger an (false) alarm. A reference value is therefore added to avoid this situation. Also note that this is done at the expense of possible excessive delays in detection.

Nevertheless, the goal of the control chart is to indicate whether the process average has shifted to this new mean, μ_1. If so, we would want to detect this shift fairly quickly. However, μ_1 is often unknown and the value of K must be determined using alternate methods. The most common method is to let $K = k\sigma$, where σ is the process standard deviation and k is some constant chosen so that a particular shift is detected. For example, if a shift from target of 1.5 standard deviations is important to detect, then $k = 1.5$, and $K = 1.5\sigma$. If the process standard deviation is unknown, it should be estimated from the sample data.

The two-sided CUSUM control chart plots the values of C_i^+ and C_i^- for each sample. If either statistic plots beyond a stated decision value (or threshold), H, the process is considered out of control. The choice of H is not arbitrary and should be chosen after careful consideration. Most commonly, $H = 5\sigma$. The choice of H, in combination with the value of K, is often made with respect to appropriate ARLs or small false alarm rates. The reader is encouraged to consult Hawkins (1993) and Woodall and Adams (1993) for the design of CUSUM control charts. We present an example using the CUSUM control chart for individuals data.

Example 10.2 Consider the failure rate data given in Example 10.1. A decrease or increase in the target average failure rate, $\mu_0 = 2.158$, of one process standard deviation is important to detect. The process standard deviation is believed to be $\sigma = 0.5$. A two-sided CUSUM control chart is to be implemented.

Solution

From the information given, the following quantities are known: $\mu_0 = 2.158$, $\sigma = 0.5$, $K = 1\sigma = 0.5$, and $H = 5\sigma = 2.5$. As a result, the cumulative sums are calculated using

$$\begin{aligned} C_i^+ &= \max[0, x_i - (\mu_0 + K) + C_{i-1}^+] \\ &= \max[0, x_i - (2.158 + 0.5) + C_{i-1}^+], \\ C_i^- &= \max[0, (\mu_0 - K) - x_i + C_{i-1}^-] \\ &= \max[0, (2.158 - 0.5) - x_i + C_{i-1}^-]. \end{aligned}$$

If the process target has shifted from the target of 2.158, the cumulative sums will become nonzero. The resulting CUSUMs are then compared with the decision interval $(-H, H) = (-2.5, 2.5)$. When comparing the CUSUMs with the decision interval, the appropriate sign is placed on any nonzero value. To illustrate, since C_i^- is a measure of a downward shift, a negative sign is placed on the value for comparing to the decision interval. If the CUSUM plots outside the decision interval, the process is considered to be out of control.

We will now illustrate how to calculate the CUSUMs. For example, consider the first observation, $x_1 = 2.21$. Initially, $C_i^+ = C_i^- = 0$, $\mu_0 = 2.158$, and $K = 0.5$. The first set of calculations is

$$\begin{aligned} C_1^+ &= \max[0, x_1 - (\mu_0 + K) + CS_0^+] \\ &= \max[0, 2.21 - (2.158 + 0.5) + 0] \\ &= \max[0, -0.448] \\ &= 0, \\ C_1^- &= \max[0, (\mu_0 - K) - x_i + CS_0^-] \\ &= \max[0, (2.158 - 0.5) - 2.21 + 0] \\ &= \max[0, -0.552] \\ &= 0. \end{aligned}$$

Table 10.2 Failure rate data for Example 10.2.

Observation	Measurement, x_i	C_i^+	C_i^-
1	2.21	0.000	0.000
2	2.31	0.000	0.000
3	1.88	0.000	−0.028
4	2.11	0.000	0.000
5	2.00	0.000	0.000
6	1.75	0.000	−0.158
7	2.05	0.000	−0.016
8	1.88	0.000	−0.044
9	2.07	0.000	0.000
10	2.15	0.000	0.000
11	2.51	0.102	0.000
12	2.43	0.124	0.000
13	2.41	0.126	0.000
14	2.18	0.000	0.000
15	1.80	0.000	−0.108
16	1.80	0.000	−0.216
17	2.33	0.000	0.000
18	2.70	0.292	0.000
19	2.78	0.664	0.000
20	3.00	1.256	0.000
21	2.97	1.818	0.000
22	3.30	2.710	0.000
23	3.22	3.522	0.000
24	3.11	4.224	0.000

The cumulative sums, with the appropriate signs, are given in Table 10.2. In examining the last two columns in Table 10.2, we see that the last three observations, starting with the 22nd month, plot outside the upper decision value, $H = 2.5$. As a result, the process appears to be out of control. The failure rate appears to be increasing over time. The CUSUM control chart for this process is given in Figure 10.3. The cumulative sums are plotted on this chart with a centerline at 0.

10.5 Multivariate control charts

Many processes being monitored involve several quality characteristics. In these situations, it is possible that there is some relationship between two or more of the characteristics. For example, the inside diameter and outside diameter of copper tubing may be two quality characteristics that must be monitored in the manufacturing process. Univariate control charts could be constructed for each quality characteristic separately, but these individual univariate charts may not capture the functional relationship(s) between the variables being measured. That is, the

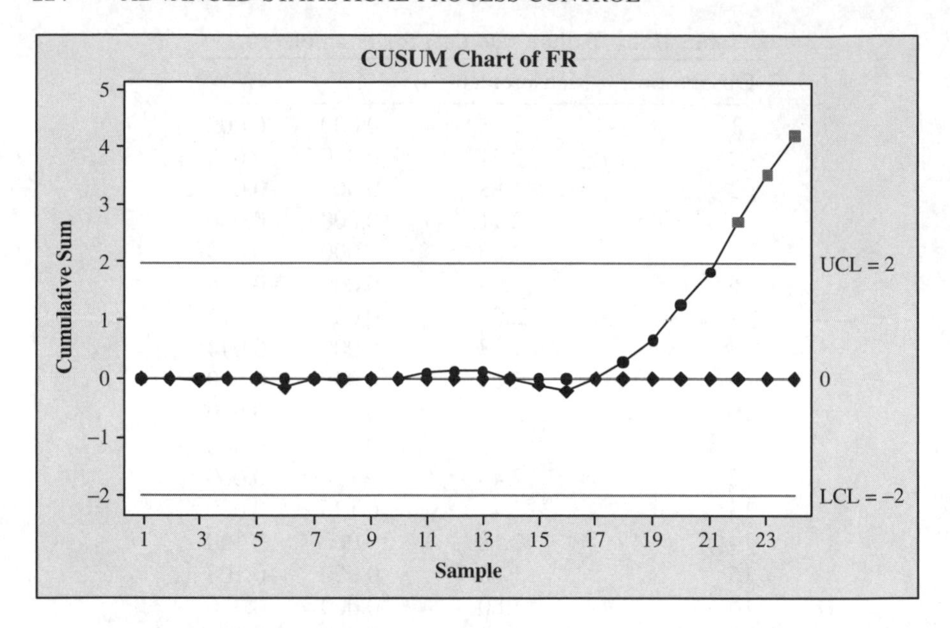

Figure 10.3 CUSUM control chart for Example 10.2.

process may actually be out of control, but the univariate charts do not indicate any out-of-control situation. This can occur when there is a correlation among two or more quality characteristics of interest, but each is monitored individually. In this section, two multivariate techniques will be discussed: Hotelling's T^2 control chart and the multivariate exponentially weighted moving average (MEWMA) control chart. The interested reader should consult Hotelling (1947), Crosier (1988), Hawkins (1993), Jackson (1985), Lowry *et al.* (1992), Lowry and Montgomery (1995), Montgomery (2005), Pignatiello and Runger (1990), and Tracy *et al.* (1992) for complete descriptions of these methods and more.

10.5.1 Hotelling T^2 control chart

The Hotelling T^2 control chart is the most common and familiar monitoring technique for multivariate data. It is can be thought of as the multivariate counterpart to the Shewhart $\overline{x}$ control chart. The Hotelling T^2 control chart can be used for either individual or subgroup data.

Suppose that $X_1, X_2, \ldots, X_p$ are p quality characteristics that are to be monitored concurrently. The statistic of interest can be written as

$$T^2 = n(\overline{\mathbf{x}} - \overline{\overline{\mathbf{x}}})'\mathbf{S}^{-1}(\overline{\mathbf{x}} - \overline{\overline{\mathbf{x}}}), \quad (10.3)$$

where $\overline{\mathbf{x}}$ is the vector of sample means, $\overline{\overline{\mathbf{x}}}$ the vector of in-control means, and $\mathbf{S}$ the averaged covariance matrix for the quality characteristics when the process is in control. If individual observations ($n = 1$) are used, the statistic to be plotted is

$$T^2 = (\mathbf{x} - \overline{\mathbf{x}})'\mathbf{S}^{-1}(\mathbf{x} - \overline{\mathbf{x}}),$$

where $\overline{\mathbf{x}}$ and $\mathbf{S}$ are as before.

Phase I

If the multivariate control chart is being implemented for a Phase I process, the appropriate control limits are a function of the F distribution. The control limits for m samples of size $n > 1$ can be written as

$$UCL = \frac{p(m-1)(n-1)}{mn - m - p + 1} F_{\alpha,p,mn-m-p+1},$$
$$LCL = 0.$$

Again, in Phase I, the control charts are being used to establish statistical control.

Phase II

In Phase II, the goal is to monitor the process to maintain statistical control. The appropriate control limits for the T^2 control chart in Phase II are

$$UCL = \frac{p(m+1)(n-1)}{mn - m - p + 1} F_{\alpha,p,mn-m-p+1},$$
$$LCL = 0.$$

Example 10.3 Variability in photovoltaic processing sequences can lead to significant increase in costs and monetary loss to a company. The variation in photovoltaic cell performance can be induced by many possible electrical metrics that are often correlated and follow a multivariate distribution. Two such electrical metrics include short-circuit current (Isc), and fill factor (FF). The reader is referred to Coleman and Nickerson (2005) for a complete definition and description of photovoltaic cell performance. Representative data for 30 randomly selected cells from a cell test station are presented in Table 10.3. Previously in Phase I, the in-control means for Isc and FF were estimated to be 3.427 and 74.285, respectively.

Solution

The Hotelling T^2 control chart for Phase II was constructed for the cell performance data and is shown in Figure 10.4. Based on the control chart, there does not appear to be any significant problem with the process being out of control.

Table 10.3 Photovoltaic data for Example 10.3.

Observation	Isc	FF
1	3.44441	74.7286
2	3.45710	74.8878
3	3.38529	74.4432
4	3.41384	73.4057
5	3.42979	75.3835
6	3.43748	74.5891
7	3.39756	72.7596
8	3.38162	73.4492
9	3.41059	73.3987
10	3.45983	75.0340
11	3.40482	73.0397
12	3.41644	74.6807
13	3.40737	73.3860
14	3.43992	74.6807
15	3.43433	73.8538
16	3.42393	75.2013
17	3.42678	75.0678
18	3.43546	74.3762
19	3.42314	75.0678
20	3.42905	75.2013
21	3.41812	74.7240
22	3.38231	74.3762
23	3.42533	75.3835
24	3.41491	74.5891
25	3.41920	74.7286
26	3.44297	74.7240
27	3.42099	74.8878
28	3.42177	75.0340
29	3.43120	73.7772
30	3.43621	74.4432

10.5.2 Multivariate EWMA control charts

The disadvantage of the Shewhart control chart for univariate processes is also a disadvantage for the T^2 control chart. That is, only the most recent observation is used in the monitoring of the process. As a result, the T^2 control chart is poor at detecting small shifts in the process mean. One alternative to the T^2 control chart as a monitoring technique is the MEWMA control chart, introduced by Lowry *et al.* (1992). These charts have been designed and evaluated using average run

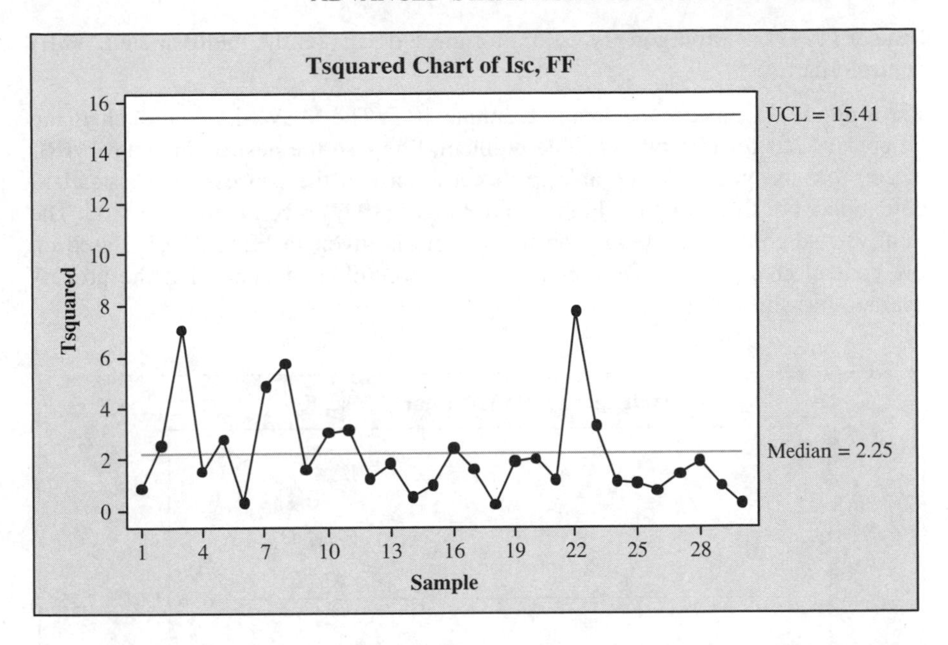

Figure 10.4 T^2 *control chart for photovoltaic cell performance involving Isc and FF.*

lengths by several authors. In general, the MEWMA statistic is

$$\mathbf{Z}_i = \lambda \mathbf{X}_i + (1 - \lambda)\mathbf{Z}_{i-1},$$

where $\mathbf{X}_i$ represents the ith observation in the dataset. For example, consider the data given in Table 10.3. For these data, $\mathbf{x}_1 = (3.4444, 74.7286)$ representing the first observation.

For the MEWMA, the quantity monitored is

$$\mathbf{T}_i^2 = \mathbf{Z}'_i \mathbf{\Sigma}_{z_i}^{-1} \mathbf{Z}_i,$$

where

$$\mathbf{\Sigma}_{z_i} = \frac{\lambda}{2 - \lambda}[1 - (1 - \lambda)^{2i}]\mathbf{\Sigma}$$

is the covariance matrix. Notice that the MEWMA statistic is the multivariate counterpart of the univariate EWMA.

The control limits and tuning parameters necessary for implementing the MEWMA control charts are dependent upon the shift that the practitioner wishes to detect, the in-control ARL he or she is willing to allow, and the number of quality characteristics to be monitored. Prabhu and Runger (1997) present tables for these parameters. The reader is referred to Lowry *et al.* (1992), Prabhu and

Runger (1997) or Montgomery (2005) for more details on the multivariate EWMA control chart.

Example 10.4 Consider the data in Example 10.3. The MEWMA control chart can be constructed for this two-variable problem. Suppose the desired in-control ARL is 200 and the goal is to be able to detect a shift in the process of 0.5 standard deviations. For this situation, Prabhu and Runger (1997) recommend $\lambda = 0.05$. The multivariate control chart was constructed and is given in Figure 10.5. Based on the control chart, the process signals out of control. It appears that the process average has shifted upward.

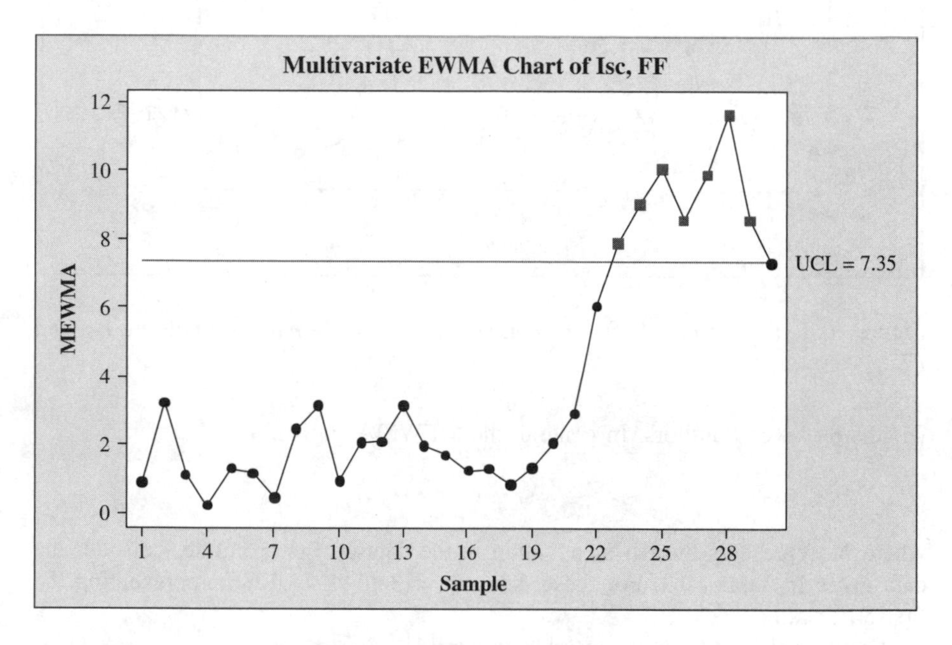

Figure 10.5 MEWMA control chart for photovoltaic cell performance involving Isc and FF.

Notice that the MEWMA control chart signaled an out-of-control situation whereas the T^2 control chart did not. It should be noted that the data monitored were in fact out of control. Therefore, the MEWMA with its weighted monitoring of the process was able to detect a shift in the process.

In this case, the multivariate control chart was much more informative than the T^2 control chart. In addition, the univariate individual control charts for Isc and FF were constructed. These charts are displayed in Figures 10.6 and 10.7. You will note that the individual control charts also did not detect the process shift. When monitoring multiple quality characteristics, multivariate control chart techniques are the most appropriate methods for identifying out of control processes. If small shifts in the process average are important to detect, then the MEWMA is more informative than the individuals control chart and even possibly the T^2 control chart.

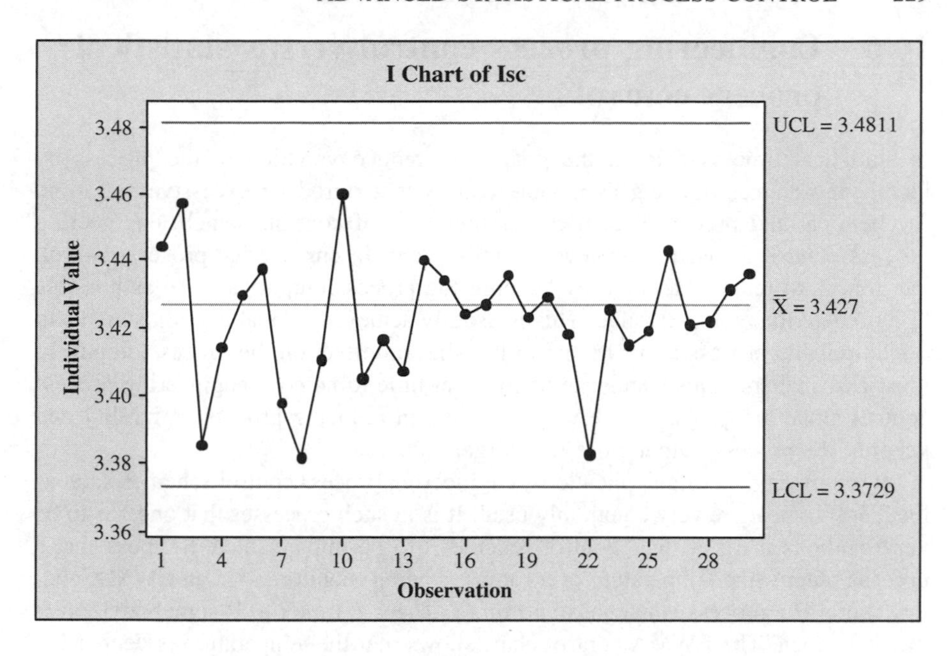

Figure 10.6 Shewhart individuals control chart for Isc.

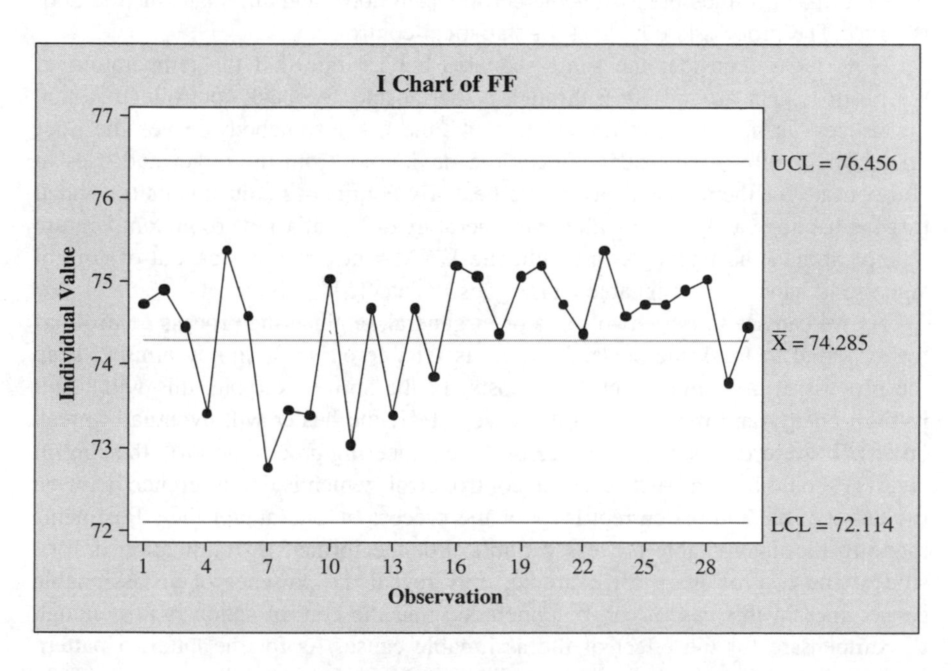

Figure 10.7 Shewhart individuals control chart for FF.

10.6 Engineering process control versus statistical process control

In statistical process control, the goal is to reduce variation in the process by identifying and eliminating assignable causes that introduce excessive variation. In chemical and process industries, another type of control, namely engineering process control, is used to achieve a similar goal. In engineering process control, the goal is to reduce variation by keeping the process output as close as possible to a predetermined target value. This is usually achieved by making adjustments to a manipulable process variable that in turn has an effect on the process output. In most cases where adjustments can be made at little to no cost, engineering process control alone has proven to be very effective in reducing process variability and keeping the process output close to a target value.

In many engineering applications, engineering process control schemes such as feedback control are very commonly used. It is in such processes that one has to be very cautious if a statistical control scheme is to be implemented. Suppose that a process output (the temperature of a room) is being monitored via an EWMA control chart. The process is in control at time t. Then, at time $t + 1$, somebody leaves the door open. The EWMA control chart shows that the temperature is decreasing (assuming that the outside temperature is colder) and eventually it signals an alarm. An investigation finds the assignable cause (open door) and eliminates it (the door is shut). The process is now back in statistical control.

Now let us consider the same scenario but assume that the temperature of the room is kept around 70 °F through a thermostat (feedback controller). Again, the process is in control at time t, and at time $t + 1$ somebody leaves the door open. The feedback controller notices the deviations from the target and tries to compensate for them by increasing the heat. By continuous adjustments (provided that the heating can keep up with the temperature differential) the room temperature is kept around the target. As a result, the EWMA never shows an out-of-control signal and hence the assignable cause goes undetected.

As we can see in these two cases or in general, engineering process control has the potential to mask the presence of an assignable cause. It simply aims to keep the process at a certain target at all costs. In the above example, this will result in high energy and repair costs (at the very least the heater will eventually break down). Therefore under the presence of the engineering process control, the control chart(s) should be constructed for the control error, which is the difference between the target value and the current level of the process output, and/or the adjustments made to the manipulable process variable. For the former, a stratification on one side of the control chart, for example, may reveal the presence of an assignable cause since in this case it can be concluded that the control action is not enough to compensate for the effect of the assignable cause. As for the latter, a pattern in the adjustments to manipulable variable, such as increased frequency or only positive adjustments, can again suggest the presence of an assignable cause. For further details, see Box and Luceño (1997).

10.7 Statistical process control with autocorrelated data

In most statistical process control applications, the construction of control charts is based on two assumptions: normality and independence. In many cases, the statistical process control schemes are robust to slight deviations from normality. The independence assumption, however, can not only be fairly often violated but also have serious impact on the performance of the control chart if indeed violated. Sampling frequency relative to the system's dynamics will in general cause the data to have serial dependence. Indeed, the consecutive observations taken not too far apart in time will be somehow positively correlated. That is, we cannot expect the temperature readings from a large industrial furnace to vary independently from one observation to another taken, for example, every 10 seconds. System dynamics and sampling frequency will force high observations to follow high ones and vice versa. This may, however, not necessarily cause the process to be out of control. In fact a (weakly) *stationary* process for which the mean, variance and cross-correlation between observations from different points in time are constant, will still be in statistical control (Bisgaard and Kulahci, 2006; Box *et al.*, 1994).

The moving range is commonly used in the estimation of the standard deviation of the observations, which in turn determines the control limits. If, as in many applications, the process exhibits positive autocorrelation, the moving ranges of successive observations are expected to be small. Hence, for these processes, the standard deviation of the observations will be underestimated and consequently the control limits will be tighter. This will lead to too many false alarms. There are generally two approaches recommended to deal with this situation: obtain a 'better' estimate of the standard deviation and calculate the control limits accordingly; and model the autocorrelation out of the data and apply the control chart to the residuals. For both cases, we strongly recommend the use of autoregressive moving average (ARMA) models to model the data. For a detailed discussion of these models, see Box *et al.* (1994) and Bisgaard and Kulahci (2005b)

We will now illustrate both approaches in the following example from Bisgaard and Kulahci (2005a).

Example 10.5 Table 10.4 shows the hourly temperature readings from an industrial furnace. Figure 10.8 shows the time series plot of the temperature readings. The EWMA control chart with $\lambda = 0.2$ in Figure 10.9 shows numerous out-of-control signals. Yet from the time series plot of the data, we observe that the successive observations tend to follow each other and not to be independent. The scatterplots of z_t vs. z_{t-1}, z_t vs. z_{t-2}, do offer simple checks for serial dependence in the data. Indeed, Figure 10.10 shows that particularly the z_t vs. z_{t-1} scatterplot shows some correlation.

We therefore proceed with fitting an ARMA model to the data. For further details of the time series modelling procedure and subsequent residual analysis, see Bisgaard and Kulahci (2005a). It turns out that a second-order autoregressive

Table 10.4 Industrial furnace temperature data (reading horizontally line-by-line).

1578.71	1578.79	1579.38	1579.36	1579.83	1580.13	1578.95
1579.18	1579.52	1579.72	1580.11	1580.41	1580.77	1580.05
1579.53	1579.00	1579.12	1579.13	1579.39	1579.73	1580.12
1580.23	1580.25	1579.80	1579.72	1579.49	1579.22	1579.03
1579.76	1580.19	1580.17	1580.22	1580.44	1580.71	1579.91
1579.48	1579.82	1580.34	1580.56	1580.05	1579.63	1578.82
1578.59	1578.56	1579.56	1579.46	1579.59	1579.66	1579.89
1580.03	1579.76	1579.84	1580.41	1580.30	1580.17	1579.81
1579.71	1579.77	1580.16	1580.38	1580.18	1579.59	1580.06
1581.21	1580.89	1580.82	1580.48	1579.97	1579.64	1580.42
1580.06	1580.12	1579.92	1579.57	1579.56	1579.40	1578.90
1578.50	1579.30	1579.93				

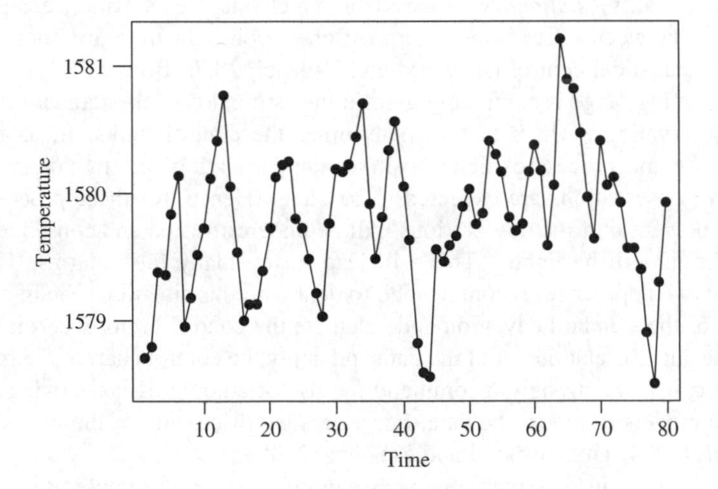

Figure 10.8 Times series plot of 80 consecutive hourly temperature observations from an industrial furnace.

model (AR(2)),

$$z_t = \delta + \phi_1 z_{t-1} + \phi_2 z_{t-2} + \varepsilon_t,$$

fits the data well; here the ε_t are independent and identically distributed errors with mean 0 and variance σ^2. Note that the AR(2) model has the form of a standard linear regression model for which the inputs are z_{t-1} and z_{t-2}, hence the term 'autoregressive'. It can be shown that for the AR(2) process, the process mean is

$$\mu = \frac{\delta}{1 - \phi_1 - \phi_2}.$$

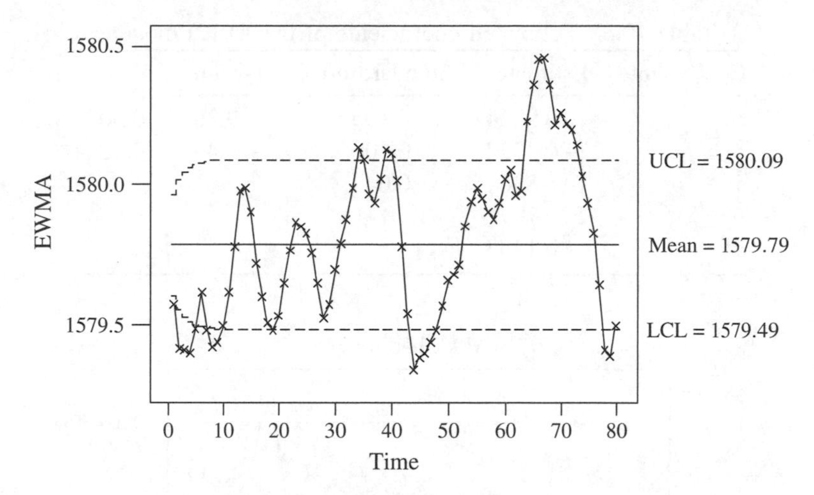

Figure 10.9 EWMA of hourly temperature observations with $\lambda = 0.2$.

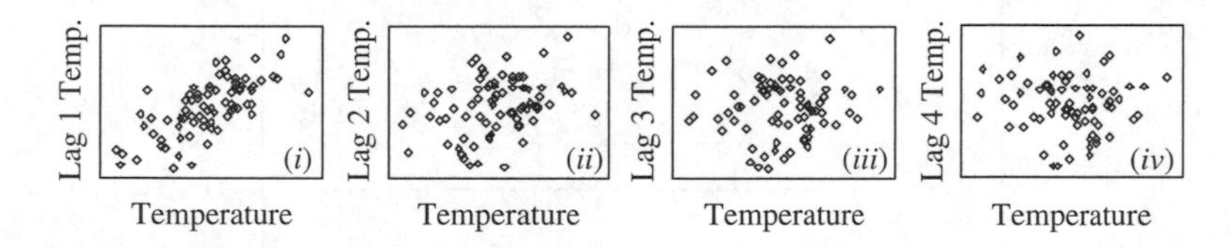

Figure 10.10 (Auto)correlation between observations (i) one time unit apart, (ii) two time units apart, (iii) three time units apart, and (iv) four time units apart.

Similarly the process variance is

$$\sigma_z^2 = \left(\frac{1-\phi_2}{1+\phi_2}\right)\frac{\sigma^2}{(1-\phi_2)^2-\phi_1^2}.$$

The parameter estimates of the AR(2) model are given in Table 10.5. We then have the estimate for the process variance given by

$$\hat{\sigma}_z^2 = \left(\frac{1+0.3722}{1-0.3722}\right)\frac{\hat{\sigma}^2}{(1+0.3722)^2-0.9824^2}$$
$$= 2.38146\hat{\sigma}^2.$$

In Figure 10.11 we show the EWMA control chart after the standard deviation is properly adjusted. It can be seen that there are now only two observations giving an out-of-control signal. Similarly, Figure 10.12 shows the EWMA control chart of the residuals of the AR(2) model. This control chart shows no out-of-control signals. We should remember that these control charts are constructed with an arbitrary choice of $\lambda = 0.2$. Since the two control charts offer different conclusions, we would recommend the investigation of the two out-of-control signals

Table 10.5 Estimated coefficients of the AR(2) model.

Coefficient	Estimate	Standard error	t-value	p-value
$\hat{\phi}_1$	0.9824	0.1062	9.25	0.000
$\hat{\phi}_2$	−0.3722	0.1066	−3.49	0.001
$\hat{\delta}$	615.836	0.042		
$\hat{\mu}$	1579.79	0.11		
$\hat{\sigma}^2$	0.1403			

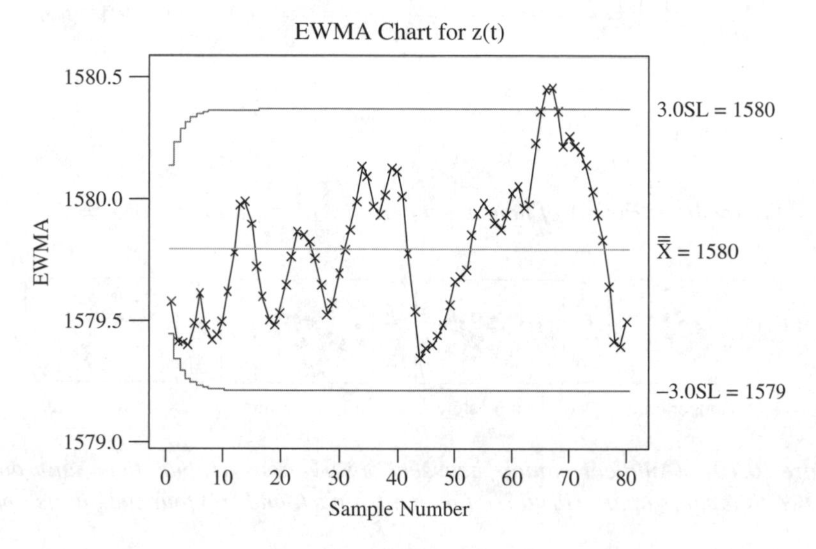

Figure 10.11 EWMA control chart of the temperature data after the variance is adjusted.

in Figure 10.11. But if no assignable cause can be identified, we will deem the process in control and proceed with our monitoring efforts.

10.8 Conclusions

In this chapter we have provided an overview on some more advanced topics in statistical process control. Control charting procedures for univariate and multivariate data have been presented as well as process monitoring for autocorrelated data. In addition, an overview of Phase I and Phase II, engineering versus statistical process control, and criteria for evaluating control charts (ARLs, false alarm rates) has been presented. The reader is encouraged to consult any of the references given here for more complete details on each topic in this chapter.

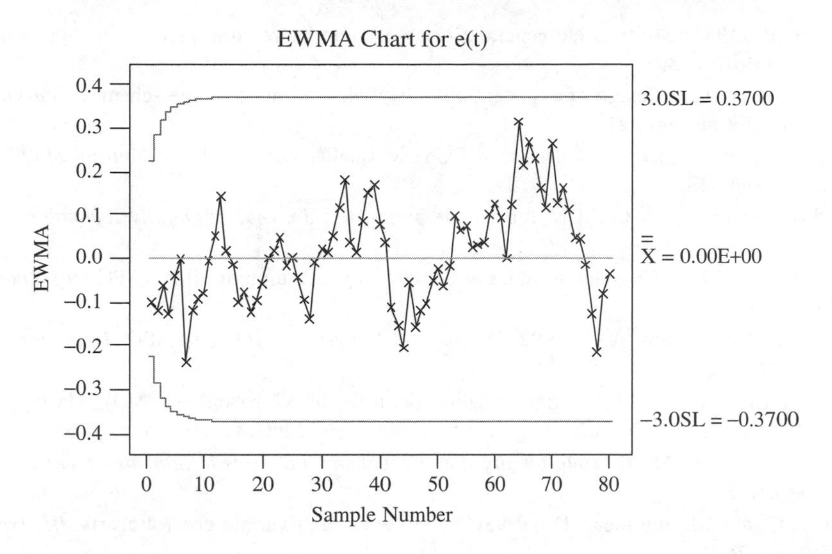

Figure 10.12 EWMA control chart of the residuals after fitting an AR(2) model to the temperature data.

References

Alt, F.B. (1985) Multivariate quality control. In S. Kotz and N.L. Johnson (eds), *Encyclopedia of Statistical Sciences*, pp. 110–122. John Wiley & Sons, Inc., New York.

Bisgaard, S. and Kulahci, M. (2005a) The effect of autocorrelation on statistical process control procedures. *Quality Engineering*, **17**(3), 481–489.

Bisgaard, S. and Kulahci, M. (2005b) Interpretation of time series models. *Quality Engineering*, **17**(4), 653–658.

Bisgaard, S. and Kulahci, M. (2006) Quality Quandaries: Studying Input–Output Relationships. *Quality Engineering*, **18**, 273–281, 405–410.

Box, G.E.P. and Luceño, A. (1997) *Statistical Control by Monitoring and Feedback Adjustment*. John Wiley & Sons, Inc., New York.

Box, G.E.P., Jenkins, G.M. and Reinsel, G. (1994) *Time Series Analysis, Forecasting and Control*. Prentice Hall, Englewood Cliffs, NJ.

Box, G.E.P., Bisgaard, S., Graves, S., Kulahci, M., Van Gilder, J., Ting, T., James, J., Marko, K., Zatorski, H. and Wu, C. (2003) Performance evaluation of dynamic monitoring systems: *The waterfall chart*. *Quality Engineering*, **16**(2), 183–191.

Chouinard, R (1997) Equipment performance monitoring: a Canadian Air Force perspective. In *Reliability and Maintainability Symposium: 1997 Proceedings, Annua*, pp. 37–43. IEEE.

Coleman, J.L. and Nickerson, J. (2005) A multivariate exponentially weighted moving average control chart for photovoltaic processes. In *Conference Record of the Thirty-First IEEE Photovoltaic Specialists Conference*, pp. 1281–1284. IEEE.

Crosier, R. (1988) Multivariate generalizations of cumulative sum quality control schemes. *Technometrics*, **30**.

Crowder, S. (1989) Design of exponentially weighted moving average schemes. *Journal of Quality Technology*, **21**.

Gan, F.F. (1991) An optimal design of CUSUM quality control charts. *Journal of Quality Technology*, **23**.

Hawkins, D. (1981) A CUSUM for a scale parameter. *Journal of Quality Technology*, **13**, 228–231.

Hawkins, D. (1993) Cumulative sum control charting: An underutililized SPC tool. *Quality Engineering*, **5**.

Hoerl, R.W. and Palm, A.C. (1992) Discussion: Integrating SPC and APC. *Technometrics*, **34**, 268–272.

Hotelling, H. (1947) Multivariate quality control. In C. Eisenhart, M.W. Hastay and W.A. Wallis (eds), *Techniques of Statistical Analysis*. McGraw-Hill, New York.

Jackson, J. (1985) Multivariate quality control. *Communications in Statistics – Theory and Methods*, **34**.

Lowry, C. and Montgomery, D. (1995) A review of multivariate control charts. *IIE Transactions*, **26**.

Lowry, C., Woodall, W., Champ, C. and Rigdon, S. (1992) A multivariate exponentially weighted moving average control chart. *Technometrics*, **34**.

Lucas, J. (1976) The design and use of cumulative sum quality control schemes. *Journal of Quality Technology*, **31**.

Lucas, J. and Saccucci, M. (1990) Exponentially weighted moving average control schemes: Properties and enhancements. *Technometrics*, **32**.

Mason, R., Tracy, N. and Young, J. (1995) Decomposition of T^2 for multivariate control chart interpretation. *Journal of Quality Technology*, **27**.

Montgomery, D. (2005) *Introduction to Statistical Quality Control* (5th edition). John Wiley & Sons, Ltd, Chichester.

Page, E. (1954) Continuous inspection schemes. *Biometrics*, **41**.

Palm, A.C. (2000) Discussion: 'Controversies and contradictions in statistical process control' by W.H. Woodall. *Journal of Quality Technology*, **32**(4), 356–360.

Prabhu, S. & Runger, G. (1997) Designing a multivariate EWMA control chart. *Journal of Quality Technology*, **29**.

Quesenberry, C. (1993) The effect of sample size on estimated limits for $\overline{X}$ and X control charts. *Journal of Quality Technology*, **29**, 237–247

Pignatiello Jr, J.J. and Runger, G.C. (1990) Comparison of multivariate CUSUM charts. *Journal of Quality Technology*, **22**, 173–186.

Roberts, S. (1959) Control chart tests based on geometric moving averages. *Technometrics*, **1**.

Testik, M.C., McCullough, B.D. and Borror, C.M. (2006) The effect of estimated parameters on Poisson EWMA control charts. *Quality Technology and Quantitative Management*, **3**, 513–527.

Tracy, N.D., Young, J.C. and Mason, R.L. (1992) Multivariate control charts for individual observations. *Journal of Quality Technology*, **24**(2), 88–95.

Wheeler, D.J. (2000) Discussion: 'Controversies and contradictions in statistical process control' by W.H. Woodall. *Journal of Quality Technology*, **32**(4), pp. 361–363.

Woodall, W.H. (2000) Controversies and contradictions in statistical process control. *Journal of Quality Technology*, **32**(4), pp. 341–350.

Woodall, W.H. and Adams, B.M. (1993) The statistical design of CUSUM charts. *Quality Engineering*, **5**(4), 559–570.

11

Measurement system analysis

Giulio Barbato, Grazia Vicario and Raffaello Levi

11.1 Introduction

Statistical methods play a key role in the analysis of measurement and testing work performed within the framework of quality systems. The traditional statistical methods are repeatability and reproducibility (R&R: see Phillips *et al.*, 1998; Smith *et al.*, 2000) and measurement system analysis (MSA: Carbone *et al.*, 2002; Senol, 2004; Automotive Industry Action Group, 2003). More recently, measurement and testing procedures have called for particular attention to be given to a fundamental quality index related to both cost and benefit, namely uncertainty associated with measurement. Statements of uncertainty are mandated by current quality standards, starting with ISO 9001:1994 and, more specifically, ISO 17025, that govern the management of testing and measurement laboratories. Recognition of the paramount importance of the subject led major international organisations concerned with metrology and standardization – such as BIPM, IEC, IFCC, IUPAC, IUPAP and OIML – to draft and publish under the aegis of the International Organisation for Standardisation a fundamental reference text, *Guide to the Expression of Uncertainty in Measurement* (ISO, 1993),[1] known as GUM and embodied into European standard ENV 13005.

Statistical procedures dictated by GUM cover a broad range of applications. Apart from definition of such a delicate undertaking as evaluation of measurement uncertainty with a clear set of universally accepted rules, by no means a minor achievement, they cater for planning measurement and testing work aimed

[1]This was followed by a number of documents concerned with applications; see Taylor and Kuyatt (1994), Nielsen (1997), Chen and Blakely (2002) and Comité Europeén de Normalisation (2004).

Statistical Practice in Business and Industry Edited by S.Y. Coleman, T. Greenfield, D.J. Stewardson and D.C. Montgomery

at specific levels of uncertainty, to avoid failure to reach mandated accuracy, and costly overdesign. These methods, covering both specific metrological work, such as calibration of sensors and instrument systems, and generic testing work, deal with three main items, namely:

- contributions to uncertainty estimated using statistical methods, referred to as 'type A'. Besides managing random errors (never absent in actual measurement) with such tools as normal and t distributions, they enable detection and estimation of systematic effects (such as with tests of normality, of linearity, and analysis of variance), and proper treatment of such outliers, typically associated with the ever-increasing diffusion of high impedance electronic instrumentation. These instrument systems are often sensitive to spurious signals from electromagnetic noise, leading to measurement accidents which need to be dealt with according to proper exclusion procedures.
- contributions to uncertainty assessed by non-statistical methods (mainly expert evaluation and practical experience), referred to as 'type B'. Their importance is underlined by GUM as cost-effective and highly robust. Introduction of type B contributions is a major step by GUM, by describing how to transform them into equivalent variance components, relying upon uniform, U-shaped or triangular distributions selected according to experience.
- composition of contributions mentioned above, and allotment of proper degrees of freedom, leading to evaluation of overall uncertainty in the form of confidence intervals, related to normal or t distributions.

Categorisation and tabulation of terms entering measurement uncertainty computation cater for assessment and comparative evaluation of contributions pertaining to individual factors, according to an iterative computation routine (PUMA method, ISO 14253-2:1999; for practical details, see Bennich, 2003). What in the past was an awkward part of the process of uncertainty budget management, namely dealing with individual contributions to be associated with every uncertainty factor, and suggesting which factors are best acted upon to cut costs and/or measurement uncertainty, may thus be made straightforward.

11.2 Historical background

The evolution of the management of testing and measurement work may be traced in the *modus operandi* currently encountered in industry, covering a broad range of attitudes. Control of instruments in some companies reduces to a formal bureaucratic procedure, covering documentation of periodic calibration and related expiry date, often coupled with reliance upon deterministic results, and disregard of scatter. Other companies choose exacting studies on measurement uncertainty instead, basing them on the latest developments, which may go well beyond practical requirements.

Such attitudes may have originated in methods for management of testing and measurement, due to intense and fruitful research work on theoretical aspects of metrology. This has led to remarkable progress, going far beyond current standards and leading to extreme consequences. Historic evolution may be summarised as follows. The maximum error of a measurement instrument was required to be just a small fraction (initially one tenth, then one fifth; now even one third is permitted even within conservative sectors in legal metrology) of the maximum error tolerated in results. This trend in admissible instrument error may be linked to the evolution from a deterministic approach to a more probabilistic approach, from sum of errors to sum of variances.

Statistical management of effects due to noise factors leads to estimation of uncertainty from scatter observed in replicated tests. Results are critically affected by what replications actually cover. Thus there is a distinction between *repeatability* (variation of measurements on identical test items under the same conditions – same operator, same apparatus, same laboratory and a short interval of time), and *reproducibility* (variation of measurements on identical test items under different conditions – different operators, different apparatus, different laboratories and/or different time).

Major limitations of the repeatability and reproducibility approach were the substantial cost (large number of tests required), and the global nature of results, where individual contributions to overall variability may be hard to extract. MSA overcomes such limitations by exploiting statistical design of experiments (typically with two or three operators, two to ten parts, and two or three replications), and related analysis, currently relying on either the average and range method or analysis of variance (ANOVA).

There has recently been a remarkable evolution in applied metrology and quality management. Traditional concepts, with rigid, inflexible schedules, such as ISO 9001:1994, point 4.11 (traceability to national standard mandatory for every measurement instrument . . .), were thrown overboard. Instead, a more flexible procedure was advocated, in terms of actual needs instead of abstract prescriptions. Thus ISO 9001:2000 mandates traceability to national standards only if necessary, each case being assessed on actual needs. Accordingly, measurement uncertainty becomes the yardstick to judge if a given instrument is appropriate for a given application. Most major international standard and metrological institutions (BIPM, IEC, IFCC, ISO, IUPAC, IUPAP, OIML) took part in the effort, leading to an universally accepted document, the GUM referred to in the Introduction.

The practical implications are remarkable, since unnecessary expenditure and complication may legitimately be avoided. We can dispense with the sacred cow of requirement of higher-class instruments for calibration, a most welcome breakthrough in industrial work. Given a set of similar instruments, we can choose the best and calibrate all the others against it. At the cost of minor, often acceptable, increase of uncertainty, major savings in costs and streamlined procedure are readily obtained.

The ISO GUM revolution ushered in a major breakthrough, with the formal introduction (or simply recognition, since the concept was already familiar to every experienced operator) of type B uncertainty terms, assessed in terms of cumulated experience and estimated in terms of accepted tolerances, for example, or expert evaluation. Given a substantial body of relevant experience, accurate assessment is readily available at no extra cost, as opposed to running an expensive and time-consuming R&R program. Treatment of effects and measurement system evaluation are not affected, but the new concept leads to substantial practical advantages.

Unbiased assessment of the current method of uncertainty evaluation does not necessarily imply that the old, traditional framework is inferior; progress is mainly in minor refinements. Let us consider the case of a calibration centre, concerned, *inter alia*, with pendulums for impact testing purposes.

Tolerance on distance between edges of specimen-supporting surfaces, nominally 10 mm, is within a few tenths of a millimetre. In the early 1990s Italy's national standards institution suggested procuring an inexpensive gauge block to be used for calibration of callipers. According to the conventional approach, the maximum allowed error of a gauge block was two orders of magnitude smaller than that of a calliper (a few tenths of a micrometre against 20 μm). No periodic calibration of gauge block was deemed necessary.

A few years later, however, ISO 9001:1994 required some sort of R&R procedure, no matter what common sense suggested; overregulation entails costly overkill.

Eventually, according to ISO 9000:2000, and evaluating uncertainty along ISO GUM taking into account type B errors, we are back at square one. The contribution to overall uncertainty due to gauge block is small enough to allow it to be disregarded.

In the light of the example discussed above, a plain, no-nonsense procedure leads efficiently to results that are at least as good as those yielded by a more complex, and substantially more expensive method. Current standards now support exploitation of type B contributions to uncertainty, matching efficiency with the necessary thoroughness, thus leading to sensible, reliable results.

11.3 Metrological background

Application of statistical methods is effective only if the field of action is sufficiently known, as clearly stressed in clause 3.4.8 of GUM:

> Although this guide provides a framework for assessing uncertainty, it cannot substitute for critical thinking, intellectual honesty and professional skill. The evaluation of uncertainty is neither a routine task nor a purely mathematical one; it depends on detailed knowledge of the nature of the measurand and of the measurement.

This underlines the need to describe what can happen during measurement. The main effects due to systematic factors (effects emerging from the noise background), enable identification of a cause-and-effect relationship. Identification of systematic factors permits correction, compensation or allowance for their effects, and control over random factors (sometimes related to systematic factors, even if they are buried within overall noise, and as such beyond association with specific cause-and-effect relationships).

Considerations of systematic and random errors lead to the concepts of bias, repeatability and reproducibility. Metrology theory also subdivided bias into linearity, hysteresis, drift and other specific errors. Some causes of random errors are investigated too, such as reading and mobility errors. Advanced metrological analysis is not required in order to relate statistics to metrology. But blind analysis of measurement performance, based only on results obtained without regard for metrological concepts, is seldom satisfactory; neither are simple, practical considerations, unsupported by basic statistical concepts.

The rapidly growing importance of measurement accidents should not be overlooked. Measuring instruments evolve with very sensitive and sophisticated electronic systems. Therefore what used to take the form of an accident, like a shock or exceptional vibration, which would be detected on mechanical devices, now correspond mainly to electromagnetic disturbances, definitely beyond detection, let alone control, of the operator. Therefore, exclusion criteria, such as the Chauvenet principle, the Grubb method or others, seem to be more and more used out of sheer necessity.

11.4 Main characteristics of a measurement complex[2]

A measurement complex comprises the measuring instrument, the operators, the working conditions and the measurand.

The terminology given in the ISO/DGuide 99999 *International Vocabulary of Basic and General Terms in Metrology* (ISO, 2004), known as VIM, may be applied usefully to the different metrological characteristics of measuring instruments and measurement complexes, as the working information is often transmitted using these characteristics. The traditional use of R&R or MSA involves mainly repeatability and reproducibility, but seldom accuracy. ISO (1993) uses all the metrological characteristics of a measurement complex.

11.4.1 Resolution

Resolution is a characteristic of the measuring instrument, or system, more than of the measurement complex, even if the effect of the complex is included in the definition given by VIM:

---

[2]Attention should be given to different concepts: *measuring instrument, measuring system* and *measuring complex*. VIM defines a measuring instrument as a 'device or combination of devices designed for measurement of quantities'. VIM also defines a measuring or measurement system as a 'set of measuring instruments and other devices or substances assembled and adapted to the measurement of quantities of specified kinds within specified intervals of values'. The concept of 'measurement complex' is involved in ISO 17025 where it stresses the importance of the effect of operator, ambient conditions and measurand in a measurement result.

> [The resolution of a measuring system is the] smallest change, in the value of a quantity being measured by a measuring system, that causes a perceptible change in the corresponding indication.
>
> NOTE The resolution of a measuring system may depend on, for example, noise (internal or external) or friction. It may also depend on the value of the quantity being measured.

This definition is not strictly connected to the reading of the measuring instrument only; for uncertainty evaluation we may prefer to distinguish different sources of contributions, and therefore use a specific definition for the reading resolution, as given by VIM:

> [The resolution of a displaying device is the] smallest difference between indications of a displaying device that can be meaningfully distinguished.

This definition is easily applied to discrete scales, that is, digital instruments, where the resolution is the count increment. With continuous scales, that is, with analogue instruments, the measured quantity is displayed by the position of an index on the measurement scale. A watch is an example of an analogue scale. Anybody reading the watch shown in Figure 11.1 would say it said 10:16 or 10:17, certainly neither 10:15 nor 10:20. We naturally interpolate when we read an analogue scale: the reading is not limited to the divisions shown on the scale, but is interpolated by dividing the smallest division into 2, 5 or 10, depending on the scale geometry.[3]

Figure 11.1 Analogue watch.

Resolution is a first contribution to uncertainty. The effect of a digital scale with the unit as resolution is easy to understand. Output possible values are, for

[3]Good analogue scales have a controlled geometry: the thickness of the index is constant and equal to the thickness of the lines separating the divisions on the scale. That thickness is made $\frac{1}{2}$, $\frac{1}{5}$ or $\frac{1}{10}$ of the dimension of the smallest division of the scale so that the user is aware of the capability of the instrument, and can interpolate within the smallest division to the nearest half, fifth or tenth.

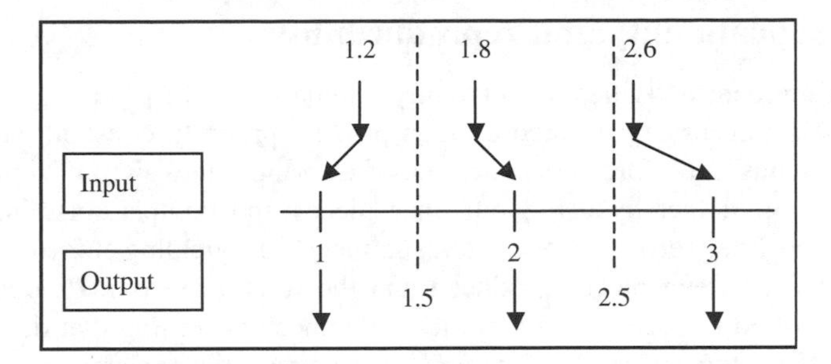

Figure 11.2 Performance of a digital indicator.

instance, 1, 2, 3, ... as represented in Figure 11.2. Input is continuous, therefore it may be 1.2, 1.8 or 2.6 as shown. Now an input of 1.2 will become an output of 1, and accordingly inputs of 1.8 and 2.6 become outputs of 2 and 3, respectively. The discriminating zones are narrow bands around the centre of resolution increments, nominally below 1.45 the output will be 1, above 1.55 it will be 2, and in between will be oscillating. Thus for an input between 2 and 3, the output will be 2 as long as the input is below 2.45, and will remain as 2 so long as input is, roughly, between 1.5 and 2.5. Thus a given output means that the input is between half a resolution step above, or below, the value obtained. Given a resolution step r, the range is between $-r/2$ and $+r/2$, and this is used to identify reading uncertainty.

With measuring instruments used in legal metrology, resolution as a rule is the largest contribution to uncertainty; some instruments belonging to the family of legal instruments, such as balances and mass comparators, behave accordingly. Therefore the contribution of resolution to the uncertainty budget is underlined as a rule, even when – as for some technical instruments – resolution is embedded in other measurement characteristics, such as repeatability or reproducibility.

11.4.2 Stability

In any measurement, a major requirement is the ability to evaluate small differences, since resolution is an important metrological characteristic. Also the reading should not fluctuate wildly, thus confusing the measurement. Ideally an instrument should yield a constant output for a constant input, as defined by VIM: stability is the 'ability of a measuring system to maintain its metrological characteristics constant with time'.

Note that readings may change owing to a zero shift or a change of sensitivity. A zero shift may be detected by applying a constant input close to the zero value. Once zero instability is taken care of, a change of sensitivity can be evaluated by applying a constant input near the maximum.

Adequate resolution and stability are sufficient metrological characteristics, whenever a measuring system is used to check a continuous production process, and what really matters is to keep a given process response constant within a control band.

11.4.3 Repeatability and reproducibility

Many measurement activities involve only comparisons. In production processes it is commonly necessary to keep a given product property constant, such as the diameter of a bush, the force characteristic of a spring, the weight of a mass. Such constancy is often met by using reference pieces and comparators, installed on the production line, zeroed on the reference piece and yielding information about the difference between pieces produced and the reference item. Production may thus be controlled in terms of an empirical tolerance concerning that difference, as opposed to the actual value of the parameter. To stress this concept, let us consider the possibility of checking ball production in a plant using a weighing scale. The reference piece is a ball that has been checked for diameter and sphericity. The density and sphericity of balls produced are as a rule inherently stable, so that periodical checks are sufficient. The constancy of the mass of balls, easily checked on all balls produced, can therefore provide an efficient means of checking size, without any length measurement.

Measurement in this case may not involve stability only, since the force on the balance, far from being kept constant, falls to zero between weighing cycles only to return to full scale whenever a new ball is loaded. It is necessary, therefore, to evaluate whether the measuring instrument or system performs correctly over loading–unloading cycles.

Two different situations may apply, namely the focus may be on the instrument only, or on the measuring complex, that is including also the effects of operators, ambient conditions and other sources of disturbances. These different situations correspond to metrological definitions of repeatability and reproducibility, connected with the measurement conditions focused on instrument, or measuring complex performances. In both instances the metrological characteristic involved is precision, so that a sound order of presentation is precision first, then repeatability and reproducibility. According to VIM, precision is defined as:

> closeness of agreement between quantity values obtained by replicate measurements of a quantity, under specified conditions.
>
> NOTE Measurement precision is usually expressed numerically by measures of imprecision, such as standard deviation, variance, or coefficient of variation under the specified conditions of measurement.

11.4.3.1 Repeatability

Repeatability is the ability of a measuring instrument to return small differences in measurement results, when repeated measurement cycles are performed under strictly controlled conditions, that is, avoiding external disturbances as much as possible. The VIM definition of repeatability is 'measurement precision under repeatability conditions of measurement', a *repeatability condition* being 'that condition of measurement in a set of conditions including the same measurement procedure, same operator, same measuring system, same operating conditions and same location, and replicated measurements over a short period of time'.

The information content of repeatability is limited, as it cannot refer to real measurement situations, where working conditions are invariably affected by external effects. Nevertheless, repeatability offers some advantages, since:

(1) it is based on a measurement condition defined as closely as possible;

(2) it requires a small amount of work to provide information on a limiting operating condition – no better results may reasonably be expected;

(3) it yields a reference value of variability, to be used as a yardstick, for instance with ANOVA, to evaluate the effects of other uncertainty factors;

(4) it is a characteristic very sensitive to the functionality of the instrument, so that, also in connection with the limited amount of work required, it provides a very efficient way to perform metrological confirmation (ISO 10012).

11.4.3.2 Reproducibility

Reproducibility carries more comprehensive information on the performance of a measurement complex, that is, the instrument, operators, working conditions and sometimes also the measurand. VIM defines reproducibility as 'measurement precision under reproducibility conditions of measurement', a *reproducibility condition* being a

> condition of measurement in a set of conditions including different locations, operators, and measuring systems.
>
> NOTES
>
> 1. The different measuring systems may use different measurement procedures.
> 2. A specification should give the conditions changed and unchanged, to the extent practical.

The main advantage of reproducibility is complete information on measurement complex performance, with reference to dispersion of measurement results. Disadvantages relate to the complexity of reproducibility evaluation: a clear and complete description of the factors involved is required, and repeatability tests entail a large amount of work.

11.4.4 Bias

Comparative measurements require a reference piece, to be measured in such a way as to obtain values compatible with measurements performed by other partners within the production system and/or the supply chain, or by inspection authorities. Agreement among measurement results entails referring the measurement scale to a uniquely defined reference, as obtained by an uninterrupted chain of calibrations from the international standards of measurement. These standards are maintained at the Bureau International des Poids et Mesures (BIPM), through the standard of measurement units, maintained by National Institutes of Metrology, and transfer measurement standards of calibration services.

The metrological concept connected with bias is defined by VIM:

> [Bias is] systematic error of indication of a measuring system.
>
> NOTE The bias of a measuring system is the average of the errors of indication that would ensue from an infinite number of measurements of the same measurand carried out under repeatability conditions.

The concept of error, frequently discussed in metrology, is defined by VIM as 'the difference of quantity value obtained by measurement and true value of the measurand'. The true value is a thorny issue: it is the

> quantity value consistent with the definition of a quantity.
>
> NOTES
>
> 1. Within the [classical approach] a unique quantity value is thought to describe the measurand. the true value would be obtainable by a perfect measurement, that is a measurement without measurement error. The true value is by nature unobtainable.
> 2. Due to definitional measurement uncertainty, there is a distribution of true values consistent with the definition of a measurand. This distribution is by nature unknowable. The concept of 'true value' is avoided in the [uncertainty approach] (D.3.5 in the GUM).

The concept of true value is hardly acceptable within the framework of operational philosophy, establishing the definition of any physical quantity strictly in terms of the measurement process; therefore no such quantity as a true value, which does not lend itself to actual measurement, may be defined.

In fact there is only a problem of a contradiction in terms between the adjective 'true' and the uncertain characteristics of every measurement result. Indeed, any physical quantity is unmeasurable, in the sense that it will always be known with an uncertainty, but no one says that it is not possible to define length, mass or other physical quantities.

The riddle is solved by considering, instead of 'true value', a 'reference value' providing the best measurable value of that quantity, according to real metrological practice. Measured quantities are at best compared with the primary reference standards, provided by international measurement standards maintained by BIPM. In this case, error is a measurable quantity, being the difference between two quantities that are known, albeit within their uncertainty, namely the reference value or standard, and the measured value.

The characteristics of systematic error connected with bias should be underlined, entailing that in its evaluation the measurement compared with the reference value should be free from random errors, that is, the average of infinite measurements; this being impossible, bias is affected by random error, the systematic part being as a rule predominant.

11.4.5 Uncertainty

Finally, but none the less important, uncertainty is defined as the opposite of accuracy. VIM defines accuracy of measurement as

> closeness of agreement between a quantity value obtained by measurement and the true value of the measurand.
>
> NOTES
>
> 1. Accuracy cannot be expressed as a numerical value.
> 2. Accuracy is inversely related to both systematic error and random error.
> 3. The term 'accuracy of measurement' should not be used for trueness of measurement and the term 'measurement precision' should not be used for 'accuracy of measurement'.

VIM defines the accuracy of a measuring system as the ability of the system

> to provide a quantity value close to the true value of a measurand.
>
> NOTES
>
> 1. Accuracy is greater when the quantity value is closer to the true value.
> 2. The term 'precision' should not be used for 'accuracy'.
> 3. This concept is related to accuracy of measurement.

Here again, a correct philosophical meaning to these definitions may be obtained by introducing the term 'reference value' as a substitute for 'true value'.

Having defined 'accuracy', VIM defines uncertainty as a 'parameter that characterises the dispersion of the quantity values that are being attributed to a measurand, based on the information used'. A clearer definition of uncertainty is provided in clause 2.2.3 of GUM: it is a parameter 'associated with the result of a measurement that characterises the dispersion of the values that could reasonably be attributed to the measurand'. This definition can be translated easily in statistical terms, by identifying 'reasonably' with the usual statistical risks of results outside the established uncertainty interval, which is the confidence interval around the measured value defined with that risk of error.

11.5 Measurement uncertainty evaluation according to GUM: Part I

Traditionally, uncertainty was evaluated by statistical analysis of measurement results. GUM refers to the uncertainty contribution determined in this way as 'type A'. As statistical analysis needs experimental data that can be costly, one of the most effective developments by GUM enabled the exploitation of all information accumulated by experience, generally known as variability intervals, such as resolution and accuracy classes of instruments, standard tolerances of measurement procedures or ambient conditions, and experience of trained operators. These contributions, known from previous experience, are referred to by GUM as 'type B'.

As type A uncertainty is widely employed in traditional methods such as R&R and MSA, we now describe it, deferring the description of type B contributions until after the presentation of the traditional method of analysis.

General concepts should be stated in terms of major aims.

The level of accuracy of an analysis should be clear, remembering that uncertainty is usually given to only one significant digit, rarely two, never more than two. Accuracy classes, for instance, are a conventional expression of the instrument uncertainty and are stated, for instance, as 1 %, 2 % or 5 % – never 1.5 % or 3.2 %. With this evaluation aim, it is not necessary to define every uncertainty contribution with great accuracy, and this is clearly stated in ISO 14253-2 under the procedure for uncertainty management (PUMA). Uncertainty evaluation has a difficult part, evaluation of contributions, and a simple part, calculation. The PUMA method suggests reasonable values of uncertainty should be introduced, avoiding many measurements or deep literature searches.

Type A contributions of uncertainty are generally obtained from a set of experimental results by statistical analysis, often as just a standard deviation.[4]

Experimental results may also exhibit constant systematic error (such as calibration errors), variable systematic errors (such as the effect of large temperature variation over replicated measurements) and measurement accidents (such as those caused by electromagnetic disturbance). Constant systematic errors may not be apparent, and should be avoided by proper instrument calibration and measurement procedures, while variable systematic errors and measurement accidents should be detected by normality tests and exclusion principles.

[4]Note that GUM introduces uncertainty contributions of type B together with an estimate of standard deviation from such contributions; therefore, to underline that it cannot be a rigorous statistical evaluation of standard deviation, GUM also introduces the term 'standard uncertainty', denoted by the symbol u.

11.5.1 Outliers: detection and exclusion

Generalised use of sensitive electronic instruments, typically with high gain and large impedance circuitry, dictated the need to take into account measurement accidents. Chauvenet's principle may be applied, by assigning to either tail of data distribution, where outliers are likely to appear, a probability $1/(4n)$, where n is the size of the experimental data set at hand. The data distribution is generally assumed to be normal, so the boundaries for identifying outliers are readily obtained. The procedure is shown in Table 11.1, which shows ten replications by five operators.

Table 11.1 Ten replicated measurements by five operators.

Oper. 1	Oper. 2	Oper. 3	Oper. 4	Oper. 5
8.0038	8.0040	8.0060	8.0092	8.0053
8.0051	8.0025	8.0041	8.0104	8.0076
8.0037	8.0093	8.0028	8.0110	8.0113
8.0017	7.9985	8.0049	8.0087	8.0112
8.0023	8.0046	8.0025	8.0047	8.0066
8.0145	8.0020	8.0030	8.0076	8.0065
8.0080	8.0046	8.0073	8.0056	8.0114
8.0058	8.0085	8.0065	8.0113	8.0064
8.0031	8.0056	8.0071	8.0068	8.0065
8.0084	8.0037	8.0064	8.0093	8.0057

Since $n = 50$, the boundaries of an outlier zone are assigned a probability $1/(4 \times 50) = 0.50\,\%$; as the grand average is 8.0063, with standard deviation 0.0031, the boundaries are computed as $8.0063 \pm 2.576 \times 0.0031$, that is, 7.9983 and 8.0143. The value 8.0145 exceeds the upper limit, and is therefore excluded as outlier. After this first exclusion $n = 49$, therefore the boundaries of the outlier zone are assigned a probability 0.51 %; the new average and standard deviation are respectively 8.0061 and 0.0029, and the updated boundaries are closer, 7.9986 and 8.0136. The value 7.9985 is now an outlier, and excluded; after this second step no further outliers are found.

Exclusion principles are based on a weak logical principle, namely that results with a low probability of occurrence are necessarily the outcome of measurement accidents. However, given that accidents are rare, by no means every rare event is necessarily an accident. Exclusion principles should therefore be preferably taken as guidance to suggest the possible presence of outliers, and values should be excluded only after careful examination of each case.

11.5.2 Detection of the presence of variable systematic errors

If the effects of several random errors are not disturbed by variable systematic errors, measurement results should be normally distributed. The presence of variable systematic errors may therefore be detected by testing the data distribution for normality, for example with normal probability plots, and for departures from normal. These plots may yield additional useful information.

Normal distributions appear on a normal probability plot as straight lines. A tendency towards an S shape indicates a *hypernormal* distribution, a situation occurring frequently with improperly trained operators. Whenever a new type of measurement is proposed to an operator, the chances are that, from experience, he will replicate the measurement at least three times to get acquainted with the usual range of results to be expected. Operators know that in the case of mistakes or gross measurement errors, results will fall outside that usual range. If that happens, competent operators take it as a warning to review the measurement process carefully, and find out whether something wrong has happened. If they find no clear evidence of a mistake, they accept that result. On the other hand, improperly trained operators often tend to reject out of hand results falling outside the usual range so that fewer results fall in either tail than ought to, leading to a hypernormal distribution.

Another situation commonly encountered is shown on a normal probability plot by a line with a slope in both tails noticeably larger than in the central part, indicating a *hyponormal* distribution. The normal probability plot shown in Figure 11.3 pertains to the case of Table 11.1 (outliers deleted). Two common causes of such an occurrence are drift or discontinuity in the measurement complex.

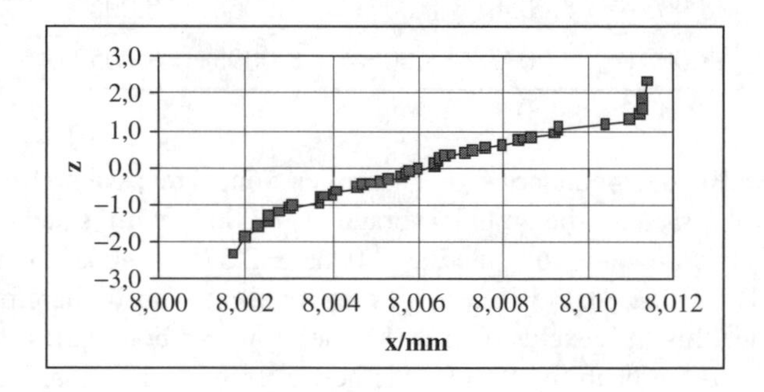

Figure 11.3 Normal probability plot of data in Table 11.1 (outliers deleted). A hyponormal condition is shown by the slope, higher in both tails and lower in the middle.

Let us examine the effect of drift. If the total time span of replicated measurements is considered as subdivided into small parts, over any one of these parts results usually exhibit a normal distribution. Drift causes a progressive shift in the averages pertaining to subsequent parts, thus leading to a broadening of the central part of the overall distribution, which takes a hyponormal shape.

When there is discontinuity of the measurement complex the corresponding distribution is multimodal. Take, for instance, the simple case of a measurement complex with two operators, with a systematic difference in results between them. Values corresponding to the average of each operator are more frequent, and less common both in between and in the tails. The resulting hyponormal distribution is bimodal.

A hyponormal condition suggests a check for drift or discontinuity. Drift evaluation simply requires plotting data in time sequence, and estimating the slope, if any, by linear regression (see Figure 11.4). A slope of about 1×10^{-4} per time unit with a standard deviation of about 2×10^{-5} is found, indicating the presence of systematic error associated with drift. Data corrected for drift are listed in Table 11.2. If we repeat the analysis on corrected data, we see no further evidence of measurement accidents, while the normal probability plot (Figure 11.5) is again of the hyponormal type.

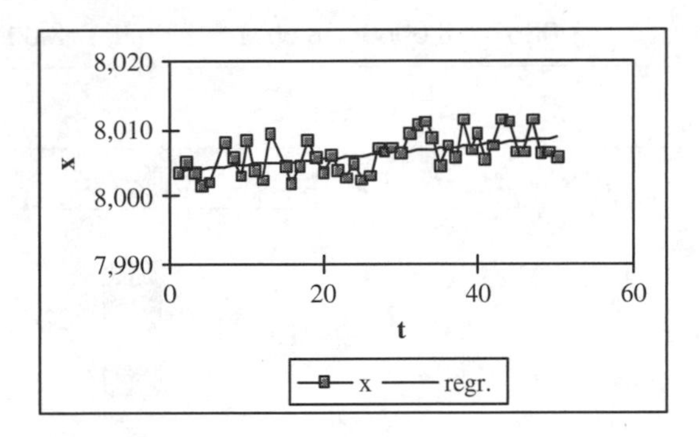

Figure 11.4 Plot of data in time sequence. Trend suggests presence of drift.

Table 11.2 Data of Table 11.1 corrected for drift.

Oper. 1	Oper. 2	Oper. 3	Oper. 4	Oper. 5
8.0063	8.0055	8.0065	8.0087	8.0038
8.0075	8.0039	8.0045	8.0098	8.0060
8.0060	8.0106	8.0031	8.0103	8.0096
8.0039	–	8.0051	8.0079	8.0094
8.0044	8.0057	8.0026	8.0038	8.0047
–	8.0030	8.0030	8.0066	8.0045
8.0099	8.0055	8.0072	8.0045	8.0093
8.0076	8.0093	8.0063	8.0101	8.0042
8.0048	8.0063	8.0068	8.0055	8.0042
8.0100	8.0043	8.0060	8.0079	8.0034

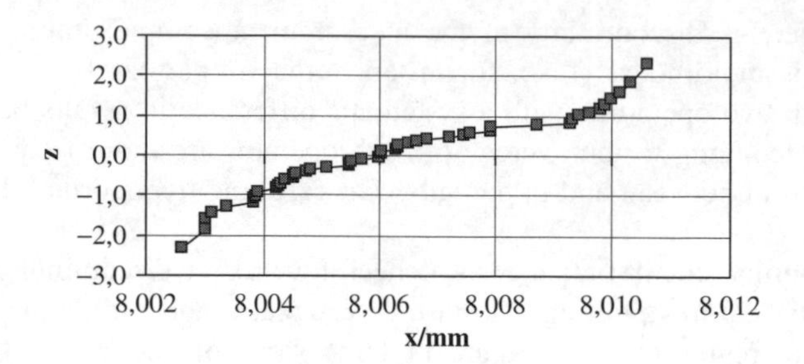

Figure 11.5 Normal probability plot for Table 11.2, showing hyponormal form.

Table 11.3 Averages obtained by different operators.

Operator	1	2	3	4	5
Average	8.0067	8.0060	8.0051	8.0075	8.0059

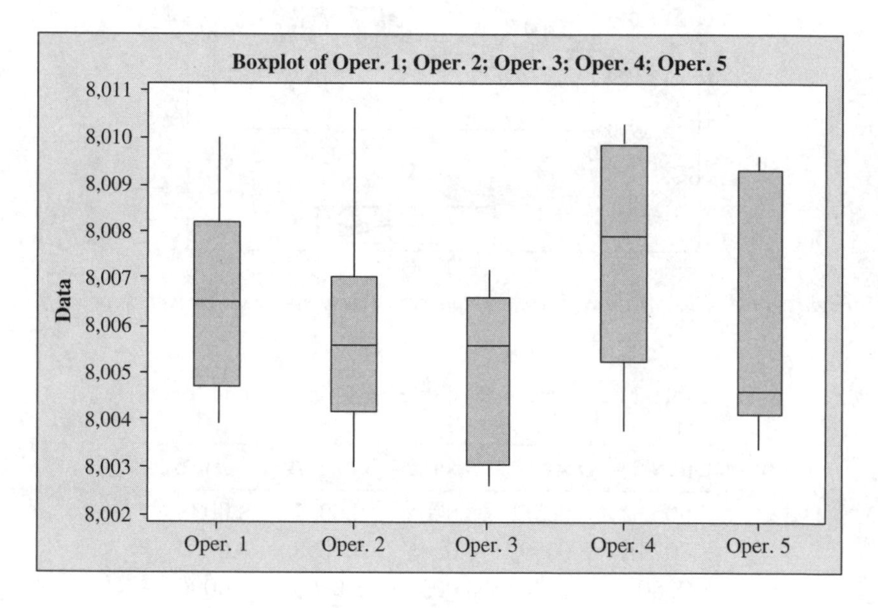

Figure 11.6 Boxplot for data of Table 11.2, showing differences between operators.

The situation becomes clearer with a closer analysis of the operators' performance. Averages for individual operators are in Table 11.3, with a grand average of 8.0062, and a 80 % confidence interval for averages ranging from 8.0053 to 8.0072. Substantial deviations observed concerning operators 3 and 4, whose averages fall outside the interval referred to above (see also the boxplot in Figure 11.6), explain the hyponormal distribution.

11.5.3 Data used for calculation of type A uncertainty contributions

The previous analysis showed management of experimental data based upon evaluation of variance and correction for effects of systematic variable errors and measurement accidents as required. Variance and standard deviation are widely used in traditional methods as R&R and MSA. For these methods, as for every method based on the analysis of experimental data, precautions concerning systematic errors and measurement accidents detection are advisable.

11.6 Measurement system analysis method

The aim of MSA is to assess the quality of a measurement system in terms of its statistical properties, through the use of a suitable set of procedures, statistical methods and metrics. The statistical properties of major interest are bias, stability, linearity and precision variability. The analysis examines procedures for quantification of total measurement variation, identification and explanation of major sources of such variation, assessment of stability of measurement process and evaluation of suitability of the measurement system for use in a particular manufacturing line or test area.

Procedures and analytical methods suitable for a wide range of MSA studies range from the traditional gauge capability study involving multiple parts and multiple operators (or appraisers), to studies involving highly automated measurement systems with only one operator and only one part. MSA studies address both variable and attribute data measurement systems and apply the average and range method and ANOVA method for variable data.

11.6.1 Responsibility for gauge studies

The timing of assessments, organisational responsibility for conducting the assessments, how they are conducted, and the responsibility for reacting to results of assessments should be delegated by the operations and/or the manufacturing manager. The person responsible should specify how data are to be gathered, recorded, and analysed. When items to be measured are selected and the environment in which the test procedures are undertaken is specified, the following recommendations apply:

- Make sure the correct forms are used.
- Get the correct number of items to measure for the study.
- Select the appropriate personnel to perform the gauge study.
- The person managing the gauge study is responsible for collecting and analysing data, documenting the results of the study and submitting the completed forms to the in-process quality assurance and/or the divisional

quality assurance group for record retention. Results of a measurement system capability study must be on file.

11.6.2 When should MSA studies be used?

Every gauge should have an MSA study before being placed in a production environment. No gauge should be used until the results of the study are acceptable.

A repeat study should be done at least once a year or as initiated by the following situations:

1. The gauge is found to be out of calibration during regular gauge maintenance.
2. A change has been made to the gauge – for example, a component replaced – which might affect gauge performance.
3. A major repair has been carried out on the gauge.
4. The gauge will be used as part of a process or equipment characterisation, a process capability study, implementation of statistical process control in a process area or running a designed experiment.
5. The gauge measurement system does not have adequate discrimination.
6. The measurement system is statistically unstable over time.
7. Measurement variation has significantly increased over time.
8. Measuring devices need to be compared.
9. A gauge is being used over a new range of a given characteristic, not covered by previous MSA studies.
10. Significantly reduced total variation in the process or product has been observed.

11.6.3 Preparation for an MSA gauge capability study

First, determine if repeated measurements can be made of the same characteristic on the same part (and at the same location within the part if applicable). Secondly, the gauge must be calibrated, repaired or adjusted before, not during, the MSA study. Thirdly, determine the sampling plan of the gauge capability study. Specifically, if multiple operators, parts and trials are required in the gauge capability design, some factors to be considered are as follows:

- Whenever possible, the operators should be chosen from among those who usually operate the gauge. If these individuals are not available, then other suitable personnel should be properly trained in the correct use of the gauge, so that they can replicate the actual daily usage of the gauge as closely as possible.

- Sample parts must be selected from the process, representing its entire operating range. This is sometimes done by taking one sample per day for several days. This is necessary because the parts will be treated in the analysis as if they represent the full range of product variation taking place in production. Any simulation carried out must be representative of the parts/process variation.
- Sample parts must be labelled and measurements must be in a random order. To avoid any possible bias, the operator who is measuring a part should not know which numbered part he is measuring.
- Criticality: critical dimensions require more parts and/or trials to have a higher degree of confidence.
- Part configuration: bulky or heavy parts may dictate fewer samples and more trials.
- Randomisation and statistical independence: the method of collecting data is important in a gauge capability study. If data are not collected randomly, there may be bias. A simple way to ensure a balanced design for a gauge study with n parts, k appraisers, and r trials is through randomisation. One common approach to randomisation is to write A(1) on a slip of paper to denote the measurement for the first appraiser on the first part. Do this up to A(n) for the measurement by the first appraiser on the nth part. Follow the same procedure for the 2nd, ..., kth appraisers, letting B(1), C(1), ... denote the measurement for 2nd, 3rd, ... appraiser on the first part. Once all $n \times k$ combinations are written, the slips of paper can be put into a bowl. One at a time, a slip of paper is drawn. These combinations (A(1), B(2), ...) identify the order in which the gauge study will be performed. Once all $n \times k$ combinations are selected, they are put back into the bowl and the procedure followed again. This is done a total of r times to determine the order of experiments for each repetition. An even better method would be to use a random number table.

11.6.4 Designing gauge capability studies: rationale for selecting different gauge capability designs

The purpose of an MSA study is to identify, estimate and understand the main sources of measurement precision, to assess the bias, linearity and stability of the measurement process, and to evaluate the suitability of the measurement system for use in a manufacturing line or test area.

A measurement process is similar to a manufacturing process. There are inputs and outputs, and several sources of variation may affect the outputs. To understand, evaluate and improve a measurement process, it is necessary to design a gauge capability study using many of the criteria required in a well-designed experiment.

The traditional standard form of gauge capability study (with two or three operators, between two and ten parts and two or three trials) has proved to be useful. However, this form may not be the best design for many measurement systems, particularly in wafer fabrication facilities. Many gauges are part of an automated measurement system, where operators do not contribute significantly to measurement error, and/or there are many more factors to study than just operators and repeatability. These more complex gauge studies require a choice of different designs to ensure that appropriate, relevant information is derived from the study.

11.6.5 Analysis of measurement systems

Distributions that can be used to describe the measurement system's variation or types of error may be characterised in terms of location and spread.

The analysis of a measurement system consists of examining the statistical properties of the system, including bias, stability, linearity and precision. Appropriate metrics are then calculated to evaluate the suitability of the measurement system for a specific use in a manufacturing line or test area. Most of the guidelines discussed in this section are for variable data.

11.6.5.1 Estimating measurement system bias

Bias is the difference between the average of observed measurements and a master average or reference value, as determined by measurement with the most precise measuring equipment available. Bias is often defined as accuracy. However, accuracy is sometimes defined as precision plus bias, so the use of bias to represent the offset from the reference value is recommended (see Figure 11.7).

The amount of bias is determined as follows. The acceptance criterion is that the observed average cannot exceed the reference value (true average). Obtain a sample

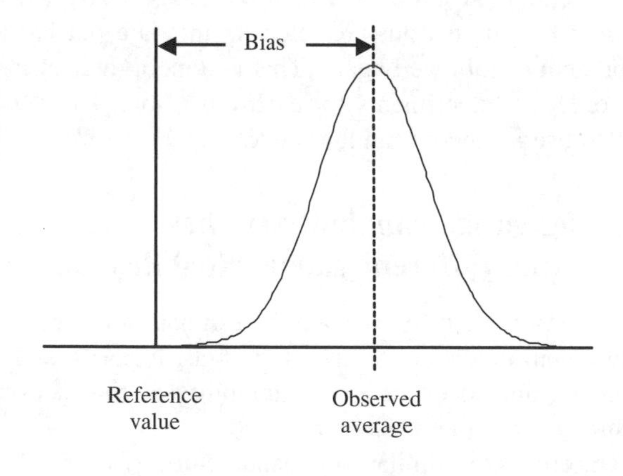

Figure 11.7 Definition of bias.

and establish its reference value(s) as described before. Have an appraiser measure the sample n times and compute the average of the n readings. Compute the bias by subtracting the reference value from this average. Set the process variation equal to the six sigma range. Then the percentage bias is the computed bias divided by the process variation.

Here is an example of a bias study. A sample of one part is measured 12 times by an appraiser and the values x_i (expressed in millimetres) are as follows:

$$0.75, 0.75, 0.80, 0.80, 0.65, 0.70, 0.80, 0.75, 0.75, 0.75, 0.80, 0.70.$$

The reference value determined by NIST is 0.80 mm, the observed average is 0.75 mm and the process variation for the part is 0.70 mm. The bias is the difference between reference value and observed average; the percentage tolerance for bias is calculated in the same way, where tolerance is substituted for process variation. Therefore, the thickness gauge to be used in the gauge R&R study has a bias of −0.05 mm. This means that the observed measurements on the average will be 0.05 mm smaller than the reference values, which is 7.1 % of the process variation.

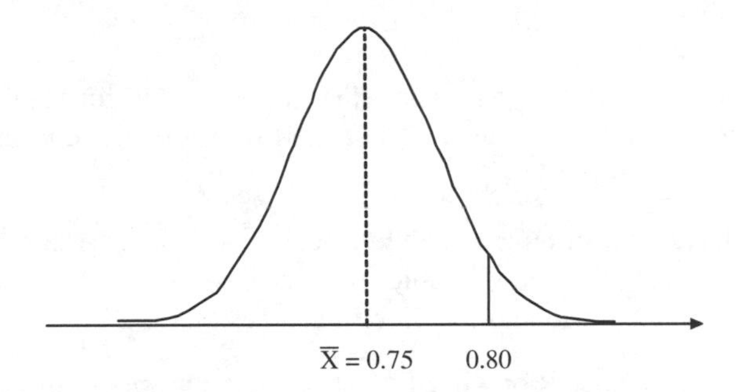

An Xbar-R chart can be used to evaluate bias as well as stability. We can compute $\overline{X}$ from the chart and then subtract the reference value from $\overline{X}$ to obtain the bias.

11.6.5.2 Evaluating gauge stability of measurement systems

Gauge stability is estimated as the difference in the averages of at least two sets of measurements taken at different times, using the same gauge on the same parts (see Figure 11.8). Gauge stability is not the same as the statistical stability of the measurement system.

The amount of total variation in the system's bias over time on a given part or reference gauge stability can be estimated in the following manner from the control charts used to monitor statistical stability.

- Obtain a sample and establish its reference value(s) relative to a traceable standard. The reference value may be obtained from a standard traceable to the relevant national standards institute (such as ISO). If one is not available, select a production part that falls in the mid-range of the production

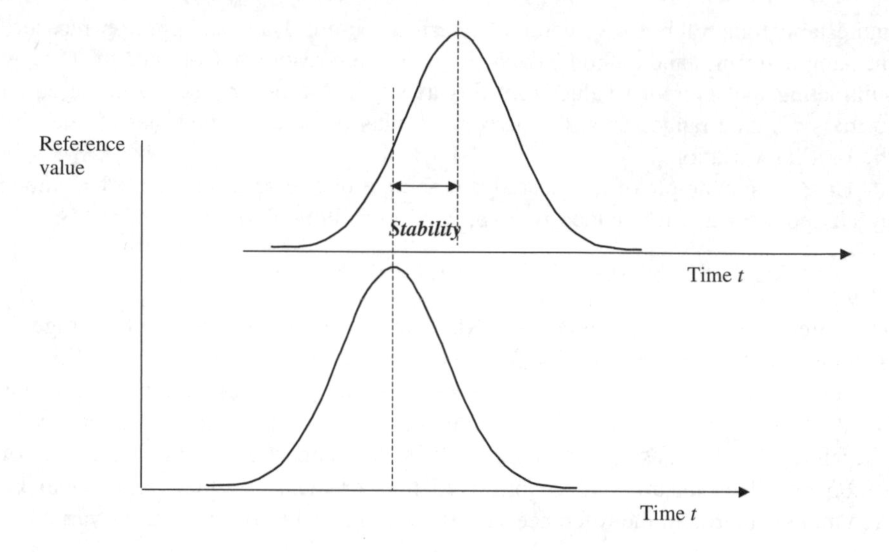

Figure 11.8 Gauge stability.

measurements and designate it as the master sample for stability analysis. (Reference samples may also be obtained from the low end or high end if applicable or desired.)

- Establish the measurement schedule (daily, weekly, ...) and frequency of sample times. Times and schedules must be based on engineering judgement concerning gauge use and recalibration times.

- Plot data on an Xbar-R or Xbar-S control chart. Out-of-control conditions on the R chart indicate an unstable repeatability and out-of-control conditions on the Xbar chart indicate the measurement system no longer measures correctly (bias has changed).

- Design of experiments or other suitable problem-solving techniques may be used to determine the cause of measurement instability. Try to determine the cause of the change and correct it.

11.6.5.3 Linearity of measurement systems

Gauge linearity is determined by doing a simple linear regression fit to gauge bias vs. reference value throughout the operating range of the measurements. The goodness of fit (R^2) of the linear regression line will determine whether the biases and reference values have a good linear relationship. The magnitude of the linearity and percentage linearity can be evaluated to determine whether linearity is acceptable.

- Select five or more parts covering the whole range of process variation.

- Establish reference values for each part.

- Select the parts at random and have each part measured 12 times by operators who usually operate the gauge.
- Calculate part average and bias average for each part. The part bias average is calculated by subtracting part average from part reference value.
- Plot bias average and reference value as shown in the linearity example (Appendix 11B).
- Linear regression equations and goodness-of-fit (R^2) equations are shown in Appendix 11B.

11.6.6 R&R studies

11.6.6.1 Reproducibility

The gauge must be previously calibrated, repaired or adjusted as required. Determine the number of operators, the number of parts and the number of trials. Among the factors that have to be considered is criticality – critical dimensions require more parts and/or trials to have a higher degree of confidence. What are the techniques for collecting data? Sometimes repeated measurements cannot be made because of the nature of measurements, such as destructive tests.

A gauge capability study is a part of an MSA, aimed at estimating and evaluating precision of a measurement system. It contributes to the evaluation of a measurement system for suitability of use in a manufacturing or test area.

11.6.6.2 Graphical methods for assessing precision and stability

Several methods can be used in a graphical analysis, to provide confirmation and further insight into the data, such as trends, cycles, interactions, and outlier detection. Among the graphs most useful for gauge capability studies are multivariate charts, control charts, interaction and residual plots.

A *multivariate chart* illustrates multiple sources of variation in a gauge capability study, providing information about non-random patterns in the study data such as time trends, shifts, clustering, cycles and outliers. It is valuable for detection of non-constant variation within each factor, such as operators, over all the levels of the other factor(s). The multivariate chart displays graphically the numerical results of the average and range or ANOVA method for gauge capability studies.

Figure 11.9 is an example of a multivari chart for a gauge capability study that used the standard form for the study design: three operators named A, B and C, ten parts and two readings per part–operator combination.

The means of the parts within each operator are connected by dashed lines. The repeatability of this gauge capability study is illustrated by the spreads of the 30 part–operator combinations. Any instability in the repeatability across different parts and/or different operators would show up clearly on this chart (see the residual plots in Figure 11.11). The reproducibility (in this example the operator-to-operator variation) is illustrated by the difference in the operator means: the solid lines. The

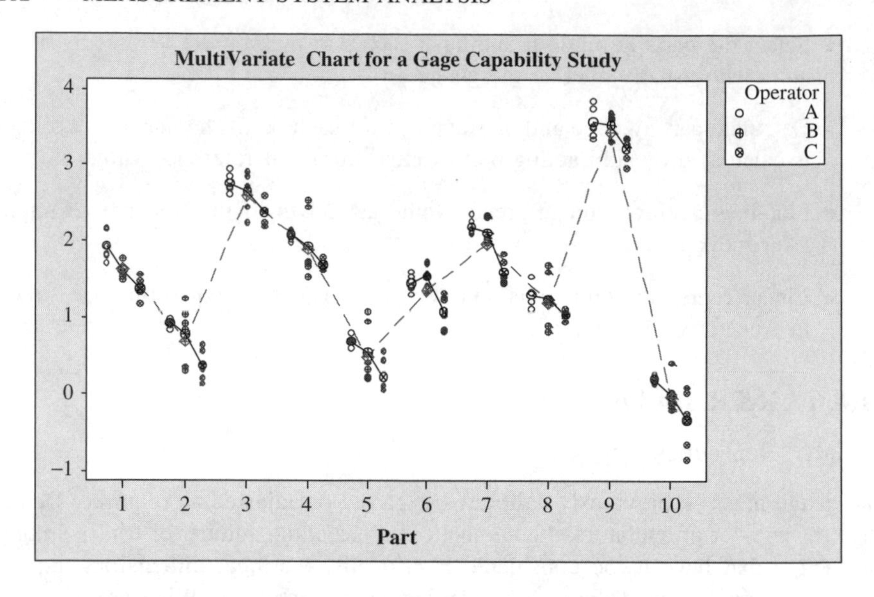

Figure 11.9 Multivari chart for a gauge capability study.

part-to-part variation is represented by the variation of the part means within each operator. If the part-to-part variation changes over the three operators, then this will be evident in the chart.

If there are more than three sources of variation to be illustrated, then multiple multivari charts can be constructed for each level or combination of levels of the additional factors.

We strongly recommend the use of multivari charts for all gauge capability studies.

Control charts are used to monitor the statistical stability of a process and to provide direction for corrective actions and continuous improvement. Since the measurement process is very similar to a manufacturing process, it is important to use control charts to monitor measurement systems, to ensure that the bias and precision of these processes remain stable and that corrective actions are taken when needed.

Another graphical method recommended when ANOVA is used is the *interaction plot*, devised to indicate whether there is a substantial interaction between two factors, such as operator and part. Figure 11.10 illustrates an interaction plot for the same example used in the multivari chart of Figure 11.9. The means of the two readings have been used.

In the part–operator example, the part means within each of the k operators are connected to form k lines. If the k lines are approximately parallel, little if any interaction is present. When the lines are definitely not parallel, an interaction may exist. This example shows the three lines associated with three operators (A, B and C). The plot indicates the presence of at least a moderate interaction between parts and operators, since there is some crossing of the lines.

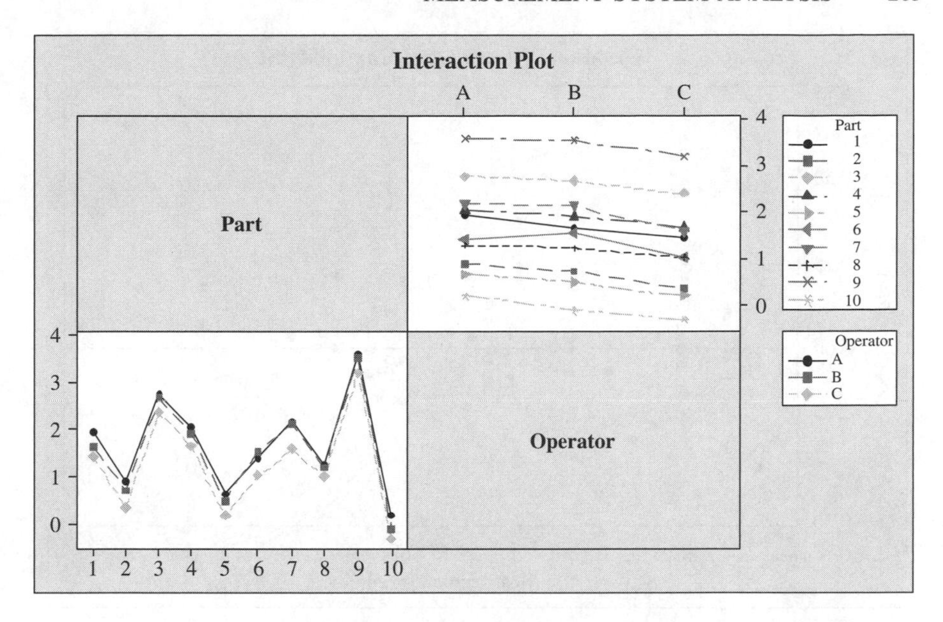

Figure 11.10 Interaction plot.

Indications suggested by graphs should be confirmed by appropriate statistical tests before making firm inferences. The causes of interaction should be eliminated whenever possible.

A *residual plot* indicates departures from assumptions on repeatability of the measurement system under study. It can be used to check for non-random trends, such as cycles and outliers, in the multiple readings that provide the estimate of repeatability.

Differences between observed readings and corresponding values predicted from the model used in the gauge study are called *residuals*. In our part–operator example, the predicted value is the mean of each part–operator combination, and the residual is the difference between each reading and its associated part–operator mean.

A residual plot is a graph of residuals against predicted values. If residuals are not randomly scattered above and below zero (horizontal reference line), assumptions may be incorrect and further investigation is suggested. Figure 11.11 is a residual plot of the part–operator example used in Figures 11.9 and 11.10. The residuals in this plot appear to be randomly distributed about the zero line. There is no evidence of trends, cycles, or outliers.

The residuals can also be used to test the assumption that the gauge repeatability (or error) is a random variable from a normal distribution. A normal probability (or quantile) plot (see Figure 11.12) can be used to test the assumption of normality of the residuals. This assumption is necessary for the application of F tests and the construction of confidence intervals from the ANOVA method.

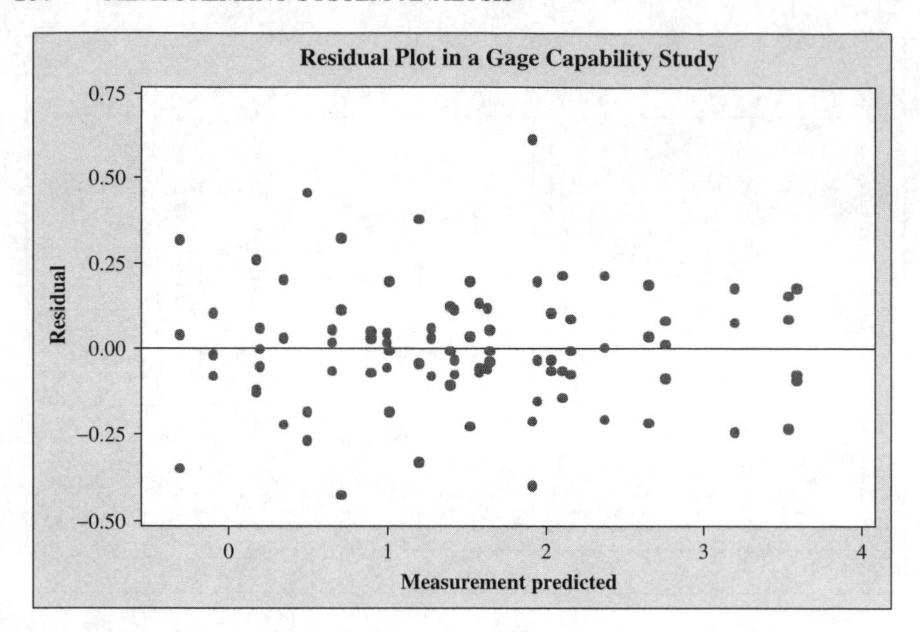

Figure 11.11 Residual plot.

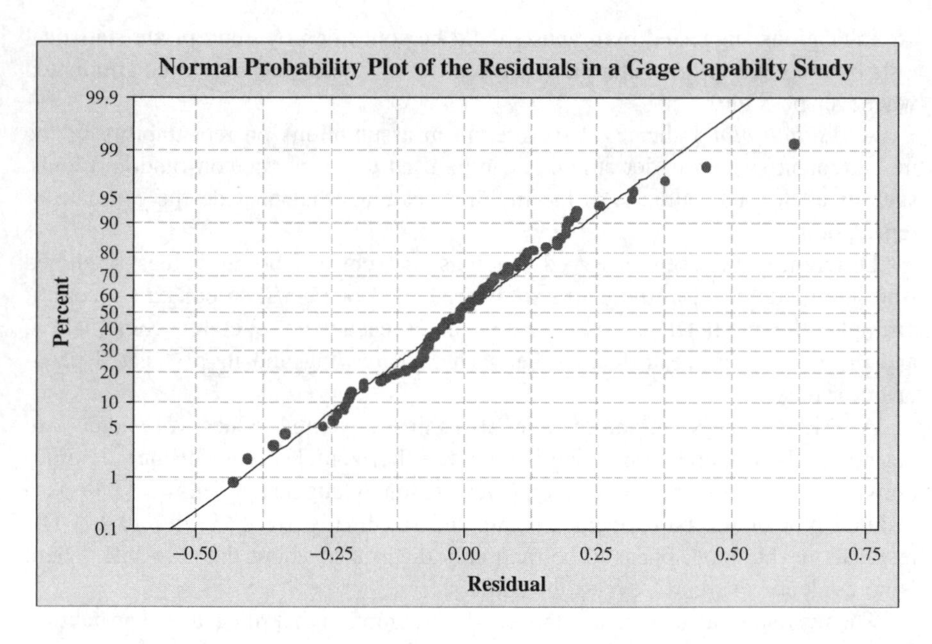

Figure 11.12 Normal probability plot.

11.6.6.3 Average and range method

The average and range method uses averages and ranges of the data from a gauge capability study to estimate total variability. This method can be applied only when the standard form is used for gauge capability study design.

1. The average and range method will determine both repeatability and reproducibility for a measurement system.

2. The standard form uses ten parts and three operators with three trials each. If these conditions cannot be met, the number of parts and trials may be reduced, to no less than five parts and three operators with two trials each. A smaller number of parts and trials will reduce the precision of estimates of repeatability, and part-to-part variation.

   2.1. Data for the average and range method are collected according to the following procedure:

      2.2.1. Select three operators A, B, C, and ten parts labelled from 1 to 10; numbers shall not be visible to operators. The parts will be completely randomised before each trial for each operator.

      2.2.2. For the first trial, let operator A measure the ten parts in random order, and have another observer enter the results in the gauge study form. Repeat with operators B and C.

      2.2.3. Operator A repeats measuring the same 10 parts for trial 2. Record the measurements. Repeat with operators B and C.

      2.2.4. Repeat the cycle for trial 3.

3. If the operators are on different shifts, operator A may measure ten parts for all three trials on shift 2 and operator C measures all three trials on shift 3.

   3.1. If parts are large or simultaneous availability of parts is not feasible, the following procedure is acceptable:

      3.1.1. Operator A measures the first part for the first trial. Then B measures the same part, then C measures the same part.

      3.1.2. Operator A measures the first part for the second trial. Repeat with operators B and C.

      3.1.3. Repeat the first part for the third trial for operators A, B and C.

      3.1.4. Repeat the cycle using parts 2 to 10 in all three trials for all three operators A, B and C.

4. *Reproducibility*. After the data are collected, the computing procedure is as follows:

   4.1 For operator data, add up the readings of the 10 parts for the first trial (add horizontally) and divide by 10 to give the average (AVG). Enter the average of the 10 readings in the AVG box.

   4.2. Repeat with the data for the second and the third trials.

   4.3. Add up the AVGs of the three trials and divide by 3 to get the average reading for operator A ($\overline{X}_A$).

   4.4. Repeat for the readings for operators B and C.

   4.5. Examine $\overline{X}_A$, $\overline{X}_B$ and $\overline{X}_C$. The smallest of the three is $\overline{X}_{min}$, the largest is $\overline{X}_{max}$.

   4.6. Calculate $\overline{X}_{diff} = \overline{X}_{max} - \overline{X}_{min}$. Enter the form as $\overline{X}_{diff}$.

5. *Part variation.*

   5.1. Calculate the total of all readings for part numbered 1 by adding (vertically) all part 1 readings from all three operators A, B and C. Enter the result in the form.

   5.2. Calculate the average reading $\overline{X}_P$ for part no. 1 by dividing the total of part 1 readings by 9 (the number of times part 1 was measured by all three operators).

   5.3. Repeat calculating $\overline{X}_P$ for the remaining nine parts.

   5.4. Calculate R_P as the range of all $\overline{X}_P$ by subtracting the smallest from the largest. Enter in the form as R_P.

6. *Repeatability.*

   6.1. In the box for operator A, determine the range of operator A's readings on part no. 1 by subtracting the smallest of the three readings from the largest. Enter the range as R.

   6.2. Repeat determining the ranges for the remaining nine parts for operator A. Enter the results horizontally.

   6.3. Calculate the average of the 10 ranges for operator A by adding up all ten ranges horizontally and divide by 10. Enter the result as R_A.

   6.4. Repeat determining the ranges and calculate average of the ranges for operators B and C. Enter the results as R_B and R_C.

6.5. Calculate the average range $\overline{R} = \frac{1}{3}(\overline{R}_A + \overline{R}_B + \overline{R}_C)$.

6.6. Calculate the upper control limit for the ranges by multiplying $\overline{R}$ with constant $D_4(n)$ which can be found in a table in the form.

7. Examine all the individual ranges of the readings (R) of each part for all operators.

7.1 If any of these ranges is greater than the UCL, it shall be marked, its cause identified and corrective action taken. Then one of the following options applies:

7.1.1. Repeat the marked readings using the same operator and parts as originally used and repeat all affected calculations.

7.1.2. Discard the marked readings, recompute average, ranges and UCL from the remaining readings.

11.6.7 Estimation of total variations

After performing all the calculations listed above, estimates of total variation within each source are obtained as follows:

1. Calculate repeatability (ev), reproducibility (av), combined R&R, part variation (pv) and total variation (tv) as per formulas in the form.
2. Calculate %ev, %av and %pv as percentage of tv. The sum of percentages corresponding to all factors will not equal to 100 %, since ev, av, pv and tv are in terms of standard deviations (σ) rather than variances since variances are additive but standard deviations are not.

11.6.8 ANOVA method for the standard form gauge study design

The analysis of variance (ANOVA) is a statistical method for partitioning total variation into components associated with specific sources of variation. It can be used to estimate and evaluate total measurement variation, and the contributing sources of variation derived from a gauge capability study. It can be used to analyse measurement error, and other sources of variability of data in a study. For a more detailed explanation. refer to Appendix 11A.

ANOVA can be applied to all gauge capability designs. This section specifically applies the ANOVA techniques to the standard form design, where ANOVA splits total variation into four main categories: parts, appraisers, interaction between parts and appraisers, and replication error within each part–operator combination due to the gauge.

The part by operator interaction is a part of the reproducibility of the measurement system. Its variation is estimated by ANOVA. It cannot be estimated by the average and range method. Estimates of individual variations from different sources, such as repeatability and reproducibility, are called variance components and are outputs of ANOVA.

11.6.9 ANOVA analysis of the standard form

The numerical analysis of the standard form can be done as shown in Table 11.4. This ANOVA table has six columns. The source column lists sources of variation. The next column lists the degrees of freedom associated with each source. The SS or sum of squares column lists squared deviations around the mean of each source. The MS or mean square column is obtained by dividing sums of squares by the corresponding degrees of freedom.

Table 11.4 ANOVA with k appraisers, n parts and r replications.

Source of variation	Degrees of freedom	Sum of squares (SS)[a]	Mean squares (MS)	Expected mean squares (EMS)	F
Appraiser	$k-1$	SS_O	$MS_O = \frac{SS_O}{k-1}$	$\tau^2 + r\gamma^2 + nr\omega^2$	$\frac{MS_O}{MS_{Op}}$
Parts	$n-1$	SS_p	$MS_p = \frac{SS_p}{n-1}$	$\tau^2 + r\gamma^2 + kr\sigma^2$	$\frac{MS_p}{MS_{Op}}$
App × Part	$(n-1)(k-1)$	SS_{Op}	$MS_{Op} = \frac{SS_{Op}}{(n-1)(k-1)}$	$\tau^2 + r\gamma^2$	$\frac{MS_{Op}}{MS_e}$
Error	$nk(r-1)$	SS_e	$MS_e = \frac{SS_e}{nk(r-1)}$	τ^2	
Total	$nkr-1$	TSS			

[a]The sums of squares in this column are calculated as follows:

$$SS_O = \frac{1}{nr}\sum_{j=1}^{k} x_{\cdot j \cdot}^2 - \frac{x_{\cdots}^2}{nkr},\ SS_p = \frac{1}{kr}\sum_{i=1}^{n} x_{i\cdot\cdot}^2 - \frac{x_{\cdots}^2}{nkr},\ SS_{Op}$$

$$= \frac{1}{r}\sum_{i=1}^{n}\sum_{j=1}^{k} x_{ij\cdot}^2 - SS_p - SS_O - \frac{x_{\cdots}^2}{nkr},$$

$$SS_e = TSS - (SS_O + SS_p + SS_{Op}),\ TSS = \sum_{i=1}^{n}\sum_{j=1}^{k}\sum_{m=1}^{r} x_{ijm}^2 - nkr\bar{x}_{\cdots}^2.$$

The estimates of the variance components for each source are given below:

Error	SS_e	$\sigma^2_{\text{Repeatability}}$
Interaction	$\frac{MS_{Op} - MS_e}{r}$	$\sigma^2_{\text{Interaction}}$

Appraiser	$\frac{MS_0 - MS_{0p}}{nr}$	$\sigma^2_{Appraiser}$
Part	$\frac{MS_p - MS_{0p}}{kr}$	σ^2_{Part}

Note: All components of variation are assumed to be random effects.

Table 11.4 decomposes total variation into four components: parts, appraisers, interaction of appraisers and parts, and replication error due to the gauge.

The EMS or expected mean square column provides the true or expected value of each MS, as a linear combination of the true variance components of each source.

The F values in the last column, obtained as mean squares ratios, allow the variation of appraisers, parts, or the interaction of parts and appraisers, to be tested for statistical significance.

Since each mean square is subject to sampling variation, and computations involve differences of mean squares, negative variance components estimates are possible. This is a problem since no variance may be negative. Therefore, any negative variance component estimate is assumed to be small, and assigned a zero value.

The square root of each variance component (standard deviation) is easier to interpret than the variance, having the same units as measurements. The total variation including 99 % of measurements for each component is given by $2 \times 2.576 = 5.15$ times its associated standard deviation estimate. For the standard form, Table 11.5 shows the 5.15 sigma spread for a measure of repeatability called equipment variation (EV) and the measure of appraiser variation (AV).

Table 11.5 Standard form of the 5.15 sigma spread.

$EV = 5.15\sigma_{Repeatability} = 5.15\,MS_e$
$AV = 5.15\sigma_{Appraiser} = 5.15\sqrt{\frac{MS_0 - MS_{0p}}{nr}}$
$I = 5.15\sigma_{Interaction} = 5.15\sqrt{\frac{MS_{0p} - MS_e}{r}}$

The interaction of part and appraiser (I) is often significant, and its estimate is provided by the ANOVA method. We have a non-additive model and therefore an estimate of its variance components is given. The reproducibility is then the sum of the appraiser variation (AV) and the interaction variation (I).

The R&R in Table 11.6 is the total measurement system variation.

Table 11.6 Standard form of the 5.15 sigma spread.

$R\&R = \sqrt{EV^2 + AV^2 + I^2}$
$PV = 5.15\sigma_{Part} = 5.15\sqrt{\frac{MS_p - MS_{0p}}{kr}}$

To determine whether any of the sources of variation correspond to statistically significant contributions, compute the F values according to the formulas in Table 11.4. Compare the computed F values with the upper percentage point of a F distribution at a given level (such as 0.95), numerator and denominator degrees of freedom being taken from the ANOVA table.

11.6.10 Metrics for gauge capability assessment

Gauge capability metrics compare the magnitude of total variation of the measurement system ($R\&R$) with either total variation (TV) or the tolerance of the measured characteristic of the part. These metrics allow for setting specific limits for acceptable levels of precision for the measurement system.

11.6.11 Guidelines for selecting a gauge capability metric

Considering the metrics

$$\%R\&R = 100\frac{R\&R}{TV} \text{ or } \%P/T = 100\frac{R\&R}{\text{Tolerance}}.$$

- Use $\%R\&R$ when the gauge is used to monitor a characteristic's variation (such as to be plotted on a control chart, or within a designed experiment). In this case both individual readings and variation from reading to reading matter.

- Use $\%R\&R$ when the tolerance is one-sided, for example if there is a minimum specification (such as strength must be greater than 3.5 N), or when the tolerance has not yet been determined, as in the case of a new process or product.

- Use $\%P/T$ when the gauge is used to measure a characteristic with a tolerance for part acceptance. In this case only individual readings matter, as opposed to relationship among readings.

In an additive model, the interaction (I) is not significant and a pooled error variance is obtained as follows: the sum of squares due to the part × appraiser interaction is added to the sum of gauge errors to give the sum of squares pooled (SS_{pool}) with $nkr - n - k + 1$ degrees of freedom. Then SS_{pool} is divided by $nkr - n - k + 1$ to give MS_{pool}. The 5.15 sigma spread limit then will be:

$$EV = 5.15\sqrt{MS_{pool}}$$

$$AV = 5.15\sqrt{\frac{MS_O - MS_{pool}}{nr}}$$

$$R\&R = \sqrt{EV^2 + AV^2}$$

$$PV = 5.15\sqrt{\frac{MS_P - MS_{pool}}{kr}}$$

To find out whether the interaction is significant, compute the variance ratio F for part $\times$ appraiser interaction. Compare this F statistic with an upper percentage point of the appropriate F distribution, numerator and denominator degrees of freedom being listed in ANOVA table. To decrease the risk of falsely concluding that there is no interaction effect, select an appropriately high significance level. Once we have decided what the R&R is, we can calculate *%R&R* in relation to process performance.

11.6.12 Comparison of average and range method with ANOVA

The advantages of ANOVA over the average and range method are as follows:

1. ANOVA is capable of handling almost any gauge capability design, not just the standard form.

2. Variance component estimates from ANOVA can provide more estimates of the contribution of each source of variation.

3. ANOVA can extract more information (such as interaction between parts and appraisers effect) from experimental data.

4. The average and range method relies on the assumption of normality of the measurements to provide estimates. ANOVA does not require the assumption of normality, unless tests of significance are involved. Even in this case, results are rather robust to non-normality thanks to the central limit theorem.

A disadvantage of ANOVA is the need for a computer and statistical software to do numerical computations. Users will also need a basic knowledge of statistical methods to design the appropriate gauge capability study, and to interpret results.

ANOVA as described in the following sections is recommended for both the standard form and for more general designs, especially if a software package is available. We discuss statistical models for more general gauge capability designs in the next section.

11.6.13 ANOVA methods for automated measurement systems

A traditional standard gauge capability study form may turn out not to be a suitable design for many measurement systems. Gauges often belong to an automated measurement system, where operators do not contribute significantly to measurement error, and/or there are many more factors to study than just operators and repeatability.

Automated mechanisms in a measurement system affect the reproducibility, minimising operator effect. Measurement variation is due mainly to factors such as loading mechanisms, part positioning, focusing, and multiple software algorithms.

There may be also a short-term time effect to be estimated and evaluated. All of these factors would be part of the reproducibility component.

Repeatability in automated systems (same interpretation as in the standard form) is estimated by taking multiple readings, trials or replications, other factors being kept constant.

In this section, we address the analysis of more general gauge capability designs, applicable to both automated and non-automated measurement systems with more than two sources of measurement variation. ANOVA provides the best approach, since it is flexible with regard to number of factors (sources of variation), interaction of those factors, and design of the study. ANOVA also yields more accurate estimates of components of variation than the average and range method.

11.6.13.1 ANOVA models

There are many gauge capability designs, of which the standard form is just one possibility. To use ANOVA and related statistical software, the ANOVA model associated with the selected design must be specified. We shall discuss only five typical models: those that represent designs appropriate for many measurement systems.

The ANOVA model partitions the total manufacturing process variation into sources or components. The first-level model partitions total variation into variation due to process or parts, and variation due to measurement:

$$\text{Total Variability} = \text{Process Variability} + \text{Measurement Variability}. \tag{11.1}$$

In gauge capability studies, process variability is usually represented as part-to-part variation (PV). Measurement variability can be further partitioned into repeatability and reproducibility:

$$\text{Measurement Variability} = \text{Reproducibility} + \text{Repeatability}. \tag{11.2}$$

Reproducibility can be further partitioned into the variation associated with each of the factors (such as operators, loading/unloading and time periods) contributing to reproducibility:

$$\text{Reproducibility} = \text{Operator} + \text{Loading/Unloading} + \text{Time Period}. \tag{11.3}$$

ANOVA is used to estimate, *inter alia*, the variance components of total variation and factors contributing to the total. For example, in model (11.1), the mathematical representation of the model is

$$\sigma^2_{\text{Total}} = \sigma^2_{\text{Process}} + \sigma^2_{\text{Measurement}},$$

and the measurement system component can be further partitioned as in model (11.2) as

$$\sigma^2_{\text{Measurement}} = \sigma^2_{\text{Repeatability}} + \sigma^2_{\text{Reproducibility}}.$$

Other models can be represented similarly as sums of variance components.

11.6.13.2 ANOVA model for the standard form

In the standard form, total variation is partitioned as the sum of part-to-part variation, reproducibility and repeatability, where reproducibility is a function of operators (or appraisers), and interaction between operators and parts. The resulting model for the total variability is:

$$\text{Total Var.} = \text{Part Var.} + \text{Operator Var.} + \text{Operator} \times \text{Part interaction} + \text{Repeatability} \quad (11.4)$$

or, equivalently,

$$TV = PV + AV + I + EV.$$

We described the estimates of the corresponding variance components and the calculation of the appropriate metrics in previous sections.

11.6.13.3 Nested and crossed designs

In the terminology of design of experiments, model (11.4) for the standard form is an example of a *two-factor crossed design*, which usually caters for presence of interaction terms. *Nested designs* are an alternative structure, applicable in many gauge capability studies. The crossed and nested effect models are illustrated in Figure 11.13.

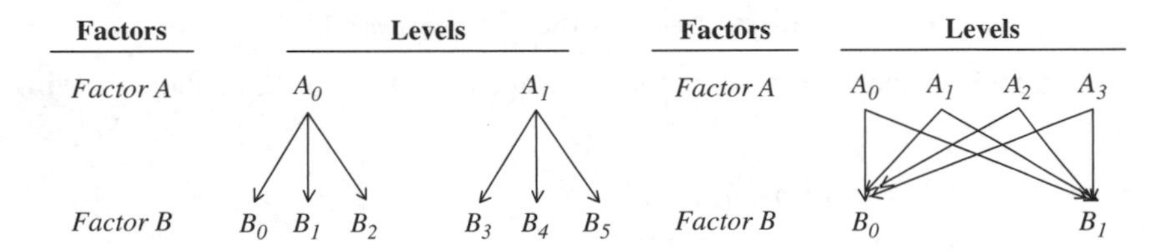

Figure 11.13 Graphical representation of nested effects model (left) and crossed effects model (right).

In the crossed effects model, factors A and B are crossed because each level of B (B_0 and B_1) appears with each level of A (A_0, A_1, A_2, A_3), and vice versa. All eight combinations are run in the experiment or gauge study. For example, the standard form might have four parts (factor A) crossed with two operators (factor B), therefore each part is measured by each operator, and each operator measures each part. Since A and B are crossed, the interaction between them ($A \times B$) can be included in the model and estimated with ANOVA.

The left-hand side of Figure 11.13 represents a *two-factor nested design*. There are six unique levels of factor B ($B_0, B_1, \ldots, B_5$), and each level of B is run with only one level of A. For example, B_3 appears only with A_1, but never with A_0, therefore factor B is said to be nested within factor A, and denoted in the model as the effect $B(A)$. No interaction is possible between A and B, because no crossing of the levels of B with the levels of A occurs.

11.6.13.4 Example of a two-factor nested design

For some automated measurement systems, the nested effects model is more appropriate for a gauge capability study than the crossed effects model. Perhaps the least complex gauge capability study for an automated measurement system might consist of a single part (such as a standard reference wafer), measured across multiple time periods (such as days), multiple cycles within each day (such as loading/unloading of a wafer on the gauge), and multiple trials or replications of the part within each cycle of a given day. An appropriate model for this design is

$$\text{Measurement Variability} = \text{Day} + \text{Cycles(Day)} + \text{Reps(Cycle, Day)}.$$

There are no part-to-part or operator effects in the model since only one part is used in the study, and operator effect is assumed to be negligible. Also, there are no interaction effects because the design is nested and not crossed. The part is assumed to be measured at one site or position only, and thus there is no estimate of process variation.

Cycles are nested within days, since each cycle appears only once in the study, and within only one day. The replication effect is nested within days and cycles, since each replication appears only once within each day–cycle combination.

With this model, ANOVA yields estimates of variance components of each effect in the model:

$$\sigma^2_{\text{Total}} = \sigma^2_{\text{Day}} + \sigma^2_{\text{Cycle}} + \sigma^2_{\text{Day}\times\text{Cycle}}.$$

Referring back to the terminology for the standard form, we have the following relationships:

$$\text{Repeatability (EV)} = 5.15\ \sigma_{\text{Replication}},$$

$$\text{Reproducibility} = 5.15\ \sqrt{\sigma^2_{\text{Day}} + \sigma^2_{\text{Cycle}}},$$

$$AV = 5.15\ \sigma_{\text{Operator}} = 0,$$

$$PV = 5.15\ \sigma_{\text{Part}} = 0,$$

$$R\&R = 5.15\ \sigma_{\text{Measurement}} = 5.15\ \sqrt{\text{Reproducibility}^2 + \text{Repeatability}^2},$$

where multiplying by 5.15 gives the spread including 99 % of measurements. The metric *%R&R* can be calculated in the usual way, given an independent estimate of total process variation (TV); $\%P/T$ can be calculated in the usual way if the part tolerance is specified.

11.6.13.5 More complex nested gauge capability designs

The two factors in the above example are days and cycles (within days). Repeated trials are often designated as originating the error term in the model; they are not among design factors, since replication effect will always be present in gauge study designs and models.

The two-factor nested example can be expanded to include other useful designs for gauge studies. Among a number of possible extensions of this model, we consider the following three:

1. *Three-factor nested.* Multiple days, multiple shifts (within days), cycles (within shifts and days) and replicate readings (within each day-shift-cycle combination). The model for this design would be

$$\text{Measurement Variability} = \text{Day} + \text{Shift(Day)} + \text{Cycles(Shift, Day)} \\ + \text{Reps(Cycle, Shift, Day)}.$$

   Shifts are nested within days, cycles are nested within shifts and days, and replications are nested within cycles, shifts and days. Note that shift effect may cover different operators on different shifts. ANOVA would provide the estimates of the variance components of each effect in the model.

2. *Three factors: two factors crossed and one factor nested.* Many gauge capability studies require a mixture of the crossed and nested effects models. A *mixed effects model* will cover both crossed and nested factors. A slight extension of the standard form (two factors crossed design) provides an example of a mixed model. Consider a design with multiple operators and multiple parts, both crossed as before, but with multiple loading/unloading of each part. Each loading cycle is within only one part–operator combination, so that cycles are nested within parts and operators. The model for the total variability of this mixed design is

$$\text{Total Var} = \text{Process Var.} + \text{Measurement Var} = \text{Part} + \text{Operator} \\ + \text{Operator} \times \text{Part} + \text{Cycle(Part, Operator)} \\ + \text{Reps(Cycle, Part, Operator)}.$$

   In this model we can obtain an estimate of process variability (PV), since there are multiple parts in the study. The measurement variability consists of operator, part by operator interaction, cycle and replications.

3. *Four factors: two factors crossed and two factors nested.* Another example of a mixed effects design would cover multiple days (or operators), crossed with multiple parts, multiple cycles within each day–part combination, and multiple sites or positions on the same part within each day–part combination. The model for this design would be

$$\text{Total Var} = \text{Day} + \text{Part} + \text{Day} \times \text{Part} + \text{Cycle(day,part)} \\ + \text{Site(Day,Part,Cycle)} + \text{Reps(Day,Part,Cycle,Site)} \ldots.$$

   Day and part are crossed, and cycle and site are nested; the combined part and site effects estimate process variation. Day, day $\times$ part, cycle and replications contribute to the measurement variation.

11.6.14 Guidelines for acceptance of measurement system precision

The metrics discussed previously can be used to evaluate the acceptability of the precision of the measurement system. The two metrics to use for this purpose are *%R&R* and %P/T (as discussed above). The criteria below for acceptance can be applied to both metrics:

- Under 10 %: present gauge system is acceptable.
- 10–30 %: may be acceptable according to the importance of the application, cost of gauge, cost of repair or other consideration. If the gauge is to be used for product acceptance, qualification or characterisation of significant process/product characteristics, it may be used only when (a) it is documented that either no better gauging method or system is available, or its cost is prohibitive, and (b) a documented action plan has been prepared and approved to improve gauge capability.
- Over 30 %: the gauge is not acceptable. The gauge may not be used for product acceptance, qualification or characterisation of any process/product characteristics, unless with the written authorisation of the responsible R&QA manager. A periodic analysis of the measuring system shall be performed to determine if improvement or replacement with a capable system can be made. The gauge may be used for non-critical engineering evaluations or studies only if result will not be affected by the imprecision of the measurement system. The gauge may be used temporarily by averaging multiple readings.

Table 11.7 Gauge linearity study data.

	Part	*1*	*2*	*3*	*4*	*5*
T	1	3.2	3.4	3.8	4.1	4.3
R	2	3.0	3.7	3.7	4.3	4.5
I	3	2.9	4.5	3.9	4.3	4.3
A	4	3.0	3.3	3.9	4.3	4.4
L	5	3.2	3.4	4.0	4.3	4.5
S	6	2.8	3.4	4.1	4.2	4.5
	7	2.9	3.4	4.4	4.3	4.5
	8	2.9	3.4	4.1	4.0	4.6
	9	3.1	3.6	4.0	4.1	4.2
	10	2.9	3.3	4.1	4.2	4.4
	Part Avg.	***2.99***	***3.54***	***4.00***	***4.21***	***4.42***
	Ref. Value	***3.00***	***3.50***	***4.00***	***4.50***	***5.00***
	Bias	***−0.01***	***0.04***	***0.00***	***−0.29***	***−0.58***
	Range	***0.4***	***1.2***	***0.4***	***0.3***	***0.3***

11.6.15 Example of gauge linearity study

The goodness of fit can be used to make an inference of the linear association between the bias and reference value. The linear model, $y = b + ax$, is fitted where x is the reference value, y the bias and a the slope. From this we can determine whether there is a linear relationship and, if so, whether it is acceptable. Generally, the lower the slope, the better the gauge linearity and, conversely, the greater the slope, the worse the gauge linearity. For the data in Table 11.7 the statistical analysis was done with MINITAB. The output is in Figure 11.14.

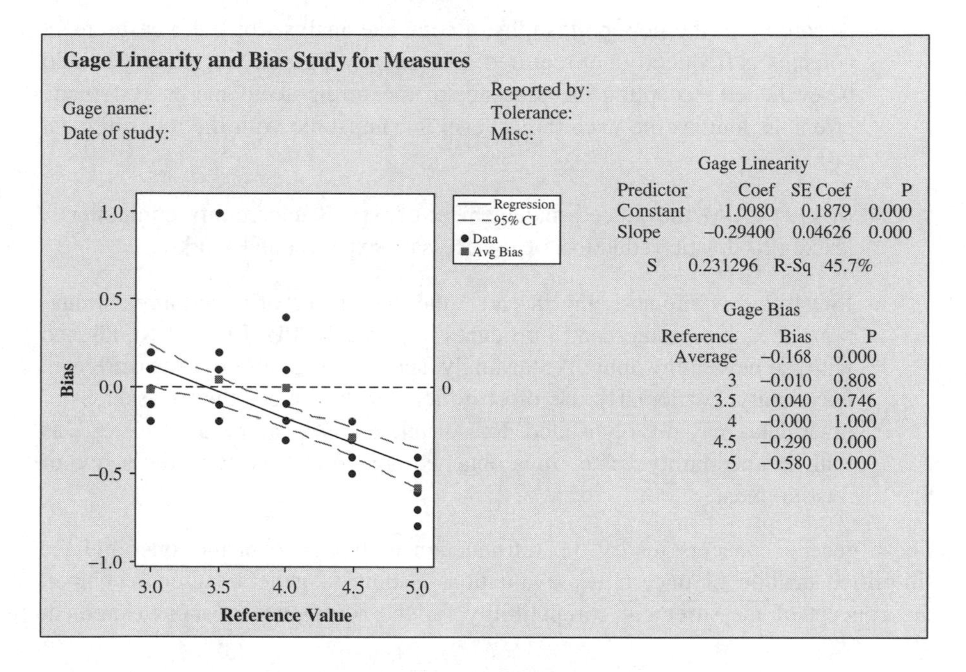

Figure 11.14 Gauge linearity study output.

11.7 Measurement uncertainty evaluation according to GUM: Part II

Traditional MSA, as we have described, requires a considerable amount of experimental work, and delves deeply into evaluation of specific causes of measurement uncertainty. Some points evidenced by practical experience should, however, be taken into account:

- A major problem with quality systems is the cost; and specifically those costs imposed by requiring unnecessarily for every instrument a certification stating its accuracy, and its traceability to national standards. The problem of costs involved by rigidity induced important changes of ISO 9001:1994, leading to the new ISO 9001:2000. A number of fixed protocols contained in the previous standard were made more flexible, and aimed at performing only those actions really needed to achieve quality objectives. Specifically, point 4.11 of ISO 9001:1994, requiring traceability of each measuring instrument, disappeared; the specific standard for measurement system management, ISO 17025:2000, is in line with the new philosophy, requiring only those actions needed for every specific measurement task.

- According to the new philosophy, a complete analysis of the measurement complex is frequently not required; instead only a global uncertainty should be evaluated, accepting the presence of measuring accidents or systematic errors, as long as the uncertainty level is compatible with the measurement tasks.

- In 1994 GUM introduced management of type B uncertainty components, leading to drastic reduction of unnecessary experimental work.

- ISO 14253, while stressing in part 1 the importance of uncertainty evaluation for each measurement, introduces in part 2 the PUMA method, aligned with the new philosophy. Accordingly, accurate evaluation of measurement uncertainty, particularly the most difficult part, namely the assessment of input data, may not be needed. Reasonable guesses can be used instead, as long as uncertainty values thus obtained are compatible with the scope of measurement.

These general concepts justify the introduction of type B contributions, and the simplified method of uncertainty evaluation, leading to general agreement upon the concept of measurement compatibility for the acceptance of a measurement system.

11.7.1 Type B contributions to uncertainty

Faced with a new measurement task, most expert measurement operators immediately check the natural range of the measurement. This habit is quite general, and rightly so: range is an important part of measurement activity, to be memorised and noted among the important points of measurement practice. A lot of useful information is thus made available: operator experience, to be relied upon as much as possible, tolerances contained in standard prescriptions, resolution of measurement instruments, resolution of numerical data from handbooks and lists of specifications, the amount of influence of quantities in common situations. For instance, if you want to examine the effect of temperature in a laboratory, you could buy a good thermometer, have it certified as traceable to national standards, measure the temperature during routine measurement work, and record the temperature along with the measurement result, thus managing the effect of temperature as a type A contribution. Alternatively, you could content yourself with observing that if people in the lab work happily without complaining about overly cold or hot conditions, the temperature should be somewhere in between 20 °C and 25 °C. In many cases, such a zero-cost solution is all that is required.

Notice that the factors examined have a substantial difference from statistical information given, for instance, by standard deviation or variance. Nevertheless the distinction made by GUM is not on the substance but on the way the information is obtained: the same standard deviation is considered of type A if directly calculated from experimental data, or of type B, on the other hand, if received in a calibration certificate calculated by someone else. A common case is the use of an instrument that has a calibration certificate with uncertainty information given by the value of standard deviation, standard uncertainty or expanded uncertainty.[5] Sometimes variance can also be received as information on measurement dispersion. In any case that information can be converted into a value for the standard deviation. The conversion is not difficult, as standard deviation is numerically equal to standard uncertainty and to the square root of variance. Only when the information arrives as expanded uncertainty is a little more attention required: generally expanded uncertainty is evaluated under the hypothesis of a normal distribution at a confidence level 95 %, that is, twice the standard deviation (the coverage factor). Sometimes, more correctly, the expanded uncertainty information is accompanied by the number of degrees of freedom. In that case, the relevant t distribution is used to calculate the coverage factor. The general rule is to apply the relationship $U = ku$, where U is the expanded uncertainty, u is the standard uncertainty and k is the coverage factor, which can be calculated using the t value for the number of degrees of freedom and the specified level of confidence.

Returning to the management of non-statistical contributions to the range, it is necessary to estimate the variance and to combine all uncertainty contributions. GUM gives the procedure for this. The connection between range and variance depends on the statistical distribution of data. GUM schematically considers three types of distributions: uniform distribution, U-shape distribution and triangular distribution.

11.7.1.1 Generic factors: uniform distribution

The uniform distribution is sometimes used because it is known to apply to the variable under evaluation; more often it is dictated by ignorance, when no specific information is available. Sometimes specific knowledge can help: factors connected with resolution, both of the instrument or of a numeric information taken from handbooks or instruction manuals. A large number of factor categories in the interval corresponding to the relevant range can justify the use of a uniform distribution.

[5] Standard deviation and standard uncertainty define an interval around the measured value that contains only about 68 % of possible results in case of normal distribution. That information can be misleading, so GUM introduced the concept of *expanded uncertainty*, defined as a confidence interval at a chosen confidence level (usually 95 %) for a normal (in the case of more than 30 data) or t distribution (if there are fewer than 30 data).

When the distribution has been specified, it is easy to determine the connection between range and variance. GUM uses one half of the range, a. That is, it considers a variability interval of $\pm a$. The variance is then evaluated as $u^2 = a^2/3$ (since GUM introduces type B, that is non-statistically evaluated uncertainty components, it avoids traditional statistical symbols such as s. Instead, it introduces the term 'standard uncertainty' and using the symbol u instead of standard deviation, so that variance is denoted by u^2).

11.7.1.2 Sinusoidal factors: U-shaped distribution

A number of natural factors have a sinusoidal pattern; vibration, oscillation, common wave propagations, daily temperature cycle. In such cases the distribution may not be assumed uniform; taking, for example, a sinusoidal pattern, extreme values occur more often than central ones, and shall accordingly have a higher probability. Evaluation of the distribution, assumed to be U-shaped, is a little more difficult, but the relevant variance has a simple formula: $u^2 = a^2/2$.

11.7.1.3 Triangular distribution

Sometimes experience suggests that, within the declared range of variability, a central value us more likely than extreme values. A triangular distribution, going to zero at the ends of the variation interval and to a maximum value at the centre, may be a reasonable guess. The variance of a triangular distribution is $u^2 = a^2/6$.

11.7.2 Physical-mathematical model of measurement and independent variables

The measurement result is not always just the number given by the measuring instrument: for instance, the section area A of a test piece with a measured diameter d is estimated by $A = \pi d^2/4$. Even in the simplest possible case, the direct measurement of a quantity, the measurement procedure should be described by a mathematical formula. For instance, the measure of the diameter d is, at least, given by $d = d_r \pm res/2$, where d_r is the output of the measuring instrument and res its resolution. Generally, for every direct measurement, we should take into account the bias and the resolution of the instrument and the reproducibility of the measurement. The mathematical model is $d = d_r - bias \pm res/2 \pm rprd$, where d_r is the output of the instrument, $bias$ its bias, res its resolution and $rprd$ the reproducibility of the measuring complex.

Our previous examples showed that a mathematical model is always necessary. Establishing a mathematical model is a challenging part of the measurement procedure. We must answer several questions:

- Which variables should be included?
- Which are more important, and must be measured carefully?
- Which are less important, and do not require special attention?

- Does the mathematical formula describe the physical relationship between the independent variables and the dependent variable, or are the independent variables correlated?

A good measurement procedure should produce an uncertainty that is lower, but not too low, than the limit required to guarantee measurement compatibility, with a cost as low as possible. We must decide the class of accuracy of the instruments and even if, for some influence quantities, it is really necessary to perform a measurement or if an educated guess is sufficient. We must decide if the measurement of some variables should be replicated and, if so, how many replications will be useful.

One of the main advantages of uncertainty evaluation is the significant help in decision-making provided by the questions.

A general form of the mathematical model, useful to visualise definitions and symbols used in general formulas, is

$$Y = G(X_1, \ldots, X_j, \ldots, X_q),$$

where Y is the dependent variable and the X_j are the independent variables, so that the ith measurement result can be expressed by the formula

$$y_i = G(x_{i1}, \ldots, x_{ij}, \ldots, x_{iq}) + \varepsilon_i.$$

11.7.3 Evaluation of standard combined uncertainty

The first calculation step combines the effect of different factors X_j to evaluate the standard uncertainty $u(y)$ of the dependent variable y, or its variance $u^2(y)$, in terms of information on variability of the independent variables X_j, given as type A or type B uncertainty contributions, and their relevant variances $u^2(x_i)$.

Let us start with the mathematical model $Y = G(X_1, \ldots, X_j, \ldots, X_q)$. The form of the function G is often regular enough to allow expansion in a Taylor series, limited to the first-degree contribution:

$$\delta y = \left(\frac{\partial G}{\partial X_1}\right)_{\overline{x}} \cdot \delta x_1 + \cdots + \left(\frac{\partial G}{\partial X_j}\right)_{\overline{x}} \cdot \delta x_j + \cdots + \left(\frac{\partial G}{\partial X_q}\right)_{\overline{x}} \cdot \delta x_q,$$

where the factors $c_j = (\partial G/\partial X_j)_{\overline{x}}$ are called sensitivity coefficients. Now, suppose we have a series of replicated measurements,

$$\begin{aligned}
\delta y_1 &= c_1 \cdot \delta x_{11} + \cdots + c_j \cdot \delta x_{1j} + \cdots + c_q \cdot \delta x_{1q} + \varepsilon_1, \\
&\ldots, \\
\delta y_i &= c_1 \cdot \delta x_{i1} + \cdots + c_j \cdot \delta x_{ij} + \cdots + c_q \cdot \delta x_{iq} + \varepsilon_i, \\
&\ldots, \\
\delta y_n &= c_1 \cdot \delta x_{n1} + \cdots + c_j \cdot \delta x_{nj} + \cdots + c_q \cdot \delta x_{nq} + \varepsilon_n,
\end{aligned}$$

which may be written in matrix form as

$$\{\delta y_i\} = [\delta x_{ij}]\{c_j\} + \{\varepsilon_j\}.$$

If we neglect[6] $\{\varepsilon_j\}$ we can evaluate the variance of dependent variable y:

$$s^2(y) = u^2(y) = \frac{\sum_{i=1}^{n}(\delta y_i)^2}{n-1} = \frac{\{\delta y_i\}^T\{\delta y_i\}}{n-1} = \{c_j\}^T\frac{[\delta x_{ij}]^T[\delta x_{ij}]}{n-1}\{c_j\},$$

where $(n-1)^{-1}[\delta x_{ij}]^T[\delta x_{ij}]$ is the variance–covariance matrix.

Now let us add the condition of independence among independent variables, frequently verified within the scope of uncertainty evaluation. Only the diagonal terms of variance–covariance matrix differ from zero, so that the variance of y is given by

$$u^2(y) = \sum_{j=1}^{q} c_j^2 s^2(x_j) = \sum_{j=1}^{q} u_j^2(y).$$

The calculation may be tabulated, as in Table 11.8, showing the individual contributions $u_j^2(y)$ to the variance. This makes clear which are the more important uncertainty components. In column x_j we write down the symbols for the independent variables appearing in the mathematical model, their values and, where appropriate, any remarks (such as to distinguish the contributions of accuracy, resolution, and reproducibility). Entries in column s_j are standard deviations for contributions of type A; and in column a_j one-half of the range for contributions of type B, as well as k_a factor of 2, 3 or 6 respectively corresponding to U-shaped, uniform or triangular distribution.

Table 11.8 Uncertainty budget in tabular form.

x_j			s_j	a_j	k_{aj}	n_{dj}	n_{rj}	$u^2(x_j)$	c_j	$u_j^2(y)$
Symb.	**Value**	**Note**								
X_1		bias								
		res								
		rprd								
...										
X_j										
...										
X_q										
Variance of y										
Standard deviation of y										

An important point concerns multiplicity related to input and output information, in connection with the difference between the variance s^2 of a population, and variance s_r^2 of averages of groups of r elements drawn from that population, $s_r^2 = s^2/r$. Accordingly, information related to an average should be dealt with in a different way from that concerning single elements.

[6]This hypothesis subtracts validity from the following derivation, but allows us to reach the same result as the rigorous proof in only two steps and in a simple mnemonical way.

The ISO standard on tensile testing machines, concerning bias on the average of three replicated readings, states: 'The average of three measurements shall not have a difference greater than 1 % from the reference value.' Such information can constitute a type B contribution, from which the variance u_m^2 of the mean of three results may be calculated. When that contribution is composed with others, related to unreplicated, single measurement, the relevant variance of the distribution of single elements u^2 is evaluated. In this case input multiplicity is $n_d = 3$ and the variance is

$$us^2 = u_m^2 \times n_d = 3u_m^2.$$

That is, the variance computed for data having an input multiplicity n_d is evaluated in the usual way, then multiplied by n_d.

Multiplicity can also apply to outputs. Thus the ISO standard for Vickers hardness measurement states that the indentation diagonal is to be measured by two operators, and the average is the result, that is, with a multiplicity $n_r = 2$ in output. The variance should be evaluated accordingly: since calculations yield the contribution pertaining to individual measurements, but the final result refers to an average, a proper formula for the variance is $u_m^2 = u^2/n_r = u^2/2$.

It is not sufficient to replicate measurements to get an effect of multiplicity. Replicate measurements must explore the variability of the factor being evaluated, to lead to an average closer to the expected value than single measurements.

For instance, let us consider evaluation of uncertainty of measurements, sensitive to temperature, performed in a non-conditioned test laboratory working continuously 24 hours a day. To assess temperature effect, total temperature variation over a working day must be considered. The normal test procedure is for the operator to take three consecutive measurements and report their mean value. If the measurements are in the afternoon, the three individual readings may correspond to a relatively high temperature, whereas in the morning they may correspond to a relatively low temperature. In such a case replications do not reflect temperature variability, so output multiplicity does not apply. If temperature effect is to be examined at all, the procedure should mandate the three measurements to be taken at intervals of about 8 hours, so that individual measurements exhibit the sinusoidal effect of temperature.

With these considerations in mind, it is possible to evaluate for every contribution the relevant variance as s_j^2, or a_j^2/k_a, as is proper for type B, due account being taken of multiplicity if any. Sensitivity coefficients may then be evaluated either by partial differentiation, or numerically, and eventually contributions $u_j^2(y)$ of variance of the dependent variable y can be calculated.

As an example, let us consider evaluation of the temperature effect due to thermal deformation of the measurand and thermal sensitivity of the instrument (see Table 11.9). The length of a part is measured at a temperature T, giving a reading L_T. The length L_{20} of the piece at the reference temperature of 20 °C is to be evaluated in terms of the part's thermal coefficient of deformation α, and the thermal sensitivity β of the instrument.

Table 11.9 Example of uncertainty budget in tabular form.

x_j			s_j	a_j	k_{aj}	n_{dj}	n_{rj}	$u^2(x_j)$	c_j	$u_j^2(y)$
Symb.	**Value**	**Note**								
L_T	800	Bias		4.0×10^{-3}	3	4	1	2.1×10^{-5}	1	2.1×10^{-5}
		Res		1.0×10^{-3}	3	1	5	6.7×10^{-8}	1	6.7×10^{-8}
		Rprd	$3.0 \cdot 10^{-3}$			1	5	1.8×10^{-6}	1	1.8×10^{-6}
α	12×10^{-6}			2.0×10^{-6}	3	1	1	1.3×10^{-12}	$-1.6 \times 10^{+3}$	3.4×10^{-6}
β	11.5×10^{-6}			2.0×10^{-7}	3	1	1	1.3×10^{-14}	$1.6 \times 10^{+3}$	3.4×10^{-8}
T	22			5	2	1	1	12.5	-4.0×10^{-4}	2.0×10^{-6}
L_{20}	799.98	± 0.011	Variance of y (L_{20})							2.9×10^{-5}
			Standard deviation of y (L_{20})							5.4×10^{-3}

The mathematical model is

$$L_{20} = (L_T \pm bias \pm res/2 \pm rprd)[1 + (\alpha - \beta)(20 - \mathrm{T})].$$

The measurement procedure requires five measurements, and their average is taken. The uncertainty connected with the bias was evaluated on previous tests, showing that the average of four measurements is within the standard tolerance of $\pm 4\,\mu m$ from the reference value. Reproducibility was tested by ten replications, obtaining a standard deviation of $3\,\mu m$. Resolution is $2\,\mu m$. The temperature coefficient of the piece is $\alpha = 12 \times 10^{-6}\,{}^{\circ}C^{-1}$ (a typical value for carbon steel) and the thermal sensitivity of the instrument is $\beta = 11.5 \times 10^{-6}\,{}^{\circ}C^{-1}$ (declared by the producer), while temperature variation over a 24-hour time span is estimated to a first approximation as $22 \pm 5\,{}^{\circ}C$.

For the first variable we start with the bias information, a type B contribution, so we put 4×10^{-3} in column a_j;[7] the bias factor does not exhibit a sinusoidal pattern, therefore $k_a = 3$. The multiplicity is 4.0 for input, as variability is declared for the average of four measurements, but only 1.0 for output, because, while the procedure requires five measurements, the measuring instrument is the same over all replications; therefore for each measurement the bias remains constant and does not reflect the variability interval declared. The sensitivity coefficient, obtained by differentiation, is nearly equal to 1.0.

The second line pertains to resolution. The uncertainty contribution, again of type B, is one-half of the resolution value in column a_j; it is not sinusoidal, so $k_a = 3$. Output multiplicity is 5.0, as results obtained by replications may fall anywhere within one or more resolution intervals. The sensitivity coefficient remains 1.0, depending only on the variable and not on the uncertainty factor.

The third line is about reproducibility; information on uncertainty is now given by the standard deviation. Again, output multiplicity is 5.0 as replicated measurements can fall anywhere in the reproducibility interval, if they contain the same disturbance effects.[8] The sensitivity coefficient is 1.0 as before.

The fourth line refers to the thermal deformation coefficient, for which only a figure derived from a handbook or a material data sheet is available, a common situation. No information is available about the variability of the numerical value. We must rely on the belief that numbers in the handbook were recorded within rules for experimental data, namely with errors confined to the least significant digit. A practical suggestion is to take ± 2 units of the least significant digit as the variability interval. This is what we did in the table. Should the factor examined appear later to be critical, further analysis would be needed. Multiplicity is not involved, as replication does not apply to the α value. The sensitivity coefficient is evaluated by partial differentiation as $L_T(20 - T)$. The fifth line, referring to thermal sensitivity of the instrument, is similar to the fourth, but for minor numerical changes.

[7]Note that measurement units are not necessary, provided that, as in every calculation, coherent units are used for each value.

[8]Note that if the reproducibility effect also contains factors of different operators or different ambient conditions, these factors are not present in the replications during the tests. In this case reproducibility will not gain from replications, that is, its multiplicity will be taken to be 1.

The last line concerns temperature. The variability interval was estimated as $\pm 5\,^\circ$C, clearly a type B contribution. Since the 24-hour variability is large enough to rule out the hypothesis of air conditioning, a natural sinusoidal pattern may be assumed for temperature, therefore $k_a = 2$. The sensitivity coefficient is evaluated by partial differentiation to be $-L_T(\alpha - \beta)$.

With many contributions, the expanded uncertainty at a 95 % confidence level is calculated with a coverage factor of 2.

The results of column $u_j^2(y)$ show how the PUMA method and the uncertainty table provide answers to general questions on the measurement complex. First of all as, following the PUMA method, some variability data were merely educated guesses, we shall check if further analysis is necessary. Variance contributions related to thermal coefficients and temperature are of the order of 10^{-6}, while the bias effect is of the order of 10^{-5}, therefore thermal effects warrant no further attention. This shows how even a very rough estimate may reduce considerably the amount of work involved in the analysis of a measuring complex and in uncertainty evaluation.

Proceeding with the examination of column $u_j^2(y)$, other considerations are as follows:

- The largest contribution is due to bias. If a smaller uncertainty is needed, bias calls for action, for instance with a better calibration of the instrument. Only after obtaining a substantial reduction of bias is better management of temperature factors justified, for instance by monitoring ambient temperature every few hours.

- In terms of cost reduction, we observe that resolution effect and reproducibility are both low, and there were five replications. Replications are expensive, since they involve every measurement operation, but do provide appropriate advantages. The question of the correct number of replications may be answered pragmatically by trying different numbers in the calculations, to establish whether or not uncertainty will increase significantly even without any replication.

This example shows how useful the PUMA method and the uncertainty table can be.

11.7.4 Evaluation of expanded uncertainty

Recall that uncertainty is defined by metrologists as one-half of the width of the interval around the measured value. This contains the reference value. The uncertainty appears to be barely adequate, as it covers only about 68 % of the measurement result distribution. GUM defines a more appropriate semi-interval U as *expanded uncertainty*, and denotes the relevant standard deviation as *standard uncertainty* u. If we consider the term ‘reasonably’ given in the uncertainty definition equivalent to ‘with a probability adequately high’, expanded uncertainty appears to be connected with the confidence interval at an appropriate level, given by $U = k \times u$, where the coverage factor k is taken in terms of the standardised normal variable z if the number of degrees of freedom justifies normal approximation, or by default Student's t. Confidence level should be selected according to what is 'reasonable' for the specific case. Thus, if safety is involved at all, high confidence levels are mandatory, while for accessory quantities lower levels are appropriate. When the application of the measurement results is not known, the 95 % confidence level is used, as conventional in metrology. At this confidence level, expanded uncertainty is calculated with $k = 2$ as coverage factor. This is justified by the large number of degrees of freedom and the assumption of a normal distribution. This assumption may not always apply, as a factor with a large uncertainty and a few degrees of freedom may strongly influence the dependent variable. Therefore, even if GUM requires the use of the t distribution only when the degrees of freedom are less than 30, a blanket use of t is often advisable. There are no ill effects because t and z are nearly equal when the degrees of freedom exceed 30. The Welch–Satterthwaithe formula is used to derive the number of degrees of freedom (ν_y) for the dependent variable (y), from the degrees of freedom (ν_j) of the independent variables (j):

$$\frac{u^4(y)}{\nu_y} = \sum \frac{u_i^4(y_i)}{\nu_j},$$

that is,

$$\nu_y = \frac{u^4(y)}{\sum (u_i^4(y)/\nu_j}.$$

Therefore two columns are added to complete the uncertainty table, one for ν_j and one for $u_j^4(y)/\nu_j$ (see Table 11.10).

Table 11.10 Uncertainty budget in tabular form. Additional columns for df management.

...	...	$u_j^2(y)$	ν_j	$\frac{u_j^4(y)}{\nu_j}$
X_1				
...				
X_q				
Variance of y		$u^2(y) = \sum u_j^2(y)$		$\frac{u^4(y)}{\nu_y} = \sum \frac{u_j^4(y)}{\nu_j}$
Standard deviation of y		$u(y) = \sqrt{u^2(y)}$		$\nu_y = \frac{u^4(y)}{\sum u_j^4(y)/\nu_j}$
Confidence level		ρ, such as 95 %		
Coverage factor		$k = t_{0.975;\nu_y}$		
Expanded uncertainty		$U = k \cdot u(y)$		

Table 11.11 Uncertainty budget in tabular form. Additional columns for df management.

x_j			s_j	a_j	k_{aj}	n_{dj}	n_{rj}	$u^2(x_j)$	c_j	$u_j^2(y)$	ν_j	$u_j^4(y)/\nu_j$
Symb.	**Value**	**Note**										
L_T	800	Bias		4.0×10^{-3}	3	4	1	2.1×10^{-5}	1	2.1×10^{-5}	5	9.1×10^{-11}
		Res		1.0×10^{-3}	3	1	5	6.7×10^{-8}	1	6.7×10^{-8}	100	4.4×10^{-17}
		Rprd	3.0×10^{-3}			1	5	1.8×10^{-6}	1	1.8×10^{-6}	9	3.6×10^{-13}
α	12×10^{-6}			2.0×10^{-6}	3	1	1	1.3×10^{-12}	$-1.6 \times 10^{+3}$	3.4×10^{-6}	30	3.9×10^{-13}
β	11.5×10^{-6}			2.0×10^{-7}	3	1	1	1.3×10^{-14}	$1.6 \times 10^{+3}$	3.4×10^{-8}	100	1.2×10^{-17}
T	22			5	2	1	1	12.5	-4.0×10^{-4}	2.0×10^{-6}	100	4.0×10^{-14}
L_{20}	799.98	± 0.012	Variance of y (L_{20})							7.4×10^{-5}	Σ	9.2×10^{-11}
			Standard deviation of y (L_{20})							8.6×10^{-3}	ν_y	8
			Confidence level							95 %		
			Coverage factor							2.3		
			Expanded uncertainty							1.2×10^{-2}		

The degrees of freedom of independent variables should be evaluated according to the following considerations:

- For type A contributions data numbers are known, and the degrees of freedom can be evaluated.

- For type B contributions information is usually fairly robust, so the number of degrees of freedom may be considered infinite (from a practical point of view 100 is large enough). The degrees of freedom may also be based on the merit of the relevant variability information: with almost certainty, for example, 100 df can be assumed; with medium certainty, assume 30 df; if low certainty, assume 15 df.

- Information on bias and accuracy, mentioned in calibration certificates and declarations of conformity to standard specifications, can be obtained from a fairly small number of tests. Therefore, if the variability information (expanded uncertainty or conformity to tolerance), is not accompanied by a statement of df., a small number of df should be assigned (such as 5).

The example discussed above may now be completed by evaluating the dependent variable's degrees of freedom, as in Table 11.11. Notice that 5 degrees of freedom were assigned to the bias contribution, since calibration needs only a few tests, 100 were assigned to resolution, usually well known, to temperature sensitivity of the instrument, clearly declared, and to the educated guess on temperature variation, because the amount of variation assigned (±5 °C) is most likely to cover the real variations.

References

Automotive Industry Action Group (2003) *Measurement System Analysis Reference Manual*, 3rd edition.

Bennich, P. (2003) Traceability and Measurement Uncertainty in Testing. MetroTrade Workshop Berlin. http://www.gpsmatrix.dk/includes/pdf_artikler/mu07.pdf

Carbone, P., Macii, D. and Petri, D. (2002) Management of measurement uncertainty for effective statistical process control. *IEEE Instrumentation and Measurement Technology Conference*, Vol. 1, pp. 629–633.

Comité Europeén de Normalisation (2004) *Guidance – Uncertainty of measurement concept in European Standards*, September. http://www.cenorm.be/boss/supporting/guidance+documents/gd063+-+uncertainty+of+measurements/index.asp (accessed October 2007).

Chen, C. and Blakely, P. (2002) Uncertainty analysis method and process for test instrumentation system measurements. *Proceedings of the International Instrumentation Symposium*, Vol. 48, pp. 97–108.

ISO (1993) *Guide to the Expression of Uncertainty in Measurement*. ISO, Geneva.

ISO (2004) *International Vocabulary of Basic and General Terms in Metrology (VIM)*, ISO/DGuide 99999, 3rd edition. ISO, Geneva.

Nielsen, H.S. (1997) Using the ISO 'Guide to the Expression of Uncertainty in Measurements' to determine calibration requirements. Paper presented at the 1997 National Conference of Standards Laboratories Workshop & Symposium. http://www.hn-metrology.com/isogum.htm (accessed October 2007).

Phillips, A.R., Jeffries, R., Schneider, J. and Frankoski, S.P. (1998) Using repeatability and reproducibility studies to evaluate a destructive test method. *Quality Engineering*, **10**(2), 283.

Senol, S. (2004) Measurement system analysis using designed experiments with minimum $\alpha-\beta$ risks and n. *Measurement: Journal of the International Measurement Confederation*, **36**(2), 131–141.

Smith, K., Preckwinke, U., Schultz, W. and He, B.B. (2000) Gauge R&R study on residual stress measurement system with area detector. *Materials Science Forum*, **347**, 166–171.

Taylor, B.N. and Kuyatt, C.E. (1994) Guidelines for evaluating and expressing the uncertainty of NIST measurement results. NIST Technical Note 1297.

Appendix 11A Analysis of variance

Analysis of variance (ANOVA) is a technique by which the variability of an observation associated with a number of defined sources may be isolated and estimated. When two or more independent sources of variation operate, the resulting variance is the sum of the separate variances. ANOVA is a method for separating the total variance of an observation into its various components, corresponding to the sources of variation that can be identified. The data must clearly contain information on any given source of variation before its contribution can be estimated from measurements which have been designed for this purpose. The ANOVA procedure will depend on the number and nature of the independent causes of variability that can be identified.

It is always possible to classify the data with respect to each source of variation and a complete classification is a necessary first step in the analysis. There are two types of classification: *cross classification* and *hierarchic classification*. The former is a classification in which each observation can be classified independently with respect to all sources of variation (see Figure 11.A1). The crossed factors occur

Factors	Levels
Factor A	A_0 A_1 A_2 A_3
Factor B	B_0 B_1

Figure 11.A1 Cross classification. Each arrow connecting two factors means that each level of each factor can be used in combination with any level of the other factors.

in an experiment when each level of each factor has a physical or fundamental property that is the same for every level of the other factors in the experiment (complete factorial experiments, by definition, involve crossed factors because each level of each factor occurs in the experiment with each level of every other factors). For example, in comparing three methods of analysis, samples from the same material were analysed by the three methods in four laboratories, named A, B, C and D; there are two sources of variation, laboratory and method of analysis, and these can be displayed in a two-way table. Every single result obtained from each laboratory can be entered in the appropriate cell in Table 11.A1. If some laboratories do not use all three methods, some cells in the table will be empty and we are dealing with missing data. This type of classification can be extended to three or more sources of variation.

Table 11.A1 General representation for a two-factor experiment: two-way cross classification.

Methods of analysis	Laboratories			
	A	B	C	D
1	y_{11}	y_{12}	y_{13}	y_{14}
2	y_{21}	y_{22}	y_{23}	y_{24}
3	y_{31}	y_{32}	y_{33}	y_{34}

Hierarchic classification, also called *nested classification*, can be represented visually by branching lines as in Figure 11.A2. This relationship is characteristic of a hierarchic classification and the pattern may be extended to three or more sources of variation. In contrast to crossed factors, nested factors have levels that differ within one or more of the other factors in the experiment.

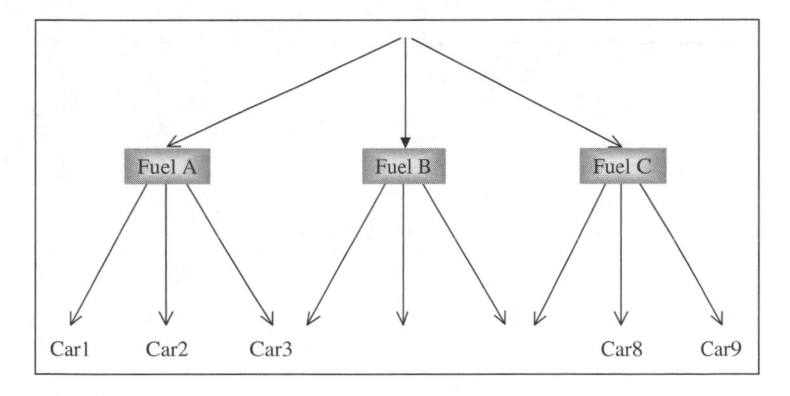

Figure 11.A2 Nested classification: cars 1, 2 and 3 are connected because they all use fuel A, but have no connection with cars 4, 5, . . . , 9.

More complicated classifications are combinations of the two basic types, cross and hierarchical classification (see Figure 11.A3).

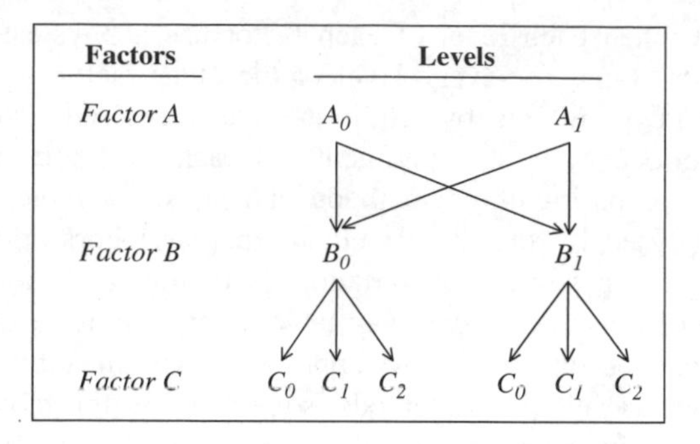

Figure 11.A3 Cross classification and hierarchical classification: the arrows connecting the factors have the meaning that each level of each factor can be used in combination with any level of the other factors.

11.A1 ANOVA for a single-factor experiment (one-way classification)

Consider a single-factor experiment (an experiment with only one controlled factor) with a different levels of the factor. Let Y_{ij} be the random variable representing the ith observation taken for factor level j, for $j = 1, 2, \ldots, a$ in the ith replication of the experiment, for $i = 1, 2, \ldots, n$ (see Table 11.A2).

Table 11.A2 General representation for a single-factor experiment; each entry Y_{ij} represents the ith observation taken under factor level j, with $j = 1, 2, \ldots, a$.

Observations					Totals
	Y_{11}	Y_{12}	$\cdot$	Y_{1a}	$Y_{1\cdot}$
	Y_{21}	Y_{22}	$\cdot$	Y_{2a}	$Y_{2\cdot}$
	$\cdot$	$\cdot$	$\cdot$	$\cdot$	
	$\cdot$	$\cdot$	$\cdot$	$\cdot$	
	Y_{n1}	Y_{n2}	$\cdot$	Y_{na}	$Y_{n\cdot}$
Totals	$Y_{\cdot 1}$	$Y_{\cdot 2}$		$Y_{\cdot a}$	$Y_{\cdot\cdot}$

To measure the total variation in the data, consider the total corrected sum of squares,

$$SS_{TC} = \sum_{i=1}^{n} \sum_{j=1}^{k} (Y_{ij} - \overline{Y}_{\cdot\cdot})^2, \qquad (11.A1)$$

where $\overline{Y}_{\cdot\cdot}$ is the total $Y_{\cdot\cdot}$ divided by na. Intuitively, this is the sample variance of the realisations y_{ij} of Y_{ij}, after dividing the SS_{TC} of (11.A1) by the appropriate number of degrees of freedom $(na - 1)$. The total sum of squares may be partitioned as

$$\sum_{i=1}^{n}\sum_{j=1}^{a}(Y_{ij} - \overline{Y}_{..})^2 = \underbrace{n\sum_{j=1}^{a}(\overline{Y}_{.j} - \overline{Y}_{..})^2}_{SS_{\text{Treatments}}} + \underbrace{\sum_{i=1}^{n}\sum_{j=1}^{a}(Y_{ij} - \overline{Y}_{.j})^2}_{SS_{\text{e}}} \qquad (11.A2)$$

The first term on the right-hand side of (11.A2) is the *sum of squares due to the different levels or treatments of the factor*, $SS_{\text{Treatments}}$, with $a - 1$ degrees of freedom. The second term is called the *sum of squares due to the error,* SS_{e}, with $a(n - 1)$ degrees of freedom. The aim of the ANOVA methodology is to decompose the total variation in the data into two terms: the sum of squares of the differences between the level averages and the grand average and the sum of squares of the differences of the observations within levels from the level average. A further possibility is to test the hypothesis that there are no differences between the means of the a populations from which the a samples (corresponding to the a levels) were extracted; but if we want to perform such a test, we need to make several assumptions. Specifically, we assume that the populations are normally distributed with common variance. Denoting the mean of the jth population by μ_j and the common variance of the a populations by σ^2, we can write

$$Y_{ij} = \mu_j + \varepsilon_{ij}, \qquad \text{for } j = 1, 2, \ldots, a \text{ and } i = 1, 2, \ldots, n. \qquad (11.A3)$$

The terms ε_{ij} represent the random component and, in accordance with the preceding assumptions, are independent, normally distributed random variables with mean zero and common variance σ^2. For uniformity with the corresponding equations for more complicated designs, μ_j is usually replaced by $\mu + \alpha_j$, where μ is the average of the μ_j and α_j is the effect of the jth level of the factor. Hence, $\sum_{j=1}^{a}\alpha_j = 0$. Using these new parameters, the model equation (11.A3) may be rewritten as

$$Y_{ij} = \mu + \alpha_j + \varepsilon_{ij}, \qquad \text{for } j = 1, 2, \ldots, a \text{ and } i = 1, 2, \ldots, n, \qquad (11.A4)$$

and the null hypothesis that the a population all have equal mean can be replaced by the null hypothesis that $\alpha_1 = \alpha_2 = \cdots = \alpha_a = 0$ against the alternative hypothesis that $\alpha_j \neq 0$ for some j.

To test the null hypothesis that the a populations means are all equal, we shall compare two estimates of σ^2, one based on the variation among the level averages and the grand average and one based on the variation within levels from the level average. Since, by assumption, each sample $\{y_{1j}, y_{2y}, \ldots, y_{nj}\}$, for $j = 1, 2, \ldots, a$, comes from a population having variance σ^2, this variance can be estimated by any one of the sample variances,

$$s_j^2 = \frac{\sum_{i=1}^{n}(Y_{ij} - \overline{Y}_{.j})^2}{n - 1}, \qquad \text{for } j = 1, 2, \ldots, a, \qquad (11.A5)$$

and a better estimate of the unknown variance σ^2 may be obtained by their mean s^2_{pool}:

$$s^2_{\text{pool}} = \hat{\sigma}^2_{\text{W}} = \frac{\sum_{j=1}^{a} s_j^2}{a} = \frac{1}{a(n-1)} \sum_{j=1}^{a} \sum_{i=1}^{n} (Y_{ij} - \overline{Y}_{.j})^2 \tag{11.A6}$$

Note that each of the sample variances s_j^2 in (11.A5) is based on $n - 1$ degrees of freedom and therefore $\hat{\sigma}^2_{\text{W}}$ is based on $a(n - 1)$ degrees of freedom.

The variance of the a sample means is given by

$$s^2_{\overline{y}} = \frac{1}{a-1} \sum_{j=1}^{a} (\overline{y}_{.j} - \overline{y}_{..})^2, \tag{11.A7}$$

and if the null hypothesis is true, $s^2_{\overline{y}}$ is an estimate of σ^2/n. Therefore, another estimate from (11.A7) of the unknown variance σ^2 (based on the differences among the sample means) is given by

$$\hat{\sigma}^2_{\text{B}} = n s^2_{\overline{y}} = \frac{n}{a-1} \sum_{j=1}^{a} (\overline{y}_{.j} - \overline{y}_{..})^2, \tag{11.A8}$$

with $a - 1$ degrees of freedom. If the null hypothesis is true, it can be demonstrated that $\hat{\sigma}^2_{\text{W}}$ and $\hat{\sigma}^2_{\text{B}}$ are two independent estimates of σ^2; it follows that $f_{\text{exp}} = \hat{\sigma}^2_{\text{B}}/\hat{\sigma}^2_{\text{W}}$ is a realisation of an F random variable with $a - 1$ and $a(n - 1)$ degrees of freedom. As the estimate $\hat{\sigma}^2_{\text{B}}$ is expected to exceed the estimate $\hat{\sigma}^2_{\text{W}}$ when the null hypothesis is not true, the null hypothesis will be rejected if f_{exp} exceeds $f_{1-\alpha}$, where $f_{1-\alpha}$ can be found from the appropriate tables of the F distribution with $a - 1$ and $a(n - 1)$ degrees of freedom.

The results presented above may be summarised in a one-way ANOVA table (see Table 11.A3).

Table 11.A3 ANOVA table with only one control factor.

Source of variation	Degrees of freedom	Sum of squares	Mean squares	f_{exp}
Among the groups (treatments)	$a - 1$	$\text{SS}_{\text{B}} = \frac{1}{n} \sum_{j=1}^{a} y_{.j}^2 - \frac{y_{..}^2}{na}$	$\text{MS}_{\text{B}} = \frac{\text{SS}_{\text{B}}}{a-1}$	$\frac{\text{MS}_{\text{B}}}{\text{MS}_{\text{W}}}$
Experimental error	$a(n - 1)$	$\text{SS}_{\text{W}} = \sum_{i=1}^{n} \sum_{j=1}^{a} y_{ij}^2 - \frac{1}{n} \sum_{j=1}^{a} y_{.j}^2$	$\text{MS}_{\text{W}} = \frac{\text{SS}_{\text{W}}}{a(n-1)}$	
Total	$na - 1$	$\text{SS}_{\text{TC}} = \sum_{i=1}^{n} \sum_{j=1}^{k} y_{ij}^2 - na\overline{y}_{..}^2$		

11.A2 ANOVA for a two-factor experiment (two-way classification)

If the experimenter has measurements for a treatments distributed over b blocks, we have a two-way cross classification as shown in Table 11.A4.

Table 11.A4 General representation for two-factor experiment. Each entry Y_{ij} represents an observation for the ith treatment and the jth block with $i = 1, 2, \ldots, b$; $j = 1, 2, \ldots, a$ (or, equivalently, for the A factor level i and the B factor level j).

	Block 1	Block 2	·	·	Block b	Totals
Treat. 1	Y_{11}	Y_{12}		·	Y_{1b}	$Y_{1\cdot}$
Treat. 2	Y_{21}	Y_{22}		·	Y_{2b}	$Y_{2\cdot}$
	·	·		·	·	
	·	·		·	·	
Treat. *a*	Y_{a1}	Y_{a2}		·	Y_{ab}	$Y_{a\cdot}$
Totals	$Y_{.1}$	$Y_{.2}$			$Y_{\cdot b}$	$Y_{..}$

The analysis of the general unreplicated two-way cross classification is by partitioning the total sum of squares into three components:

$$\sum_{i=1}^{b}\sum_{j=1}^{a}(y_{ij} - \overline{y}_{..})^2 = b\underbrace{\sum_{j=1}^{a}(\overline{y}_{.j} - \overline{y}_{..})^2}_{SS_A} + a\underbrace{\sum_{i=1}^{b}(\overline{y}_{i\cdot} - \overline{y}_{..})^2}_{SS_B}$$

$$+\underbrace{\sum_{j=1}^{a}\sum_{i=1}^{b}(y_{ij} - \overline{y}_{.j} - \overline{y}_{i\cdot} + \overline{y}_{..})^2}_{SS_e}. \qquad (11.A9)$$

The left-hand side of (11.A9) still represents the total variation. The terms on the right-hand side are respectively: the block sum of squares; the treatment sum of squares: and the error sum of squares (or, equivalently: the sum of squares due to the factor A; the sum of squares due to the factor B; and the error sum of squares).

The underlying model is given by

$$Y_{ij} = \mu + \alpha_i + \beta_j + \varepsilon_{ij}, \qquad \text{for } i = 1, 2, \ldots, a \text{ and } j = 1, 2, \ldots, b, \quad (11.A10)$$

where μ is the grand mean, α_i is the effect of the ith treatment (or of the ith level of the factor A), β_j is the effect of the jth block (or of the jth level of the factor B), and the ε_{ij} represent the random component and are independent, normally distributed random variables with mean zero and common variance σ^2. Similar to the model for the one-way classification, we have the constraints $\sum_{i=1}^{a}\alpha_i = 0$ and $\sum_{j=1}^{b}\beta_j = 0$.

Now, the aim is to test two null hypotheses:

$$H_0 : \alpha_1 = \alpha_2 = \cdots = \alpha_a = 0 \text{ against the alternative hypothesis that } \alpha_i \neq 0 \text{ for some } i \qquad (11.A11)$$

and

$$H_0 : \beta_1 = \beta_2 = \cdots = \beta_b = 0 \text{ against the alternative hypothesis that } \beta_j \neq 0 \text{ for some } j. \qquad (11.A12)$$

Testing the null hypothesis H_0 (11.A11) means testing whether the treatments have been effective (or testing the effectiveness of factor A), and testing the H_0 of (11.A12) means testing whether the blocking has been effective (or testing the effectiveness of factor B). As in the case of the one-way ANOVA, the significance tests are by comparison of different estimates of σ^2: one based on the variation among the treatments; one based on the variation among the blocks; and one measuring the experimental error. Only the latter is always a correct estimate of σ^2 (either of the two null hypotheses holds or does not), while the first two estimates are correct only under the (11.A11) and (11.A12) null hypothesis, respectively.

The null hypothesis (11.A11) has to be rejected at significance level α if

$$\frac{SS_A/(a-1)}{SS_{exp}/(a-1)(b-1)} > f_{1-\alpha},$$

where $f_{1-\alpha}$ can be found from tables of the F distribution with $a-1$ and $(a-1)(b-1)$ degrees of freedom. The null hypothesis (11.A12) has to be rejected at significance level α if

$$\frac{SS_B/(b-1)}{SS_{exp}/(a-1)(b-1)} > f_{1-\alpha},$$

where $f_{1-\alpha}$ can be found from tables of the F distribution with $b-1$ and $(a-1)(b-1)$ degrees of freedom.

The corresponding two-way ANOVA table is in Table 11.A5.

Table 11.A5 ANOVA table with two controlled factors.

Source of variation	Degrees of freedom	Sum of squares	Mean squares	f_{exp}
Among treatments (factor A)	$a-1$	$SS_A = \frac{1}{b}\sum_{j=1}^{a} y_{\cdot j}^2 - \frac{y_{\cdot\cdot}^2}{ab}$	$MS_A = \frac{SS_A}{a-1}$	$\frac{MS_A}{MS_e}$
Among blocks (factor B)	$b-1$	$SS_B = \frac{1}{a}\sum_{i=1}^{b} y_{i\cdot}{}^2 - \frac{y_{\cdot\cdot}^2}{ab}$	$MS_B = \frac{SS_B}{b-1}$	$\frac{MS_B}{MS_e}$
Experimental error	$(a-1)(a-1)$	$SS_e = SS_{TC} - SS_A - SS_B$	$MS_e = \frac{SS_e}{(a-1)(b-1)}$	
Total	$ab-1$	$SS_{TC} = \sum_{i=1}^{a}\sum_{j=1}^{b} y_{ij}^2 - ab\overline{y}_{\cdot\cdot}^2$		

If replications of measurements are available, the classical layout of a two-way cross classification with replications is like the one shown in Table 11.A4, but in the (i, j)th cell there are n observations Y_{ijk} for $i = 1, 2, \ldots, b$; $j = 1, 2, \ldots, a$; and $k = 1, 2, \ldots, n$. In such a case, it is possible to estimate the interaction between the two factors A and B. In fact, the total sum of squares can be partitioned into four components, one more than in (11.A9):

$$\sum_{i=1}^{b}\sum_{j=1}^{a}\sum_{k=1}^{n}(y_{ijk} - \bar{y}_{\ldots})^2 = \underbrace{nb\sum_{j=1}^{a}(\bar{y}_{.j.} - \bar{y}_{\ldots})^2}_{SS_A} + \underbrace{na\sum_{i=1}^{b}(\bar{y}_{i..} - \bar{y}_{\ldots})^2}_{SS_B}$$

$$+ \underbrace{n\sum_{j=1}^{a}\sum_{i=1}^{b}(y_{ij.} - \bar{y}_{.j.} - \bar{y}_{i..} + \bar{y}_{\ldots})^2}_{SS_{A\times B}} + \sum_{i=1}^{a}\sum_{j=1}^{b}\sum_{k=1}^{n}(y_{ijk} - \bar{y}_{ij.})^2. \quad (11.A13)$$

The extra term is the contribution to the variation of the interaction between the two factors A and B.

The underlying model when there are replications available is given by:

$$Y_{ikj} = \mu + \alpha_i + \beta_j + (\alpha\beta)_{ij} + \varepsilon_{ijk}, \quad \text{for } i = 1, 2, \ldots, a; j = 1, 2, \ldots, b; k = 1, 2, \ldots, n, \quad (11.A14)$$

where the terms μ, α_i and β_j have the same meaning as in (11.A10), $(\alpha\beta)_{ij}$ is the effect of the ith level of factor A when factor B is at the jth level (interaction between A and B), and the ε_{ijk} represent the random component (independent, normally distributed with mean zero and common variance σ^2). In addition to the constraints on (11.A10), we have $\sum_{i=1}^{a}(\alpha\beta)_{ij} = 0$ for every $j = 1, 2, \ldots, b$ and $\sum_{j=1}^{b}(\alpha\beta)_{ij} = 0$ for every $i = 1, 2, \ldots, a$. In addition to the two null hypotheses (11.A11) and (11.A12), the null hypothesis that there is no interaction between A and B can be tested. The null hypothesis that all the $(\alpha\beta)_{ij}$ are equal to zero is rejected at significance level α if

$$\frac{SS_{A\times B}/(a-1)(b-1)}{SS_{\text{exp}}/ab(n-1)} > f_{1-\alpha},$$

where $f_{1-\alpha}$ can be found from tables of the F distribution with $(a-1)(b-1)$ and $ab(n-1)$ degrees of freedom.

The corresponding ANOVA table is given in Table 11.A6.

11.A3 ANOVA of hierarchic (nested) data

In some experiments, the levels for a given factor differ for different levels of the other factors. Sometimes the levels are chosen at random for each factor, but this is

Table 11.A6 ANOVA with two factors at a and b levels and n replications.

Source of variation	Degrees of freedom	Sum of squares	Mean squares	f_{exp}
Factor A	$a-1$	$SS_A = \frac{1}{bn}\sum_{j=1}^{a} y_{\cdot j\cdot}^2 - \frac{y_{\dots}^2}{nab}$	$MS_A = \frac{SS_A}{a-1}$	$\frac{MS_A}{MS_e}$
Factor B	$b-1$	$SS_B = \frac{1}{an}\sum_{i=1}^{b} y_{i\cdot\cdot}{}^2 - \frac{y_{\dots}^2}{nab}$	$MS_B = \frac{SS_B}{b-1}$	$\frac{MS_B}{MS_e}$
Interaction $A \times B$	$(a-1)(b-1)$	$SS_{A\times B} = \frac{1}{n}\sum_{i=1}^{a}\sum_{j=1}^{b} y_{ij\cdot}^2 - SS_A - SS_B - \frac{y_{\dots}^2}{nab}$.	$MS_{A\times B} = \frac{SS_{A\times B}}{(a-1)(b-1)}$	$\frac{MS_{A\times B}}{MS_e}$
Experimental error	$ab(n-1)$	$SS_e = SS_{TC} - SS_A - SS_B - SS_{A\times B}$	$MS_e = \frac{SS_e}{ab(n-1)}$	
Total	$nab-1$	$SS_{TC} = \sum_{i=1}^{a}\sum_{j=1}^{b}\sum_{l=1}^{n} y_{ilj}^2 - nab\overline{y}_{\dots}^2$		

not a strictly necessary condition for being a nested (hierarchical) design; in fact, the models used for such designs can be random, mixed or fixed. The ANOVA models for nested data can be written in a form that appears to be identical to those for cross classification. There are, however, important differences in the assumptions that accompany the analysis of nested data. The model for four factors X nested in W, nested in V that is nested in the main factor U is

$$Y_{tijkl} = \mu + U_t + V_{(t)i} + W_{(ti)j} + X_{(tij)k} + \varepsilon_{tijkl}, \tag{11.A14}$$

for $t = 1, 2, \dots, h$; $i = 1, 2, \dots, m$; $j = 1, 2, \dots, p$; $k = 1, 2, \dots, q$; $l = 1, 2, \dots, n$. Here μ is the overall mean, U_t is the effect due to the tth level of the main factor, $V_{(t)i}$ is the effect due to the ith level of the *sub*[1] *factor* (the factor nested in the tth level of the main factor) where the subscripts in parentheses denote the factor levels within which the leading factor level is nested, $W_{(ti)j}$ is the effect due to the jth level of the *sub*[2] *factor* nested in the ith level of the sub[1] factor and in the tth level of the *main factor*, and so on; ε_{tijkl} is the error component assumed to be independently and identically normally distributed with mean zero and common variance σ^2. The use of the Latin letters $U, V, \dots$ is for random effects and U_t, $V_{(t)i}$, $W_{(ti)j}$ and $X_{(tij)k}$ are independently and identically normally distributed with zero mean and variance σ_U^2, σ_V^2, σ_W^2 and σ_X^2, respectively.

The responses in the cross classification models (models (11.A4), (11.A10), (11.A14)) are normally distributed with mean, in the case for example of model (11.A14), $\mu_{ik} = \mu + \alpha_i + \beta_j + (\alpha\beta)_{ij}$ and standard deviation σ; in contrast, the responses in the nested models have common normal distribution with mean equal μ and standard deviation $\sigma_Y = \sqrt{\sigma^2 + \sigma_U^2 + \sigma_V^2 + \sigma_W^2 + \sigma_X^2}$. The variance terms in this formula are called variance components because the variance of the response σ_Y^2 is the sum for the individual variances of the factors, main and sub, and for

Table 11.A7 ANOVA for nested design with one main factor and four sub factors (model (11.A14)). Here $SS_M = y^2_{.....}/hmpqn$.

Source of variation	Degrees of freedom	Sum of squares
Main	$h-1$	$SS_{Main.Fact.} = \frac{1}{mpqn}\sum_{t=1}^{h} y^2_{.j.} - SS_M$
Sub[1] factor	$h(m-1)$	$SS_{Sub^1 Fact.} = \frac{1}{pqn}\sum_{t,i} y^2_{ti...} - SS_{MainFact.} - SS_M$
Sub[2] factor	$hm(p-1)$	$SS_{Sub^2 Fact.} = \frac{1}{qn}\sum_{t,i,j} y^2_{tij..} - SS_{MainFact.} - SS_{Sub^1 Fact.} - SS_M$
Sub[3] factor[2]	$hmp(q-1)$	$SS_{Sub^3 Fact.} = \frac{1}{n}\sum_{t,i,j,k} y^2_{tijk.} - SS_{MainFact.} - SS_{Sub^1 Fact.} - SS_{Sub^2 Fact.} - SS_M$
Experimental error	$hmpq(n-1)$	$SS_e = SS_{TC} - SS_{MainFact.} - SS_{Sub^1 Fact.} - SS_{Sub^2 Fact.} - SS_{Sub^3 Fact.}$
Total	$nkp-1$	$SS_{TC} = \frac{1}{hmpqn}\sum_{t,i,j,k,l} y^2_{tijkl} - SS_M$

the uncontrolled error σ^2. Therefore, in contrast to cross classification, the nested factors contribute to the overall variation observable in the response variable, not to differences in factor level means.

In such a situation there are as many null hypotheses as there are factors (main plus sub of any degree) and they are of the sort H_0: the factor (either main or sub) is not significant; the ANOVA allows the variances $\sigma^2_U, \sigma^2_V, \sigma^2_W$ and σ^2_X to be estimated to test each variance with the one of the factors at a higher level. After computing the sums of squares of each factor according to the scheme in Table 11.A7 and the corresponding degrees of freedom, the mean squares are the estimates of the variance of the main factor, partitioned into the sum of the variances of the sub factors, after multiplying them by appropriate constants related to the number of the groups in which the observed values were divided:

$$\begin{aligned} MS_{MainFact.} &= \hat{\sigma}^2 + q\hat{\sigma}^2_X + pq\hat{\sigma}^2_W + mqp\hat{\sigma}^2_V + hqpm\hat{\sigma}^2_U \\ MS_{Sub^1 Fact.} &= \hat{\sigma}^2 + q\hat{\sigma}^2_X + pq\hat{\sigma}^2_W + mqp\hat{\sigma}^2_V \\ MS_{Sub^2 Fact.} &= \hat{\sigma}^2 + q\hat{\sigma}^2_X + pq\hat{\sigma}^2_W \\ MS_{Sub^3 Fact.} &= \hat{\sigma}^2 + q\hat{\sigma}^2_X \\ MS_{exp.} &= \hat{\sigma}^2 \end{aligned} \tag{11.A15}$$

Comparing the values of the computed ratio between the variances to the theoretical ones from the table of the F distribution at a fixed significance level, the null hypothesis can be rejected or accepted:

$$\frac{MS_{Sub^3 Fact}}{MS_{exp}} \text{ to test } H_0 : \text{the effect of the sub}^3\text{factor is null,}$$

$$\frac{\mathrm{MS}_{\mathrm{Sub}^2\mathrm{Fact}}}{\mathrm{MS}_{\mathrm{Sub}^3\mathrm{Fact}}} \text{ to test } H_0 : \text{the effect of the sub}^2\text{factor is null,}$$

$$\frac{\mathrm{MS}_{\mathrm{Sub}^1\mathrm{Fact}}}{\mathrm{MS}_{\mathrm{Sub}^2\mathrm{Fact}}} \text{ to test } H_0 : \text{the effect of the sub}^1\text{factor is null,}$$

$$\frac{\mathrm{MS}_{\mathrm{MainFact}}}{\mathrm{MS}_{\mathrm{Sub}^1\mathrm{Fact}}} \text{ to test } H_0 : \text{the effect of the main factor is null.}$$

Moreover, an estimation of the variances σ^2, σ_U^2, σ_V^2, σ_W^2 and σ_X^2 can be obtained from (11.A15):

$$\begin{aligned}
\hat{\sigma}^2 &= \mathrm{MS}_{\mathrm{exp}} \\
\hat{\sigma}_X^2 &= \frac{1}{q}(\mathrm{MS}_{\mathrm{Sub}^3\mathrm{Fact.}} - \hat{\sigma}^2), \\
\hat{\sigma}_W^2 &= \frac{1}{pq}(\mathrm{MS}_{\mathrm{Sub}^2\mathrm{Fact.}} - q\hat{\sigma}_X^2 - \hat{\sigma}^2), \\
\hat{\sigma}_V^2 &= \frac{1}{mpq}(\mathrm{MS}_{\mathrm{Sub}^1\mathrm{Fact.}} - pq\hat{\sigma}_W^2 - q\hat{\sigma}_X^2 - \hat{\sigma}^2), \\
\hat{\sigma}_U^2 &= \frac{1}{hmpq}(\mathrm{MS}_{\mathrm{MainFact.}} - mpq\hat{\sigma}_V^2 - pq\hat{\sigma}_W^2 - q\hat{\sigma}_X^2 - \hat{\sigma}^2).
\end{aligned} \tag{11.A16}$$

Appendix 11B Linear relationships

Chemical and physical laws (sometimes derived empirically) can usually be expressed in mathematical form. Each law states a relationship between a number of variables and can be used to estimate the value of any variable given the values of the others. The variables are said to be functionally related if a unique relationship between the variable exists or can be hypothesised. Not all the related variables are functionally related; but if it is known that the two variables y and x are functionally related, then an approximation to the form of this relationship over given ranges of the variables can be estimated empirically from measurements of the corresponding values of y and x. For simplicity, we shall assume that the two variables y and x are linearly related (more complex relationships may be considered, but not in this appendix), even if in practice this may be only a rough approximation. Since measurements of the variable Y (a capital letter is used to denote the random variable that has to be predicted from the data and a small letter is used to denote an observed value) are subject to error, the plotted points will not therefore fall exactly on a straight line but will vary randomly about one, the graph of the functional relation.

11.B1 Linear regression

In the study of the relationship between variables, the graph of the mean values of one variable for given values of the other variable, when referred to the whole population, is called the regression curve or simply the regression of Y on x. When the regression curve is linear, the mean of the Ys is given by $\beta_0 + \beta_1 x$. In general, Y differs from this mean and the difference is denoted by ε,

$$\mathrm{Y} = \beta_0 + \beta_1 \mathrm{x} + \varepsilon, \tag{11.B1}$$

where ε is a random variable and β_0 can always be chosen so that the mean of the distribution of ε is equal to zero. The value of ε for any given observation depends on a possible error of measurement and on the values of variables other than x which may have an influence on Y.

We wish to estimate the values of the parameters β_0 and β_1 in (11.B1). The actual method of estimating a linear regression, usually referred to as fitting a straight line to the data, is straightforward. There are a number of methods of estimating β_0 and β_1, but the most popular is the method of least squares. This consists of finding the values of β_0 and β_1 that minimise the sum of squares of the deviations of the observed values from the line ($\sum_{i=1}^{n} \varepsilon_i^2$, where $\varepsilon_i = Y_i - \beta_0 + \beta_1 x_i$). Denoting by b_0 and b_1 the estimated values of β_0 and β_1 obtained from the observations (x_i, y_i) for $i = 1, 2, \ldots, n$, we have

$$b_0 = \overline{y} - b_1 \overline{x},$$
$$b_1 = \frac{\sum_{i=1}^{n} (x_i - \overline{x})(y_i - \overline{y})}{\sum_{\mathrm{i}=1}^{n} (x_i - \overline{x})^2}, \tag{11.B2}$$

with $\overline{y} = n^{-1} \sum_{i=1}^{n} y_i$, or, in an equivalent form which is more suited to computing,

$$b_0 = \overline{y} - b_1 \overline{x},$$
$$b_1 = \frac{\sum_{i=1}^{n} x_{i y_i - n\overline{x}\overline{y}}}{\sum_{i=1}^{n} x_i^2 - n\overline{x}^2}. \tag{11.B3}$$

A simpler notation is to let

$$S_{xx} = \sum_{i=1}^{n} (x_i - \overline{x})^2 = \sum_{i=1}^{n} x_i^2 - n\overline{x}^2 \text{ and } S_{xy} = \sum_{i=1}^{n} (x_i - \overline{x})(y_i - \overline{y})$$
$$= \sum_{i=1}^{n} x_i y_i - n\overline{x}\overline{y},$$

hence

$$b = \frac{S_{xy}}{S_{xx}}. \tag{11.B4}$$

It is impossible to make an exact statement about how good an estimate like the previous ones is unless some assumptions are made about the underlying distribution of the random variables ε_i. Fortunately, the Gauss–Markov theorem states that the least squares estimators are the most reliable in the sense that they are subject to the smallest chance of variations. Under the additional assumptions that the ε_i are independently normally distributed with mean zero and common variance σ^2, we can work out the distribution of the estimators B_0 and B_1 of the regression coefficients β_0 and β_1. Both the statistics

$$\frac{B_0 - \beta_0}{\sqrt{\text{var}[\hat{\beta}_0]}} = \frac{B_0 - \beta_0}{s_e\sqrt{nS_{xx}/(S_{xx} + n\bar{x}^2)}} \tag{11.B5}$$

and

$$\frac{B_1 - \beta_1}{\sqrt{\text{var}[\hat{\beta}_1]}} = \frac{B_1 - \beta_1}{s_e\sqrt{S_{xx}}} \tag{11.B6}$$

are values of random variables having a t distribution with $n - 2$ degrees of freedom, and $s_e = \sqrt{s_e^2}$ with

$$s_e^2 = \frac{1}{n-2}\sum_{i=1}^{n}(y_i - \hat{y}_i)^2 = \frac{S_{yy} - S_{xy}^2/S_{xx}}{n-2},$$

where $\hat{y}_i = b_0 + b_1 x_i$. The term s_e^2 is usually referred to as the standard error of estimate and it is an estimate of $\sigma^2 : s_e^2 = \hat{\sigma}^2$. The term $(n-2)s_e^2 = \sum_{i=1}^{n}(y_i - \hat{y}_i)^2$ is referred to as the residual sum of squares.

It is possible to construct confidence intervals for the regression coefficients β_0 and β_1 substituting for the middle term of $-t_{n-2,1-\alpha/2} < t < t_{n-2,1-\alpha/2}$ ($t_{n-2,1-\alpha/2}$ is the $1 - \alpha/2$ percentile of the t distribution with $n - 2$ degrees of freedom) the appropriate statistics (11.B5) and (11.B6), using simple algebra to obtain the following $1 - \alpha$ confidence intervals for β_0 and β_1, respectively:

$$\left(b_0 - t_{n-2,1-\alpha/}s_e\sqrt{\frac{1}{n} + \frac{\bar{x}^2}{S_{xx}}},\, b_0 + t_{n-2,1-\alpha/2}s_e\sqrt{\frac{1}{n} + \frac{\bar{x}^2}{S_{xx}}}\right), \tag{11.B7}$$

$$\left(b_1 - t_{n-2,1-\alpha/2}s_e\sqrt{\frac{1}{S_{xx}}},\, b_1 + t_{n-2,1-\alpha/2}s_e\sqrt{\frac{1}{S_{xx}}}\right). \tag{11.B8}$$

It also possible to perform tests of hypotheses concerning the regression coefficients β_0 and β_1; those concerning β_1 are of special importance because β_1 is the slope of the regression line – the change in the mean of Y corresponding to a unit increase in x. If $\beta_1 = 0$, the regression line is horizontal and, much more important, the mean of Y does not depend linearly on x.

Another problem of interest is that of estimating $\beta_0 + \beta_1 x$, namely, the mean of the distribution of Y for a given value of x. If x is held fixed at x^*, the quantity we want to estimate is $\beta_0 + \beta_1 x^*$ and it seems reasonable to use $b_0 + b_1 x^*$, where b_0 and b_1 are the estimates obtained with the least squares methods. It can be shown that the estimator $\hat{Y}$ of $\beta_0 + \beta_1 x^*$ is unbiased and has variance

$$\sigma^2 \left(\frac{1}{n} + \frac{(x^* - \overline{x})^2}{S_{xx}} \right), \tag{11.B9}$$

and substituting σ^2 with its estimate s_e^2, the $1 - \alpha$ confidence interval for $\beta_0 + \beta_1 x^*$ is given by

$$\left((b_0 + b_1 x_k) - t_{n-2,1-\alpha/2} s_e \sqrt{\frac{1}{n} + \frac{(x^* - \overline{x})^2}{S_{xx}}}, (b_0 + b_1 x_k) + t_{n-2,1-\alpha/2} s_e \sqrt{\frac{1}{n} + \frac{(x^* - \overline{x})^2}{S_{xx}}} \right). \tag{11.B10}$$

Note that the confidence interval (11.B10) has minimum size when $x^* = \overline{x}$ and maximum size at the extremities of the range of experimentation.

Having performed the aforementioned tests, it is important to check the adequacy of the assumed model. An important step is the examination of the residual using residual plots and other regression diagnostics to assess if the residuals appear random and normally distributed. Most statistical programs provide a number of residual plots (normal score plots, charts of individual residuals, histograms of residuals, plots of fits versus residuals on the same graph or not, ...) and regression diagnostics for identifying outliers or unusual observations. These observations may have a significant influence upon the regression results. The experimenter should check whether the unusual observations are correct; if they are, he needs to find out why they are unusual and consider what is their effect on the regression equation, or he might wish to examine how sensitive the regression results are to the outliers being present. Outliers can suggest inadequacies in the model or a need for additional information.

11.B2 Correlation

So far, the independent variable x has been assumed to be known without error (or at least negligible with respect to the error of the variable Y); but there are a number of problems where both the xs and the ys are values assumed by random variables. There is also a class of problems in descriptive statistics involving two or more related variables where the main interest is in checking the extent of the association between them, and for this purpose there are proper correlation methods.

Usually, the measurements supply the experimenter with a mass of data that will mean very little unless it is presented in a condensed form to show up the essential features. As is well known, any one observation of a variable can be completely described by means of a distribution (histogram, boxplot, ...) and by the two statistics mean and standard deviation, if the distribution is of the normal form.

A convenient method for exhibiting the relationship is to plot the observations of the two variables, one against the other, in the form of a scatter plot or scatter diagram. A linear association or correlation between the variables X and Y is reflected in the tendency to plot a diagonal line. Large values of the X variable tend to be associated with large values of the Y variable and vice versa, but for any given value of one variable there is a sizeable spread in the other.

If we consider any two random variables X and Y and assume that both are normally distributed, which for measurement errors is usually true, even though each variable can be separately represented by its mean and standard deviation, these four figures are not able to give a complete description of the data because no information is supplied on the relationship between the two measurements. There exists a normal distribution for two variables, just as for one; it depends on four parameters, the means and the standard deviations of the two variables, and on an additional parameter, usually denoted with the Greek letter ρ, which measures the association between the two variables. ρ is called the correlation coefficient and is also estimable from the data. First of all, it is necessary to estimate another statistic, the covariance, given by

$$\text{cov}(x, y) = \frac{1}{n-1}\sum_{i=1}^{n}(x_i - \overline{x})(y_i - \overline{y}) = \frac{1}{n-1}S_{xy}, \tag{11.B11}$$

that is, the sum of the products of the deviations of each observation x of the random variable X from the mean $\overline{x}$ and the deviation of the corresponding observation y of the random variable Y from the mean $\overline{y}$, divided by the number of degrees of freedom $n - 1$. The correlation coefficient ρ is then estimated by

$$r = \frac{\sum_{i=1}^{n}(x_i - \overline{x})(y_i - \overline{y})}{\sqrt{\sum_{i=1}^{n}(x_i - \overline{x})^2 \sum_{i=1}^{n}(y_i - \overline{y})^2}} = \frac{S_{xy}}{\sqrt{S_{xx}S_{yy}}}. \tag{11.B12}$$

The correlation can be positive or negative: a positive correlation means that one variable tends to increase as the other increases; a negative one means that one variable tends to decrease as the other increases. Moreover, the correlation coefficient cannot take values outside the limits -1 and $+1$. A large absolute value of r means a close linear relationship and a small one a less definite relationship. When the absolute value of r is the unity, the points in scatter plot fall exactly on a straight line and the linear relationship is perfect. When $r = 0$, the points scatter in all directions and we say that there is no correlation (linear relationship or association) between the two random variables. In other words, we can also state that r^2 is an estimate of the proportion of the variation of Y that can be attributed to the linear relationship with X. Since the value of r is only an estimate of the true value of ρ, we can perform a test of significance or to construct a confidence interval. To make inferences, a useful transformation due to R.A. Fisher may be used:

$$x = \tanh^{-1} r = \frac{1}{2} \ln \frac{1+r}{1-r}. \tag{11.B13}$$

The statistic x in (11.B13) is very nearly normally distributed with mean $\mu = \frac{1}{2}\ln((1+\rho)/(1-\rho))$ and standard deviation $\sigma = 1/\sqrt{n-3}$. It can thus be standardised, yielding a random variable having approximately the standard normal distribution. Thus, we are able to compute approximate confidence limits for a correlation coefficient and also for a difference between two correlation coefficients.

We stress the relation between the coefficients of correlation and the coefficients of regression. From (11.B4) and (11.B12) it follows that

$$b_1 = r\frac{S_{yy}}{S_{xx}}, \tag{11.B14}$$

and the regression equation may also be written as

$$\frac{y-\overline{y}}{S_{yy}} = r\frac{x-\overline{x}}{S_{xx}}. \tag{11.B15}$$

In other words, the correlation coefficient is numerically equal to the regression coefficient when variables are expressed in standard measure.

12

Safety and reliability

12.I

Reliability engineering: The state of the art

Chris McCollin

12.I.1 Introduction

In this part of this chapter I review engineering design reliability from my own background in aerospace with comments on where research may arise. The areas of design which mainly affect reliability are reliability specification; the initial systems design process incorporating feasibility and concept design; the development process incorporating the main computer-aided design/engineering activities including failure mode and effects analysis (FMEA), circuit design and simulation packages; the development process incorporating prototype build; and components and training.

Recent papers on design for reliability are described in Section 12.I.2. This is followed by a description of general approaches to reliability development covering costing, management and specification (Section 12.I.3), design failure investigation techniques (DFITs: Section 12.I.4), development testing (Section 12.I.5) and failure reporting (Section 12.I.6). In Section 12.I.7 some further ideas for research are presented, and in Section 12.I.8 current training and future needs are addressed.

Statistical Practice in Business and Industry Edited by S.Y. Coleman, T. Greenfield, D.J. Stewardson and D.C. Montgomery

12.I.2 The reliability engineering literature

A definition of reliability is the maintenance (over time) of system performance within specification bands (such as contained and controlled variability) across diverse environmental and operating conditions. Work on reducing variability within design has been discussed by Morrison (1957, 2000), and these ideas have been used in Drake (1999: Chapter chap13). Papers by Sexton *et al.* (2001), who have analysed design variables for hydraulic gear pumps, Thompson (2000), who discussed an approach which outlines a scoring criterion for design variables, and Swift *et al.* (1999) on design capability all come under the above definition.

The work by Andrews and Dugan (1999) and Bartlett and Andrews (1999) on binary decision diagrams (BDDs) is an extra addition to the growing pool of fault tree tools. These tools aid the engineer in lateral thinking when it comes to possible combinations of failures and should be taught to undergraduate engineers.

These research papers concentrate on engineering design improvement. The purpose of this chapter is to review the main reliability activities within each phase of a project and highlight areas where work has been done and where there are opportunities for further work.

12.I.3 Management, specification and costing issues

Typically, customer requirements are delivered into the company via the marketing department and are passed to a designated project manager. A team of engineers will be assigned to the project. A reliability engineer may be assigned to the project if the project manager is aware of any customer reliability requirements. Alternatively, a reliability manager (if there is one) should get copies of all relevant customer documentation from the marketing department; that is the statement of work, the contract and the customer's system specification and liaise with the project manager on the support required. Within these documents, there will be information relevant to reliability engineering activities which the project reliability engineer will address in terms of costing and work to be done. For example, in the contract there may be statements on limited liability, warranty and indemnity, which may require some form of analysis (life cycle costing, safety analysis) to be carried out. The specification requirements for the design of a system may be laid out in a compliance matrix with each clause being classed as to be met by development test (1), met by analysis (2), met by test (3) or met by inspection (4). Examples from these four criteria are:

1. one-off test of fungal growth for a system operating in the Tropics (carried out on a prototype unit);
2. finite-element analysis of a mechanical part;
3. the production test before delivery of the power supply unit of a PC using a voltmeter;
4. visual inspection of a TV screen, again during a production test.

Within the specification, apart from the reliability requirement, there may be information on operating regimes, built-in-test requirements, maintainability and testability requirements, which need integrating into any proposed design configuration. Each of these requirements is discussed by the customer, the project manager and the quality/reliability manager to determine the appropriate method of meeting the criteria. The statement of work details reliability activities such as reliability plans; analyses such as FMEA, preliminary hazards analysis (PHA), Markov analysis and fault tree analysis (FTA); testing, for example environmental stress screening and reliability development testing; and any applicable reports, test procedures.

The project manager's attitude to reliability is highly influenced by his/her age, project experience, background (mechanical/electrical/software/other), attitude to reliability activities, and managerial style (autocratic/democratic/laissez-faire). In fact, these attributes have a great influence on system reliability. Monitoring of projects by taping important meetings would provide examples of good and bad managerial practice leading to improved manager training on engineering issues such as reliability concerns. Each of these attributes affects the decision-making process of a project manager. The author can find no research on the relationship of these attributes to the success or failure of a project in terms of cost/time overruns and unreliability; however, it is the decisions of a project manager that decide a project's success. In most cases, a project manager starts a new development project after his/her previous system's project has gone into production and so he/she does not necessarily see the effect of some of his/her decisions in terms of field failures. A recent relevant paper in this context, which considers human error in preventing business disruption, is by Gall (1999). O'Connor (2000) points out that 'the achievement of reliability is essentially a management task'.

The main phase of the project where the expertise and experience/knowledge of the reliability engineer can have a significant effect is in preliminary design as it is in the early design meetings where the configuration of the system – including testing, safety/reliability and power supply requirements for the proposed system – is decided. Knowledge of the reliability of previous similar systems and a current understanding of requirements can be used to aid the project manager in the proposed system configuration. Knowledge of the management style of the project manager and how to influence him/her, and attending meetings which affect reliability, are paramount to improving reliability. For example, a meeting between the project manager, chief buyer and project procurement manager to discuss costs will invariably involve discussion of reliability since the more reliable a component is designated to be, the more testing it has been through and hence the more expensive it will be. Hence, a report of what was discussed in the above meeting should be fed through to the reliability engineer so that he/she can comment and give feedback. This feedback is especially important when commercial off-the-shelf (COTS) components are involved since it is what the COTS device has not been designed for in its usage within a project which has to be monitored. Feedback from meetings should be made available through the company intranet and be readily editable for safety case documentation. It is interesting to note 'no member of the Safety, Reliability and Quality Assurance staff was present at the teleconference between Marshall Space Flight Center and Morton Thiokol engineers on the night before the launch' of the ill-fated Space Shuttle Challenger. 'The reason for their absence? No one thought to invite them' (Reason, 1997).

Reliability is specified by a prospective customer. For example, the minimum failure operating period (MFOP) for aircraft is specified by the aerospace sector (Cini and Griffith, 1999) and this provides designers with a performance goal which can be fed through to intermediate levels of development. The goal requires that the equipment performance does not fail or degrade below a specified level in a given time (with a certain risk).

An excellent quality function deployment (QFD) approach to this feed-through of reliability requirements is described in Akao (1990). The paper by Reunen *et al.* (1989) gives a comprehensive conceptual design approach to reliability incorporating structured analysis and hazardous operations (HAZOP) analysis.

At the proposal stage, reliability activities are costed. The author can find little evidence of methods (such as generic formulae) of costing design or reliability activities in the literature. The only difference in cost from one contractor to the next would be the overhead if generic formulae were used. Some managers ask engineers for initially estimated development costs to be reduced at the proposal stage. A standard formula would show the manager whether this is possible.

Implementation of reliability techniques (the DFITs) which detail recommended actions that are carried out and make a significant impact on the design should reduce overhead. If recommended corrective actions are not carried out in a timely fashion, then actual failures may occur which will increase overhead for later costing exercises. When carrying out a DFIT, recommended corrective actions should be listed with associated costs. Some of these recommended actions are improvements which, if implemented, may save the company millions in lawsuits and product recall. So, in effect, the cost of the DFIT can be seen as negative as long as it is used as a continuous improvement tool with all recommended corrective actions implemented on time. Research is required on DFITs on how to improve the responsiveness of corrective action implementation.

For many suppliers, reliability tasks are seen as one-off analyses to be carried out by new or placement student engineers or consultants. Once a reliability task has been carried out to meet a customer requirement, then the paperwork sits on a shelf and the corrective actions that are associated with the tasks may not be implemented because they were not expected or costed into the analysis. Integration of tools such as DFITs and actual failure reporting requires more treatment within the research literature.

The problems are exacerbated by the use by large companies of generic specifications, which may not be relevant to the supplier's components, and it is only fear of losing a lucrative contract that makes the supplier carry out an analysis which does not carry through its recommended corrective actions.

12.I.4 DFITs

Failure modes can then be identified using DFITs to address equipment failure effects which may be alleviated by pre-emptive corrective action.

Automation of a DFIT (such as FMEA) would reduce the time to carry it out and thus provide the time and opportunity to implement corrective actions. Three examples of this automation have resulted in embellishments to design packages such as Autosteve, SPICE and MENTOR. Since DFITs should highlight possible causes and factors, the automated application of designed experiments (utilising simulation with bootstrapping/jackknifing) within these computer-aided design/engineering packages would improve reliability integration with design. However, there has been little research work on helping small to medium enterprises (SMEs) create reliable designs as they cannot afford this type of complex computer-aided design software.

Two major disasters (Piper Alpha, Flixborough) have occurred due to the failure of a permit to work (PTW) system. When an FMEA is carried out, engineers detail administrative systems such as PTW as corrective actions. Engineers assume systems are in place; however, experience with ISO 9001 certification shows that paperwork in some companies is only up to date during the certification period of three days. Other types of corrective action should be used. Corrective actions such as the use of poka yoke devices and designed experiments to optimise output do not tend to be used as alternatives or additions to improvements.

Recent research into FMEA by Braglia (2000) and Sankar and Bantwal (2001) have addressed ways of looking at the risk priority number within FMEA. Main problems associated with FMEA which are not addressed by these papers are that engineers are very conservative in their use of scoring of the likelihood, severity and detectability of a failure mode. Feynman (1989) also noticed that the job that one does affects one's subjective judgement of risk, managers being much more optimistic about outcomes than engineers. Risk is a dynamic variable affected by common incidents (such as human forgetfulness) which changes over time, and this is not allowed for in DFITs. Incorrect assumptions during analysis may change an unlikely event into one which is very likely (Gastwirth, 1991). Other comments on FMEA are presented in McCollin (1999).

12.I.5 Development testing

After the concept phase design review where the selected design configuration is justified (to the customer), detailed design starts on the various subassemblies. The reliability engineer's job may include derating, prediction and various DFITs. During this phase, designers use the amendment record scheme to implement any design changes needed. A design meeting at regular intervals discusses each amendment record. There may typically be 8000–12 000 amendments during a development phase. Once prototype units are developed then the failure reporting and corrective action system (FRACAS) and the work-in-progress (WIP) system are initiated. The WIP system records any changes required in manufacture or production planning. Each of these systems will generate records on a regular basis, and determining their effectiveness is one method of improving reliability at the design stage. Monitoring these records in terms of counts or as streams of individual records will give an indication of where extra work may be required and areas where there may be major reliability problems. Statistical methods of monitoring the counts over time using explanatory variables such as subsystem, personnel ID, cumulative calendar time, time since last review, may include discrete proportional hazards modelling, cumulative sum or time series techniques. A method of monitoring the stream of binary data over time (Harvey, 1993) could be applied to minor or major design amendments. Further work is required on these areas and also multi-level analysis (Goldstein *et al.*, 1998; Gray *et al.*, 1988).

After manufacturing a number of prototypes, each undergoes a test programme to meet customer environmental requirements. Development testing on a prototype system (such as Highly Accelerated Life Test (HALT)) should identify any failures whose effects that cannot be alleviated to assert themselves, which then provides an indication of whether the reliability requirement can be met.

A PHA highlighting the effect of these environments on the designed systems should have been carried out so that no unforeseen circumstances can arise without adequate safety precautions. For instance, how should a gimballed system react when it is first switched on since the system does not 'know' where it is? Also, when a system is first run in an environmental chamber, what is the expected effect of the chamber environment on the critical components? What is expected outcome if critical components fail catastrophically? A fault seeding within a prototype may be needed to simulate possible problem identification before entering the system into any environmental testing. Examples of faults to be seeded may be highlighted from a previously performed DFIT.

A question relating to MFOP is 'How long will this newly developed system last until first failure?'. Given no prior operating time information on the system in question, how can the probability that the item will last for a certain time be evaluated? Silverman (2006) suggests that the use of mean time between failure may be obsolete when using HALT. This test area is still developing and Bayesian approaches to Duane analysis (Ke and Shen, 1999) and intensity modelling with covariates may provide aids to identifying failed design modifications which can then be eliminated by readdressing a previous design standard and providing indications of the expected MFOP.

12.I.6 Failure reporting

FRACAS identifies any failure effects in the factory or in service use which may not have been identified by the previous methods and should be fed forward to improve the next generation equipment design. This feed-forward process should also address these possible failures in the next generation DFIT and test and hence improve reliability requirements for this next generation.

Customer requirements do not usually tie up with the reliability problems that are met in industry. For instance, a major problem with electronic systems is that a fault in the system is missed. This may be due to a number of reasons: the fault was intermittent due to dry wetting of a soldered joint, the fault was unconfirmed on retest, a high temperature problem with solder, an interface problem between two interconnecting units or a problem with traceability of the fault. However, not finding a fault may have been decided by the customer to be non-relevant for reliability calculation purposes. A customer may not be satisfied by the supplier's equipment reliability but the supplier is only offering what has been tendered for.

An example of a root cause analysis for a large system follows. This highlights problems of traceability of root cause. The customer's specification states that there will be no intermittent faults due to supplier's units. An actual intermittent fault is isolated to a sub-assembly containing hundreds of components. The sub-assembly is removed for repair. Suppose a power amplifier circuit fails because a pull-down resistor goes open-circuit (or pull-up resistor goes short-circuit) on a current-limiting transistor feeding the power amplifier. The production engineer replaces the power amplifier because it looks burnt and after switching on the power amplifier fails again. He/she now replaces the transistor and resistor in circuit and the power amplifier again.

He/she lists the components replaced on his failure report and these get recorded on a component failure database. What does the database tell us about the original fault of the resistor, given that usually a non-technical person is typing out the details of the defect report onto the database? It may be seen that the power amplifier is the main problem when actually it is not. The supplier of the power amplifier may be informed of the problem when the fault is not theirs. The main cause of failure may have been a voltage transient affecting the whole circuit and the resistor was just the first to get damaged. So a root cause solution for this problem should involve checking to see how intermittent environmental conditions are covered by test in the design phase in the customer's company. There are many problems like this which are incorrectly diagnosed (Edwards, 2000).

Reviews of root cause analysis are given in the review document available from the European Network for Business and Industrial Statistics (http:///www.enbis.org) and Goodacre (2000). There is a diverse amount of information on the subject and a unique statistical approach may involve a combination of a flowchart question based approach to FRACAS and BDDs and identifying variation within data across numerous sources.

12.I.7 Further ideas for research

Some further research ideas are presented in this section. The main reliability activity of reducing the number and type of prospective failures requires the design team (as an aim) to determine root causes at the earliest opportunity. Many companies usually only carry out reliability activities if they are specified contractually and then only to meet contractual requirements. If the system is not safety-critical, then reliability activities are usually an extra burden to a design engineer who has other more pressing tasks to carry out. Hence the integration of reliability activities does not take place and problems recur from one generation of design to the next. Only the large multinational companies can attempt to bring about an integration of activities as outlined in Figures 4 and 6 of Strutt (2000). There is a need to provide an integration methodology, which will suit SMEs because at the moment reliability techniques are seen as stand-alone.

If an engineer wants to increase his wage, then the only way is to become a manager – and an excellent engineer does not necessarily become a good manager. As engineers move from one company to another, knowledge of a technique goes with them. Intranet descriptions of previous projects and their problems and solutions may keep important knowledge within a company. This could be extended as follows. An overall equipment manufacturer (OEM) managed project of many suppliers would identify diverse common problems identified in specifications, FMEA, tests and warranty reports. The simplest template for such a tool across diverse organisations (OEMs, suppliers) would be an intranet-based tool as most companies have access to the web.

The specification compliance matrix may be converted to a House of Quality to take into account tasks to be carried out to meet the various requirements. A scoring method can be incorporated into this matrix to take account of whether each requirement has been met to the customer's satisfaction, not been met, or there is a conflict. An overall weighted unreliability score for the grid can then be calculated and monitored over time by a P chart and used as guidance for continuous improvement at design review meetings. This can also be extended. The overall dependability status of a company can be assessed by developing and identifying reliability maturity levels in a similar manner to the capability maturity model (Paulk *et al.*, 1995).

Many documents, such as specifications, test procedures and drawings, may be improved by the use of Fagan inspection (Fagan, 1976). This is a software technique which is applicable to any sort of documentation. It may be carried out before the start of a new project – for example, design amendments/drawing reviews for feedback to next-generation designs (perhaps leading to a database reviewing facility for identification of problem areas). The difference between what the customer wants and is going to get would also be highlighted with a Fagan inspection of the specification.

12.I.8 Training in reliability

Two recent papers concerning the needs of engineers highlighted some areas of reliability.

Based on a speech by Richard Parry-Jones to the Royal Academy of Engineering (Parry-Jones, 1999), Davies (2001) proposes a 12-section syllabus based around realistic engineering scenarios for the design and manufacture of a fuel filler cap. Within the syllabus are exploratory data analysis, designed experiments, statistical process control, regression, response surface methodology and Weibull analysis. Volunteer university engineering departments are being invited to look at the material with a view to running the course during the next two years. This puts some reliability back into an engineering syllabus, but reliability must not been seen as just statistics.

The results of the Engineering Quality Forum survey to establish ongoing requirements for education and competency in the field of quality management (Woodall *et al.*, 1998) listed skills and knowledge required in engineering for new graduates and experienced engineers. The 'concepts' needs targeted by/for new graduates were mainly quality systems, teams and process improvement. The top ten 'techniques' needs targeted by/for new graduates included design of experiments, statistical process control, FMEA, Weibull analysis, QFD, FTA and Duane analysis in the top ten list. Surprisingly, out of 178 responses there was no mention of root cause analysis, although the 7 old and new quality tools were listed.

There was very little response in the above questionnaire for elements which are in the Engineering Council Quality and Reliability Engineering syllabus, namely:

- reliability and maintainability (R&M) planning – identification and selection of R&M tasks appropriate to a development programme and assistance in the generation of an R&M programme plan;
- design analysis – creation of logic diagrams and application of PHA and HAZOPs;
- R&M prediction and assessment – R&M prediction using proprietary and in-house software and assessment through the analysis of trials data, and combination of parts and subsystem level data to the system level using block diagrams;
- FRACAS – set-up and operation of a data recording and corrective action system using in-house software.

This despite the fact that these are the main elements of a reliability engineer's job. There is too much literature on reliability statistics and not enough on reliability engineering activities and how to improve them. There is a growing body of professionals who believe that engineers should teach engineering statistics. It may be that, from this viewpoint, reliability will not be seen as just statistics but as a valid improvement activity during preliminary and full-scale development.

The review in this chapter has highlighted areas where the author thinks improve ments in reliability practice can be made. These include integration of reliability activities, costing and feedback/feed-forward of information to aid future projects. There needs to be a survey of reliability practitioners to determine their background, their respective needs and why reliability activities are not seen as core tools for engineering design for customer satisfaction. It may be that the expertise for reliability tasks will always lie in the hands of consultancies since management sees reliability as an add-on activity and many companies are not prepared to pay for the creation of in-house systems for reliability improvement.

References

Akao, Y. (1990) *Quality Function Deployment: Integrating Customer Requirements into Product Design*. Productivity Press, Cambridge, MA.

Andrews, J.D. and Dugan, J.B. (1999) Advances in fault tree analysis. In B.W. Robinson (ed.), *Advances in Safety and Reliability: Proceedings of the SARSS*, pp. 10/1–10/12. Safety and Reliability Society, Manchester.

Bartlett, L.M. and Andrews, J.D. (1999) Comparison of variable ordering heuristics/algorithms for binary decision diagrams. In B.W. Robinson (ed.), *Advances in Safety and Reliability: Proceedings of the SARSS*, pp. 12/1–12/15. Safety and Reliability Society, Manchester.

Braglia, M. (2000) MAFMA: Multi-attribute failure mode analysis. *International Journal of Quality and Reliability Management*, **17**(9), 1017–1033.

Cini, P.F. and Griffith, P. (1999) Designing for MFOP: Towards the autonomous aircraft. *Journal of Quality in Maintenance Engineering*, **5**(4), 296–306.

Davies, N. (2001) A new way to teach statistics to engineers. *MSOR Connections Newsletter*, May. http://www.mathstore.ac.uk/newsletter/may2001/pdf/engstats.pdf.

Drake Jr., P. (1999) *Dimensioning and Tolerancing Handbook*. McGraw-Hill, New York.

Edwards, I.J. (2000) The impact of variable hazard-rates on in-service reliability. Paper presented to the 14th ARTS Conference, Manchester.

Fagan, M.E. (1976) Design and code inspections to reduce errors in program development. *IBM Systems Journal*, **15**(3), 182–211.

Feynman, R.P, (1989) *What Do You Care What Other People Think?* Unwin Hyman, London.

Gall, B. (1999) An advanced method for preventing business disruption. In B.W. Robinson (ed.), *Advances in Safety and Reliability: Proceedings of the SARSS*, pp. 1/1–1/12. Safety and Reliability Society, Manchester.

Gastwirth, J.L. (1991) The potential effect of unchecked statistical assumptions. *Journal of the Royal Statistical Society, Series A*, **154**, 121–123.

Goldstein, H., Rasbash, J., Plewis, I., Draper, D., Browne, W., Yang, M., Woodhouse, G. and Healy, M.J.R. (1998) *A User's Guide to MLWin*. Institute of Education, London.

Goodacre, J. (2000) Identifying current industrial needs from root cause analysis activity. 14th ARTS Conference, Manchester, 28–30 November.

Gray, C., Harris, N., Bendell, A. and Walker, E.V. (1988) The reliability analysis of weapon systems. *Reliability Engineering and System Safety*, **21**, 245–269.

Harvey, A.C. (1993) *Time Series Models*, 2nd edition. Harvester Wheatsheaf, Hemel Hempstead.

Ke, H. and Shen, F. (1999) Integrated Bayesian reliability assessment during equipment development. *International Journal of Quality and Reliability Management*, **16**(9), 892–902.

McCollin, C. (1999) Working around failure. *Manufacturing Engineer*, **78**(1), 37–40.

Morrison, S.J. (1957) Variability in engineering design. *Applied Statistics*, **6**, 133–138.

Morrison, S.J. (2000) Statistical engineering design. *TQM Magazine*, **12**(1), 26–30.

O'Connor, P.D.T. (2000) Commentary: reliability – past, present and future. *IEEE Transactions on Reliability*, **49**(4), 335–341.

Parry-Jones, R. (1999) *Engineering for corporate success in the new millennium*. Speech to the Royal Academy of Engineering, London, 10 May.

Paulk, M., Weber, C.V., Curtis, B. and Chrissis, M.B. (eds) (1995) *The Capability Maturity Model: Guidelines for Improving the Software Process*. Addison-Wesley, Reading, MA.

Reason, J. (1997) *Managing the Risks of Organisational Accidents*. Ashgate, Aldershot.

Reunen, M., Heikkila, J. and Hanninen, S. (1989) On the safety and reliability engineering during conceptual design phase of mechatronic products. In T. Aven (ed.), *Reliability Achievement: The Commercial Incentive*. Elsevier Applied Science, London.

Sankar, N.R. and Bantwal, S.P. (2001) Modified approach for prioritization of failures in a system failure mode and effects analysis. *International Journal of Quality and Reliability Management*, **18**(3), 324–335.

Sexton, C.J., Lewis, S.M. and Please, C.P. (2001) Experiments for derived factors with application to hydraulic gear pumps. *Applied Statistics*, **50**, 155–170.

Silverman, M. (2006) Why HALT cannot produce a meaningful MTBF number and why this should not be a concern. http://www.qualmark.com/hh/01MTBFpaper.htm/

Strutt, J.E. (2000) Design for reliability: A key risk management strategy in product development. Paper presented to the 14th ARTS Conference, Manchester.

Swift, K.G., Raines, M and Booker, J.D. (1999) Designing for reliability: A probabilistic approach. In B.W. Robinson (ed.), *Advances in Safety and Reliability: Proceedings of the SARSS*, pp. 3/1–3/12. Safety and Reliability Society, Manchester.

Thompson, G.A. (2000) Multi-objective approach to design for reliability. Paper presented to the 14th ARTS Conference, Manchester.

Woodall, A., Bendell, A. and McCollin, C. (1998) Results of the Engineering Quality Forum Survey to establish ongoing requirements for education and competency for engineers in the field of quality management. Available from the author.

12.II

Stochastics for the quality movement: An integrated approach to reliability and safety

M F Ramalhoto

12.II.1 Introduction

It is almost 300 years since the appearance in 1713 of Jacob Bernoulli's book *Ars Conjectandi*. Starting from an investigation of games of chance, Bernoulli discovered the new science of uncertainty, which he named *stochastics*. He anticipated that this new science would have an enormous impact on every aspect of human society. After the publication of Bernoulli's book the greatest contemporary scientists, such as Laplace, recognised the importance of the new science and advocated including it in school education at any level. Nevertheless, traditionally, engineering sciences start from the assumption that the world is deterministic by nature. The overwhelming success of technology seemed to confirm this assumption. Thus education in stochastics (here meaning statistics, probability, stochastic processes and all related quantitative issues) is generally very much neglected, even today, in engineering education, and in some engineering sciences. However, nowadays it is common knowledge that natural as well as physical and engineering sciences have reached limits which cannot be crossed using deterministic models.

Indeed, the issue of quality has traditionally opposed determinism. Even the best-trained personnel and the best-designed machines cannot guarantee a completely homogeneous quality. Stochastics plays a crucial role in business and industry since uncertainty is one of the major factors in the real world.

The main objective of a reliability study is to provide information as a basis for decisions. Reliability analysis has a potentially wide range of applications. For instance, reliability analysis is a well-established part of most risk and safety studies. The causal part of a risk analysis is usually accomplished by reliability techniques, such as failure mode and effects analysis and fault tree analysis. Reliability analysis is often applied in risk analysis to evaluate the availability and applicability of safety systems, ranging from single-component safety barriers (such as valves) to complex computer-based process safety.

The concepts of quality and reliability are closed connected. Nowadays, some see reliability also as an extension of quality, for example in the mean proportion of time the item is functioning. Reliability may in some respects be considered a quality characteristic. In some cases perhaps it is the most important characteristic.

12.II.2 Reliability and survival analysis

There is a rich literature on reliability that very often appears linked to that on survival analysis. Reliability and survivability theory have an important common component. They are both concerned with the measurement of length of life, whether it be the life of a mechanism, a system, an animal, or a human being.

There are of course differences. An important problem in reliability theory is the optimisation of system reliability in multi-component systems. Such optimisation depends on the relation between the number and the location of components in the system. No analogous problem exists in survivability theory since the system in this case is living and functioning, which makes the rearrangement of components (organs) an impractical alternative, at least for the time being. On the other hand, a very important problem in survivability theory is the comparison of survival times of patients or experimental animals receiving two or more different treatments.

Similar problems arise in reliability theory when it is necessary to compare life distributions before and after engineering changes. By and large, however, they are not as important as the comparable survivability problems. Many parametric models are used in both reliability and survivability problems, namely, estimation of the requisite parameters for censored and non-censored data from exponential, gamma, Weibull, Rayleigh, inverse Gaussian and lognormal distributions.

In addition to the parametric survival distributions and models, non-parametric procedures are also used – for example, graphs of observed hazard rates, Kaplan–Meyer estimates for censored data, censored two-sample Wilcoxon tests, non-parametric graphical procedures for estimating the cumulative hazard rate, and life-table methods for analysing survival data.

The statistical procedures for censoring of data have been utilised extensively in reliability applications. In most patient survival studies the patients arrive for treatment during the course of their illness; thus the length of follow-up after entry to the study is usually different for each patient. In industry the situation would be analogous if the date of installation of equipment varied for each piece of equipment: after a period of time we could determine the failure time for the items that have failed and the length of time they have run or the survival time for those still surviving.

The Cox regression model is the standard regression tool for survival analysis in most applications. However, often the model provides only a rough summary of the effects of some covariates. It does not allow the covariate to have a time-varying effect. The additive-multiplicative (Cox–Aalen) model can handle the time-varying effect easily. When the aim is to give a detailed description of covariate effects and thus to calculate predicted probabilities, Scheike and Zhang (2002) suggest a flexible extension of the Cox regression model, which aims to extend the Cox model only for those covariates where additional flexibility is needed. For some further developments, see also Scheike and Zhang (2003).

The international journal *Lifetime Data Analysis* is devoted to methods and applications for time-to-event data. Lawless (2003) is also a good reference for the interested reader.

12.II.3 Reliability and data quality and handling

There are several data books and data banks; several computer codes have been developed to aid the analyst using them. Data quality is of course also of great concern. Despite a significant effort devoted to the collection and processing of reliability data, the quality of the data available is still not good enough.

The following are two examples of problems with reliability related to reliability data, brought to my attention by Chris McCollin.

The first is that when failure and repair data are collected in the field, the information is fed back to Accounts so that the repair engineer is paid, back to Stores so that replacement parts can be ordered, and then back to the customer from Accounts so that the customer can be charged for the work. However, the information is not typically fed back to Design so that systematic problems can be highlighted. Computer networks exacerbate the problem where a direct link between the Maintenance and Design functions does not exist and, if it does, it goes through Accounts and gets lost.

The second problem arises when Marketing requests information from Warranty databases, for example for presentations of new generation products and how previous products fared for reliability. Since Marketing does not usually get involved with the design of databases (even though they are asked) then no attempt can be made to answer their questions because the information that they require has never be collected. Thus work is required for both problems on how and what reliability information feeds to and from all areas of a company so that adequate planning can go into the next-generation failure databases in the light of future network design and implementation.

We draw the reader's attention to the practicability of reliability engineering, and the danger that poor-data quality as well as unreflective data analysis will lead to misleading conclusions. For further data handling problems and the identification of its challenges to the statistical community, see Ramalhoto and Goeb (2005, 2006) and the references therein.

12.II.4 Reliability and maintenance analysis and management

Maintenance is carried out to prevent system failures as well as to restore the system function when a failure has occurred. One of the prime objectives of maintenance is thus to maintain or improve the system reliability and production/operation regularity. A source of research papers with interesting models and methods, including stochastic Petri nets modelling among several others, is the annual proceedings of the Reliability and Maintainability Symposium.

12.II.4.1 Maintenance and imperfect repair models

A device is repaired at failure. Sometimes the repair is 'perfect', and the item is returned to a good-as-new state. At other times the item is only 'minimally repaired', that is, the item is brought back to a working condition, and it is as good then as it was just before it failed.

Brown and Proschan (1983) introduced the imperfect repair model to model such a situation in which an item, upon failure, is replaced with a new one with probability β, and is minimally repaired with probability $1 - \beta$. There are several imperfect repair models and extensions thereof in the current academic literature. Some of them equip the imperfect repair model with preventive maintenance, and give maintenance comparisons for the number of failures under different policies via a point-process approach.

There are two types of preventive maintenance policies that are widely used in reliability modelling: age replacement, where a component is replaced at failure (unplanned replacement) or is replaced when its age reaches some fixed value (planned replacement); and block replacement, where the component is replaced at failure and also at times according to the block replacement schedule.

Incorporating age or block preventive maintenance into imperfect repair models gives two general repair models: imperfect repair with age replacement, in which a component is imperfectly repaired at failure and is also replaced when its virtual age reaches some fixed value T; and the imperfect repair with block replacement, where the component is imperfectly repaired at failure and is also replaced at fixed times of a block replacement schedule. Authors are interested in finding optimal block or age intervals which minimise the long-run expected cost per unit time, and in comparisons of the effectiveness of these various maintenance policies.

To compare the effectiveness of the various maintenance policies, stochastic comparisons of the number of unplanned replacements under age, block and other policies have been studied in the literature. Thousands of maintenance models have been published, and several variations of replacement policies have been introduced and studied. Baxter *et al.* (1996) use a point-process model for the reliability of a maintained system subjected to general repair. Studies on maintenance for multi-component systems with and without economic dependence have also been studied by several authors.

It also has been shown that it is often better to have simultaneous replacements of components than replacing each component independently of the others. And preventive maintenance only makes sense for components whose lifetimes are better new than used.

The main goals of maintenance policies and strategies are to reduce failure rate and to maximise system availability. Clearly it is important to minimise the costs relating to maintenance, and the resources of maintenance, and to maximise the quality of maintenance. It should be stressed that preventive maintenance is not always effective. Often external corrective maintenance is needed, which might lead to extra problems and longer periods of unavailability.

However, it has been shown, for instance in Linderman *et al.* (2005), that a combined statistical process control (SPC) and preventive maintenance can reduce the cost of operation significantly more than either policy in isolation

Maintenance might strongly affect reliability. Industries such as nuclear power, aviation, offshore industries and shipping have realised the important connection between maintenance and reliability and implemented reliability-centred maintenance methodology. This methodology aims to improve the cost-effectiveness and control of maintenance and hence to improve availability and safety. Reliability-centred maintenance deserves a lot of attention from the academic research community.

Many industries have also integrated a reliability programme into the design process. Such integration may be accomplished through the concept of concurrent engineering which focuses on the total product perspective from inception through to product delivery.

12.II.4.2 Maintenance: a key component in productivity and competitiveness

For many quality movement analysts maintenance is a key component in enhancing industrial productivity and competitiveness. It is largely unrecognised that a major contribution to the continued economic success of Western economies will be the productivity gains available through efficient equipment reliability and maintenance planning.

Many authors who offer strategies for improving industrial productivity frequently overlook the fact that equipment downtime is substantial and can be significantly reduced. In many organisations, the maintenance function is viewed as not adding value to the product. Therefore, for them the best maintenance is the least-cost maintenance. The result of this cost cutting is often increasing cost and reduced productivity.

12.II.4.3 An example of a reliability, quality and maintenance programme

Park (1999) presents a reliability management system consisting of a reliability policy, an equipment maintenance programme, optimisation of equipment reliability, continuous effort for reliability improvement, small-group activity to enhance reliability, reliability education, information systems for reliability, and company organisation for reliability. The paper discusses how to construct such a reliability management system, how to operate it efficiently, and how to set up a typical education programme for it. It also explains its relationship to SPC and quality engineering.

12.II.5 The Bayesian approach

The Bayesian approach to reliability, maintenance and quality control is highly relevant, with its view that the ultimate purpose of reliability analysis is to take decisions relating to engineering systems in the face of uncertainty – that is, any system of concern is prone to failure no matter which one of the conceivable design solutions is implemented, and system failure (success) will cause a loss (reward) to be incurred (cashed). The aim of the book by Barlow *et al.* (1993) is to supply the reliability community with a coherent Bayesian predictive view on reliability theory and practice (see especially Section 2–4 of that book).

Gregory (2005) provides a clear exposition of the underlying concepts and a large number of worked examples. Background material is provided in appendices and supporting Mathematica notebooks are available.

12.II.6 Safety

The safety issue is a very broad and often very complex subject. However, safety might be substantially improved in many situations if we understood better the stochastic aspects of reliability, maintenance, control, and its interactions usually present in the equipment's failure that caused the lack of safety. It is advisable to support and supplement the development of methods taking an analytical rather than judgmental attitude to failure and safety, and also to set up systems for process control and accident mitigation, as well as for maintenance.

As George Box pointed out, no model is absolutely correct. In particular situations, however, some models are more useful than others. In reliability and safety studies of technical systems one always has to work with models of the systems. A mathematical model is necessary to enable one to bring in data and use mathematical and statistical methods to estimate reliability, safety, or risk parameters. For such models two conflicting interests always apply: on the one hand, the model should be sufficiently simple to be handled by available mathematical and statistical methods; on the other hand, it should be sufficiently 'realistic' that the results of applying it are of practical relevance.

Thus, it is occasionally said that the use of 'probabilistic reliability analysis' has only bounded validity and is of little practical use. And when something goes wrong, it is usually attributed to human error. But even when human error *is* to blame, a different and more constructive approach could be considered. In principle, there is nothing to prevent key persons from counting as 'components' of a system in the way that technical components do (numerical estimates for the probability of human errors in a given situation would be difficult to derive, but that is another issue).

Under the auspices of the European Union a number of new directives have been issued. Among these are the machine safety directive, the product safety directive, major hazards directive and the product liability directive. The producers of equipment must, according to these directives, verify that their equipment complies with the requirements. Reliability analysis and reliability demonstrative testing will be necessary tools in the verification process. In the European Round Table on Maintenance Management, held in Lisbon on 15–18 June 1997 and organised by Ramalhoto, it was reported that in the USA court consultancy on these matters is becoming a booming business.

The growing dependence of working environments on complex technology has created many challenges and led to a large number of accidents. Although the quality of organisation and management within the work environment plays an important role in these accidents, the significance of individual human action (as a direct cause and as a mitigating factor) is undeniable. This has created a need for modern, integrated approaches to accident analysis and risk assessment. For instance, Hollnagel (1998) presents an error taxonomy which integrates individual, technological and organisational factors based on cognitive engineering principles. It provides the necessary theoretical foundation and a step-by-step description of how the taxonomy can be applied to analyse and predict performance using a context-dependent cognitive method. In particular it shows how to identify tasks that require human cognition, how to determine conditions where cognitive reliability and ensuing risk may be reduced, and how to provide an appraisal of the consequences of human performance on system safety which can be used in probability safety assessment.

12.II.7 Software reliability

When something goes wrong, it is usually attributed to human error. However, when we look more closely at the human factors we realise that some accidents are caused by problems with equipment, including software as in the Chinook 1994 crash case, or even the lack of the required education or training.

The Chinook 1994 crash case illustrates a situation where the software reliability issue seems to have been crucial. According to Burns and Halligan (1997), Tony Collins, executive editor of *Computer Weekly* and an aviation software expert, said that evidence given in the Chinook 1994 crash case raises serious concerns about the quality of the software, and that he personally would have been worried about flying in a Chinook with these concerns. Burns and Halligan show, among other things, the increasing importance of software reliability and software maintenance in some crucial equipment and products, as well as the importance of reliability research into questions of human–machine interaction.

Software reliability has been an important issue in the software industry since it can provide critical information for developers and testing staff during the testing/debugging phase. As an essential issue in the software development process, software reliability must balance the requirements between user satisfaction and software testing costs. In order to manage the cost in the software testing/debugging phase the software testing staff have to understand the variation of software reliability and testing costs at any time in the process. Therefore, a software development process concerned with reliability, cost, and release time should be closely monitored.

By studying the repercussions, testing and debugging costs in dealing with software faults investigation can be employed to develop software reliability models to predict the behaviour of failure occurrences and the fault content of a software product. For the last two decades, several stochastic models have been proposed to assess software reliability. Most of these models for describing the software failure phenomenon are based on the non-homogeneous Poisson process. Although these models are fairly effective in describing the error-detection process with a time-dependent error-detection rate, they assume perfect debugging (that each time an error occurs the fault that caused it can be immediately removed). Recent developments can be seen in Bai *et al.* (2005), Huang (2005), Jeske and Zhang (2005) and Zhang and Pham (2006). Kenett and Baker (1999) is also a useful reference for the interested reader.

Madsen *et al.* (2006) argue that fuzzy logic can address and capture uncertainties in software reliability, and describe some fuzzy approaches to model uncertainties during software testing and debugging connected to detectability, risk and software reliability. However, to apply their fuzzy logic approach adequate databases are required and processed based on inferential procedures.

For further information, the reader is referred to the resources available on the internet and in international standards. The Goddard Space Flight Center Software Assurance Technology Center has some interesting web pages, as does the NASA Office of Safety and Mission Assurance (http://satc.gsfc.nasa.gov/). The websites of computer-aided software engineering centres and software suppliers are also often useful to refer to, for example:

- http://www.qucis,queensu.ca/Software-Engineering/vendor.html
- http://www.osiris.sunderland.ac.uk/sst/casehome.html
- http://www.cat.syr.edu
- City University Centre for Software Reliability, http://www.csr.city.ac.uk
- Software Engineering Institute, http://www.sei.cmu.edu.

The following standards are particularly relevant:

- ANSI/IEEE Std. 729-1983 (1983). *IEEE Standard Glossary of Software Engineering Terminology*. IEEE, New York.
- ANSI/IEEE Standard 1045-1992 (1993). *Standard for Software Productivity Metrics*. IEEE Standards Office.
- ANSI/IEEE 982.1. *Standard Dictionary of Measures to Produce Reliable Software*. IEEE Standards Office.
- IEEE Standard 1044-1993 (1993). *Software Quality Measures*. IEEE Standards Office.
- ANSI/IEEE Std 1008-1987 (1993). *IEEE Standards for Software Unit Testing*. IEEE, New York.
- ISO/IEC Standard 9126 (1992). *Software Product Evaluation: Quality Characteristics and Guidelines for Their Use*. International Organisation for Standardisation, Geneva.

According to the last standard in the above list, ISO/IEC 9126, reliability relates to the frequency and severity of program errors and recovery attributes. The question is, how much can the user rely on the program results?

12.II.8 Signature analysis and preventive maintenance

It is highly advantageous if future wear and failure can be detected before occurring. This enables preventive action such as maintenance to be carried out, with resulting cost savings and the avoidance of possibly dangerous failure.

The detection of faults by monitoring a dynamic characteristic such as noise, vibration or electrical output is known as signature analysis. Ideally every failure mode will have its own special signature. Signal processing methods are typically used, such as time frequency plots using wavelet analysis.

Di Bucchianico *et al.* (2004) use signature analysis for reuse prediction of parts in the context of a modern product life cycle that includes recycling.

12.II.9 Academic research and industry needs

The author has been setting up joint international and European conferences, workshops, round tables and publications, concerned with research and advanced engineering and management education, since 1992. From these activities the following lists of industry needs and drawbacks, as well as of academic fears, have emerged. In this section the word 'industry' should be taken to include business as well as the public sector and government.

We begin with industry needs:

- Most industry analysts claim that the only enterprises that will survive in the new millennium will be the ones that have the flexibility to tailor production to fluctuating demand.
- A great deal of waste of resources of all kinds, including raw materials, energy, as well as human time and effort, can be observed in the way a product (or a service) is designed, made and sold.
- Globalization and competitiveness demand that processes and resources of all kinds should be aligned with and responsive to customer fluctuations; that equipment be ready to produce without failure; and that every product be of the highest quality at the lowest price.
- Society rightly demands that everything be planned under the basic principle that machines, software, technologies (namely those based on genetic manipulation, chemicals and the like), must serve human and ecological needs and not the other way around.
- Essentially industry needs to produce 'good quality', reliable and safe products, at competitive prices, in a 'reasonable' amount of time, well adjusted to fluctuating demands, in a human- and environment-friendly manner, and, when applicable, with good after-sale maintenance. An extra undertaking to surprise and delight the customer is of course an advantage.

There are a number of drawbacks hindering the fulfilment of these needs:

- Sharing data is a taboo in many important industries. Companies often do have the data but they are with different people in different functions and the information is locked away.
- Data are neither correctly collected nor sufficiently understood.
- Industrialists are usually too busy fire-fighting in their plants.
- Many enterprises spend €10–15 million a year on maintenance in the plant that is not conditioned on anything and no research is done on its effectiveness. However, almost any division of these companies will spend €10 000 on advertising promotion and then about 15 % more researching its effectiveness.

- Technology is not necessarily enough in all cases. At least at present there is no miraculous software or hardware solution that is ready to use wherever or whenever it is needed.

- The consultant firm is usually not able to provide ready-made solutions that can be implemented without a lot of effort on the part of the client organisation and its staff. Very often, they have to carry out all the really important and definitive activities to implement the recommended solution. Usually the client organisation and their staff have not jointly created that solution nor even have any feeling for it. That very often leads to the consultancy work being a disaster or at least a massive waste of time and money.

Turning to academic research, it is recognised worldwide that in almost all knowledge fields:

- On average, it takes more than 30 years for a solution to a problem in one field to appear in another different field (clearly that must improve).

- Regarding the value system for academic journals and conferences, only academics seem to be preaching it.

- The biggest reward for a researcher is to see his/her intellectual creation (directly or indirectly) implemented for the 'good' of all.

- Addressing the needs of industry leads to big pay-offs: it brings that knowledge into the industry (if we give them the higher knowledge, not only the basics), and brings people back with an interest in utilizing resources.

- A lot of important targeted research is now being done outside the universities, behind closed doors. This is a serious threat to universities as one of the most important modern institutions that have been shaping our civilization for many centuries. It is our duty as members of that great institution not to allow universities to diminish in importance or to die altogether.

- Universities really have been an interesting globalization model that industry should perhaps better understand and benefit from.

As remarked by Brown and Duguid (2000): 'Become a member of a community, engage in its practices, and you can acquire and make use of its knowledge and information. Remain an outsider, and these will remain indigestible.' Perhaps everything would be different if academics spent sabbaticals in industry and if practitioners were encouraged to do academic research for specific periods of time, at the company's convenience. I wonder if future EU Frameworks could consider this suggestion.

The good news is:

- The quality movement in Europe has forced companies to think about how they put questions to their suppliers. For instance, they now write availability demands into most contracts and not demands about mean time to failure.
- In Europe, there are also highly relevant examples of important industries, such as the aircraft and automotive industries, that have good reliability and maintenance management programmes.

But there is also some not so good news:

- Getting in touch with small and medium size companies remains a problem – they are difficult to reach.
- Some areas are very backward and others are high-tech, and some of the latter come with their problems already cooked.

With some exceptions, communications between industry and academia have been very difficult indeed – different cultures, and even different visions and missions. Now, and perhaps much more so in the future – due to, for instance, higher productivity demands and globalization, competitiveness 'quality movement' gurus, and recent and future developments in telecommunications and information technology – industry and academia visions and missions might grow closer than ever.

We have to have the right mix of industry and academics trading ideas with each other. Whether this is a dream or a nightmare future, the only thing we are sure about is that it poses new research challenges to the existing academic theories and methodologies of reliability, maintenance, quality control, risk and safety, and its related fields.

Joint assessment is needed:

- What are the techniques currently being used?
- What further existing knowledge is available and needs to be translated for engineers and explained to managers?
- What are the applied research priorities for specific industries and business?

In any organisation, be it industry, business, academia or any other, there are always too many signals, concerns and wish lists. Prioritization and focus are extremely useful. The stochastics for the quality movement (SQM) concept addresses these issues in the context of reliability, quality and maintenance for managing risk and safety.

12.II.10 The SQM concept

The quality movement has many approaches and gurus. Here the following are considered: six sigma, quality, excellence, total quality management, lean production, concurrent engineering, re-engineering, supply chain management, learning organisation, knowledge management, benchmarking, balanced scoreboard, radical innovation, business intelligence, and all the other buzzwords, as part of the so-called quality movement, as well as value analysis, value engineering and value management.

An 'extended quality movement' is needed, as technology is not enough. Organisation members must weave new knowledge into the fabric of the company's formal and informal systems, such as the way they develop and implement products and business strategies or the way they organise themselves to accomplish tasks. A collaborative innovation of new work settings that are ecologically sound and economically sustainable and that bring out the best in human beings is required, as well as helping people to get better at working and planning together. It is important to focus people's interactions on what is working well or on kernels of possibilities, as opposed to only lists of things that have gone wrong.

It goes almost without saying that the most important part of a company is the place where manufacturing activities are conducted (in a factory) as well as the place where employees have direct contact with customers (in the service sector). Managers focus too much on results and not on the process of what achieves or delivers those results. Cost reduction is not synonymous with cost cutting – cost reduction should come as a result of better cost management.

Some senior managers already realise that they also have to achieve very high availability, reliable machines and maintenance under control, and this has to be done in an integrated manner. Maintenance is a domain in which the opportunities to generate wealth are now being seen to be enormous.

For stochastics to serve the 'extended quality movement' better, a more trans-disciplinary/integrated/merging approach is required. More than ever before, this demands accountability, interaction and joint research work among all players (statisticians, engineers, computer science experts), and the creation of open channels of communication among them and with the industry/business senior managers and the workforce, and in some cases including the government/public sector staff.

Indeed, the quality movement is fragmented at least three ways, First, quality-related endeavours in academia and industry have few relationships with respect to organisations, personnel, and exchange of information. Second, in academia, quality is a subject within several disciplines such as management science, engineering, and statistics. Finally, in industrial companies, quality control and quality assurance are local phenomena in departments such as R&D, manufacturing, sales, and maintenance, often without adequate links and without an integrating global framework. The SQM concept is an approach to overcome fragmentation in the quality movement (in the context of stochastics) and to integrate particular local endeavours into a global scheme.

12.II.10.1 Vision and mission of SQM

SQM is conceived as a cross-fertilization of the traditional disciplines – SPC, reliability and maintenance, inventory, supply chain – that might also have some features to be rediscovered and some interactions to be studied, as well as other relevant disciplines such as Petri and neural networks, pattern recognition, data mining, Bayesian networks, expert systems, fuzzy logic, to name just a few. The idea behind the SQM is to bridge all these research areas and, when applicable, bring about useful interactions among them (taking into account economic, as well as ecological and human factors and demands, whenever possible). This encourages us to focus on these issues as a whole, rather than in isolation as is usually the case in academic research.

The mission of SQM is to come up with more realistic models and better solutions for (or ways to cope with) real-life problems concerning safety, reliability and maintenance management issues, under an economic as well as ecological and human well-being viewpoint, as waste-free as possible, and to avoid problems proactively rather than reactively.

Essentially, SQM is supposed to provide industrial applications for the solutions already found by university researchers, and to promote university research into industry's tougher unsolved problems. That is to say, we have to jointly assess the techniques currently being used in industry, what further existing knowledge is available and needs to be translated for engineers and explained to managers, and what are the research priorities in SQM for the different areas of industry considered.

12.II.10.2 Benefits of SQM

The benefits of SQM come mainly from its mission, its transdisciplinary nature, its holistic corporate viewpoint, and its potential to destroy the existing self-imposed boundaries among the professionals of the different research areas involved, by imposing instead a joint commitment to the SQM research area mission.

However, in any real-life problem we know that on average solutions will be 20 % methodology, 10 % common sense and 70 % engineering. SQM is a comprehensive way of thinking and acting which combines information, opinion and common sense to achieve better products, systems, strategies and decision-making, where the accent also has to be on communication, teamwork and problem-solving (not forgetting, in accordance with the ecological and human needs principle, with as little waste as possible).

SQM will lead us to uncover the important problems, which we can then structure and put into an academic form before giving them the necessary intellectual effort. Continuous feedback from 'interested companies' will do the rest.

Part of the methodology behind SQM has been inspired by the medical experience. To learn from the medical experience means studying specific cases over and over again until some kind of library is built up. Then, once a problem is recognised to be in a certain class, we go to the library and see what to do. If the problem is not in the library then we do research on it with industry feedback, and once the problem is understood and solved, or coped with, it will be placed in the library, and the whole process restarts.

The SQM concept was first introduced in Ramalhoto (1999), and further discussed and developed within the Pro-ENBIS project; see, for example, Kenett *et al.* (2003), Ramalhoto *et al.* (2003, 2004), Di Bucchianico *et al.* (2004), Ramalhoto and Goeb (2005, 2006a, 2006b) and Goeb *et al.* (2006).

Kenett *et al.* (2003) describe a framework for an SQM library that uses profiles and pattern models and presents a database in the context of an e-library organised to provide diagnostic tools and solutions for SQM problems. Ramalhoto *et al.* (2003) present a scheme for industry–academia interaction to enhance research and development issues, through the so-called ‘industrial visit model’, described in terms of quality improvement principles.

Another very important activity of SQM is to facilitate and do research on novel transdisciplinary/integrated issues, usually not much discussed in the current literature, but of recognised importance for industry. Ramalhoto *et al.* (2004) and Goeb *et al.* (2006) established a framework for process maintenance that integrates SPC and reliability analysis. Ramalhoto and Goeb (2005, 2006b) discuss further problems resulting from the present fragmented situation and suggest methods of solution: the idea of an SQM library as an effective contribution to distribution of expert knowledge among all agents in the quality movement; the situation of industrial data management as an obstacle as well as a challenge for SQM research; discussion of variation as one of the central topics of SQM research.

Ramalhoto and Goeb (2006a) discuss industrial aspects of data quality in terms of the SQM concept. They introduce and discuss ten industrial aspects of data quality, review the most relevant aspects of reliability data and their links to warranty data, concluding with some guidelines for the construction of a holistic data quality framework.

12.II.11 Some challenges for SQM in the maritime industries and sciences

From our investigation of the marine industry we learned the following:

(a) The US Navy has been requiring quantitative reliability, availability and maintainability (RAM) assessments in ship design, operation and regulation for more than thirty years.

(b) Although around the world the commercial marine industry has been reluctant to adopt these techniques, fierce competition in the international shipping industry is forcing ship owners/operators to make efforts to systematically optimise reliability, safety and cost-effectiveness.

(c) New trends in the merchant marine industry demand basic knowledge in reliability, maintainability and risk assessment for successful implementation and safety.

12.II.11.1 Examples of research opportunities in SQM

1. The European Maritime Safety Agency is based in Lisbon. Its objectives are to guarantee a uniform and efficient high level in maritime safety and pollution prevention from ships. It is supposed to provide EU member states and the Commission the necessary technical and scientific support and a high level of specialization, in order to assist them in the correct application of EU legislation in those domains, as well as in the control of its application and in the evaluation of the efficiency of the present rules.

There has existed since 2002 an EU Directive on Maritime Traffic Safety that obliges each country to install an automatic identification system to follow the maritime traffic (including vessels' maintenance conditions, load levels, sea and coastal pollution and ecological threats) in European waters, much as has long been in place with aircraft. However, its implementation inside EU waters is still far from satisfactory, and needs urgent efficient action.

Many ENBIS/Pro-ENBIS members have know-how in air and land transportation issues and related matters that could be transferred to the sea matters.

Furthermore the Agency will coordinate:

- the fight against marine pollution;
- the protection/security of maritime transportation and ports;
- the accreditation of competence certificates of the maritime staff of third countries.

There are several industries and activities linked to maritime issues that should be looked into closely, as they are very promising areas that have stagnated for some time and there is a lot to be done in terms of application of SQM methodology. Some examples are presented below.

2. On the basis of current trends, international trade may grow to the equivalent of 30 % of world output by 2010 from its current level of around 15 %. It is recognised that the ports play a very crucial role in this international trade. Europe has some of the best high-tech ports in the world. The port of Sines (close to the ports of Lisbon and Setúbal) has been recognised as one of the best natural seaports in the world. Therefore, to build a network of trans-European seaports with adequate partnerships with the most important African, American and Asian seaports makes a lot of sense, and might be a challenge for all European member states. It also represents a great source of new jobs and opportunities for cooperation between European universities and industry. Efficient network management and logistics of such a trans-European network platform needs SQM, and the pro-ENBIS group know-how and ongoing research could be its seed.

3. Europe has one of the world's largest and most competent naval companies for the specialised repair and maintenance of large ships, located in several countries. It also builds ships specially tailored for the transportation of specific items, such as chemical materials and ferries. For example, in 2001 the Portuguese company Lisnave received the prestigious *International Ship-Repair News* annual award for its performance, excellence and innovation in naval repair. Linked to that there are more than 100 SMEs, operating with several material technologies and dedicated to several market types. Clearly, the SQM approach to all those enterprises makes a lot of sense.

4. Due to globalization and oil-supply problems, transportation by sea is more than ever crucial to world commerce. There is a lot of SQM knowledge transfer in land and air transportation sectors that should be properly adapted to sea transportation. Sea transportation is also a huge service industry in itself.

5. Another important issue is to manage all the integrated transportation (sea, land, air) and reduce costs and pollution by means of a well-designed network of all the transport and facilities needed to become available, and be properly maintained and innovated. The marketing and all the customer technology of the service quality industry has an important application here too. The SQM approach has the potential to make a difference of millions of euros.

6. SQM knowledge in aerospace and automotive industries, railways, land and air transportation and linked issues (cars, trains and aircrafts repair and maintenance) by most of the ENBIS/Pro-ENBIS members could enable a comparative study in terms of common problems and solutions and how they relate to the corresponding sea issues. This in itself would be an interesting piece of work as part of the SQM library.

12.II.11.2 The maritime section of the SQM library

The aim is to build up a large number of case histories, generalise them and put them in the library. Once somebody has a problem that can be identified as being in a certain class, then it is known that such and such a procedure will be done. Up to a point, what is advocated here is an empirical approach, that is, a case-oriented knowledge base, or case-based reasoning, after the artificial intelligence (AI) paradigm.

It is becoming possible for AI to solve real-life problems through advances in hardware (that is much higher speed processors, huge memory and band capacity) and software (probabilistic models, neural networks, and so on, that is SQM methodology).

If a problem is not available in the SQM library then it will need to be the subject of research by the SQM Library Think Tank Group, who will pick the right researchers and the right practitioners to find its solution. Once that is achieved and successfully implemented it will be part of the SQM Library, in its section targeted to maritime issues, and a kind of press release will be issued across the whole SQM library in order to quickly transfer knowledge from maritime to any other area on the SQM library.

12.II.11.2.1 The European contribution to the quality movement

The European contribution to the world quality movement, as far as I know, has so far been very modest. The SQM concept clearly has a potential to break that tradition if taken seriously by the EC, at least in its next Framework.

12.II.12 Acknowledgement

I would like to thank all the wonderful and hardworking people in the Pro-ENBIS project group, as well as the European Commission for funding my research.

References

Bai, C.G., Hu, Q.P., Xie, M. and Ng, S.H. (2005) Software failure prediction based on a Markov Bayesian network model. *Journal of Systems and Software*, **74**, 275–282.

Barlow, R.E., Clarotti, C.A. and Spizzichino, F. (eds) (1993) *Reliability and Decision Making*. Chapman & Hall, Lonfon.

Baxter, L.A., Kijima, M. and Tortorella, M.(1996) A point process model for the reliability of a maintained system subjected to general repair. *Stochastic Models*, **12**, 37–65.

Bernoulli, J. (1713) *Ars Conjectandi*. Basle.

Brown, J.S. and Duguid, P. (2000) *The Social Life of Information*. Harvard Business School Press, Boston.

Brown, M. and Proschan, F. (1983) Imperfect repair. *Journal of Applied Probability*, **20**, 851–859.

Burns, J. and Halligan, L. (1997) Chinook software trouble admitted. *Financial Times*, 20 November, p. 11.

Di Bucchianico, A., Figarella, T., Hulsken, G., Jansen, M.H. and Wynn, H.P. (2004) A multi-scale approach to functional signature analysis for product end-of-life management. *Quality Reliability Engineering International*, **20**, 457–467 .

Goeb, R., Ramalhoto, M.F. and Pievatolo, A. (2006) Variable sampling intervals in Shewhart charts based on stochastic failure time modelling. *Qualitative Technology and Quantitative Management*, **3**(3), 361–381.

Gregory, P.C. (2005) *Bayesian Logical Data Analysis for the Physical Sciences.* Cambridge University Press, Cambridge.

Hollnagel, E. (1998) *Cognitive Reliability and Error Analysis Method*. Elsevier, Oxford.

Huang, C.-Y. (2005) Performance analysis of software reliability growth models with testing-effort and change-point. *Journal of Systems and Software*, **76**, 181–194.

Jeske, D.R. and Zhang, X. (2005) Some successful approaches to software reliability modelling in industry. *Journal of Systems and Software*, **74**, 85–99.

Kenett, R. and Baker, R (1999). *Software Process Quality: Management and Control*. Marcel Dekker, New York.

Kenett R., Ramalhoto M.F., Shade J. (2003) A proposal for management knowledge of stochastics in the quality movement. In T. Bedford and P.H.A.J.M. van Gelder (eds), *Safety and Reliability*, Vol. 1, pp. 881–888. A.A. Balkema, Lisse, Netherlands.

Lawless, J.F. (2003) *Statistical Models and Methods for Lifetime Data*, 2nd edition. Hoboken, NJ.

Linderman, K., McKone-Sweet, K. and Anderson, J. (2005). An integrated system approach to process control and maintenance. *European Journal of Operations Research*, **164**, 324–340.

Madsen, H., Thyregod, P., Burtschy, B., Albenu, G. and Popentiu, F. (2006) A fuzzy logic approach to software testing and debugging. In C. Guedes Soares and E. Zio (eds), *Safety and Reliability for Managing Risk*, Vol. 2, pp. 1435–1442. Taylor & Francis, London.

Park, H.S. (1999) Reliability management and education industries. *European Journal of Engineering Education*, **24**(3), 291–297.

Ramalhoto M.F. (1999) A way of addressing some of the new challenges of quality management. In I. Schueller and P. Kafka (eds) *Safety and Reliability*, Vol. 2, pp. 1077–1082. A.A. Balkema, Rotterdam.

Ramalhoto, M.F. and Goeb, R. (2005) An innovative strategy to put integrated maintenance, reliability and quality improvement concepts into action. In K. Kolowrocki (ed.), *Safety and Reliability*, Vol. 2, pp. 1655–1660. A.A. Balkema, Rotterdam.

Ramalhoto, M.F. and Goeb, R. (2006a) Industrial aspects of data quality. In C. Guedes Soares and E. Zio (eds), *Safety and Reliability for Managing Risk*, Vol. 2, pp. 949–955. Taylor & Francis, London..

Ramalhoto, M.F. and Goeb, R. (2006b) An innovative strategy to put integrated maintenance, reliability and quality improvement concepts into action. *International Journal of Materials & Structural Reliability*, **4**(2), 207–223.

Ramalhoto, M.F., Goeb, R., Pievatolo, A., Øystein, E. and Salmikukka, J. (2003) A scheme for industry-academia interaction to enhance research and development issues. In T. Bedford and P.H.A.J.M. van Gelder (eds), *Safety and Reliability*, Vol. 2, pp. 1289–1294. A.A. Balkema, Lisse, Netherlands.

Ramalhoto, M.F., Goeb, R., Pievatolo, A., Øystein, E and McCollin C. (2004) Statistical process control procedures in relation with reliability engineering. In C. Spitzer, U. Schmocker and V.N. Dang (eds), *Probabilistic Safety Assessment and Management*, Vol. 5, pp. 3048–3059. Springer-Verlag, London.

Scheike, T.H. and Zang, M.-J. (2002) An additive-multiplicative Cox–Aalen regression model. *Scandinavian Journal of Statistics*, **29**, 75–88.

Scheike, T.H. and Zhang, M.-J. (2003) Extensions and applications of the Cox–Aalen survival model. *Biometrics*, **59**, 1036–1045.

Zhang, X. and Pham, H. (2006) Software field failure rate prediction before software deployment. *Journal of Systems and Software*, **79**, 291–300.

13

Multivariate and multiscale data analysis

Marco P Seabra dos Reis and Pedro M Saraiva

13.1 Introduction

The last four decades have witnessed a considerable increase in the complexity of processes along with the development of new, more reliable and less expensive measurement systems and a proliferation of computational hardware, with very appealing performance/cost ratios. These factors have led to new demands for improved methods for process/product design and optimisation, process monitoring and control, in a new context in which competition in the global market is very tight and the issues of product quality, safety operation and environmental awareness are key factors for success. There has thus been significant pressure on both academia and industry to come up with improved methodologies and new solutions to cope with the problems raised by this new reality. In particular, the increasing number of variables acquired and the higher sampling rates used soon lead to databases of considerable size, bringing together huge quantities of records originating from several sources within the system (such as the industrial processing unit) and its 'surroundings' (such as market trends and raw materials data). The very notion of what constitutes a 'large' data set has itself evolved at a rate of an order of magnitude per decade: in the 1970s a data set was considered 'large' if it had more than 20 variables, whereas nowadays it has to contain in excess of 100 000–1 000 000 variables (Wold *et al.*, 2002). Therefore, techniques tailored to handle problems raised by the high dimensionality of data sets as well as the extensive use of computer power play a central role in the toolboxes and activities

Statistical Practice in Business and Industry Edited by S.Y. Coleman, T. Greenfield,
D.J. Stewardson and D.C. Montgomery

of today's data analysts (statisticians, engineers), and terms such as 'multivariate analysis' are giving way to 'megavariate analysis' (Eriksson *et al.*, 2001).

However, the number of variables is by no means the only relevant factor in characterising data sets. Another source of complexity may also be present in the time dimension, as it often happens that the dynamic behaviour of variables collected in time is the cumulative result of events occurring at different locations in the process, characterised by different operating time-scales and leaving their own typical time-frequency signatures in the time series. Therefore, the ability to analyse what is occurring simultaneously at different scales may constitute a decisive development in the analysis of variables that exhibit so-called multiscale phenomena, relative to current single-scale approaches (Bakshi, 1999).

In this chapter we will focus on the methodologies available to cope with the multi- or megavariate nature of data sets and with the multiscale features that may be present. Just as in the natural sciences, where the aim is to extract regularities present in observed phenomena and to put them in some adequate form for future use (typically a mathematical formulation), the goal of the present chapter is to present adequate statistical methodologies for extracting regularities hidden in data tables, be they across the different variables present (covariance), or among the values collected at different times (dynamics or autocorrelation), or even both. Several approaches and examples will be presented regarding how their desired integration can be achieved in order to obtain improved results in practical application contexts.

In Section 13.2 we describe a general framework widely used for handling the multivariate nature of large data sets, namely that built upon a latent variable model, and show why it is so often appropriate in practice. We go on to mention several methodologies for estimating representatives from such classes of models and their main features. Several examples illustrate their usefulness in real-life contexts – multivariate calibration, development of soft sensors, experimental design, process monitoring, among others. In Section 13.3 we focus on the multiscale description of data, with special emphasis on their major supporting tool, the wavelet transform, and its properties. Several applications will also underline its utility in different tasks, such as signal denoising and compression or system identification. Furthermore, methodologies where multivariate and multiscale techniques appear integrated in order to handle relevant problems (like process monitoring and regression modelling) are also addressed. We close with some final conclusions and comments on current trends in these fields.

13.2 Multivariate and megavariate data analysis

When performing exploratory data analysis (EDA), one may check the correlation matrix and scatter plot matrix or, if one happens to have some subject-matter knowledge about the process underlying the generation of data, look more carefully into the meanings of variables and the relationships between them. Either way will quite often lead to the conclusion that there are redundancies among different groups of variables. This basically means that the dimensionality of data sets (the number of variables) is larger (usually much larger) than the true dimensionality of the underlying process, that is, the number of independent sources of variability that structure the overall dispersion of values observed. These independent sources of variability in industrial data are usually related to raw material variability, several process disturbances and other perturbations that might be introduced through other means, such as operator interventions, being typically of the order of magnitude of a dozen. The observed covariance between variables may have different origins:

(1) dependencies caused by the nature of underlying phenomena, where variables are constrained by the relevant conservation laws (mass and energy);

(2) control loops;

(3) use of redundant instrumentation;

(4) the nature of the measuring devices employed. For instance, if the measurement device is a spectrometer, the variables will be something like the 'absorbance', 'transmittance' or 'reflectance' at a set of frequencies or wavelengths. In such a situation, the variables have a natural ordering (frequency or wavelength), and the correlation arises from the natural relationships existing between, for example, the transmittance at various wavelengths and the components of the mixtures analysed.

In such cases, it is useful to consider the process as being driven by p non-observable *latent variables*, t, with the $m(m \geq p)$ measured variables, x, being the visible 'outer' part of such an 'inner' set of variation sources. These are related by

$$x = t\mathbf{P} + \varepsilon, \tag{13.1}$$

where x and t are respectively $1 \times m$ and $1 \times p$ row vectors, $\mathbf{P}$ is the $p \times m$ matrix of coefficients and ε is the $1 \times m$ row vector of random errors, including uncontrollable sources of variability such as measurement error, sampling variation and unknown process disturbances (Burnham *et al.*, 1999). Equation (13.1) is the *latent variable model*. For n observations, the $n \times m$ data table, $\mathbf{X}$, that consists of n rows for m different variables (each row vector x stacked upon the next), can be written as

$$\mathbf{X} = \mathbf{TP} + \mathbf{E}, \tag{13.2}$$

where $\mathbf{T}$ is the $n \times p$ matrix of latent variables (each row corresponds to a different vector t, representing a different observation in the p latent variables) and $\mathbf{E}$ is the $n \times m$ matrix that also results from stacking all the n rows regarding vectors ε, each one relative to a given (multivariate) observation. Sometimes it is useful to separate the variables into two groups, $\mathbf{X}$ and $\mathbf{Y}$, for instance in order to use the model in future situations where the $\mathbf{Y}$ variables are not available or their prediction is required, with the knowledge of only variables from the **X** block. In such circumstances we can write (13.2) in the form

$$\begin{aligned} \mathbf{X} &= \mathbf{TP} + \mathbf{E}, \\ \mathbf{Y} &= \mathbf{TQ} + \mathbf{F}, \end{aligned} \qquad (13.3)$$

which can be directly obtained from (13.2) after a rearrangement with the following variable grouping: [**XY**]= **T**[**PQ**]+[**EF**]. From (13.3) it is clear that there is no strong causality assumption linking variables belonging to the two blocks **X** and **Y** in the latent variable model. In fact, non-causality is also a characteristic of historical databases and normal operation data, situations where this model is so often applied (MacGregor and Kourti, 1998). Therefore, blocks **X** and **Y** share a symmetrical role regarding the underlying latent variables, and the separation of variables is only decided on the basis of its intended final use.

Model (13.2) can be estimated using factor analysis, whereas (13.3) is often estimated using principal components regression (PCR) or partial least squares (PLS, also known as projection to latent structures); in Appendix 13A we briefly describe these two techniques. When the error structures are more complex, other techniques can be used, such as maximum likelihood principal components analysis (MLPCA) regarding the estimation of model (13.2) (Wentzell *et al.*, 1997a).

Some useful features of these estimation techniques, apart from being able to deal with the presence of multicollinearity in a natural and coherent way, are their ability to handle the presence of moderate amounts of missing data by taking advantage of the estimated correlation structure among variables (Nelson *et al.*, 1996; Walczak and Massart, 2001; Kresta *et al.*, 1994), their flexibility to cope with situations where there are more variables than observations, and the availability of diagnostic tools that portray information regarding the suitability of the estimated models to explain the behaviour of new collected data. For instance, in PLS it is possible to calculate the distance between a new observation in the **X**-space and its projection onto the latent variable space (the space spanned by the latent variables, as defined by the **P** matrix), as well as to see whether this projection falls inside the domain where the model was estimated. Such features enable a quick way to check whether a new observation can be adequately described by the latent variable model, and, furthermore, whether it falls within the region used to build the model, therefore avoiding extrapolation problems in the prediction of variables that belong to the **Y** block. Another useful characteristic of these approaches is that several variables in the **Y** block can be handled simultaneously. In order to further illustrate the flexibility and potential underlying this class of approaches based upon latent variables, in what follows several different application scenarios where they can be used will be presented.

13.2.1 Multivariate calibration

In multivariate calibration, we aim to build a model using available multivariate spectral data (**X** block) and concentrations of the solutions used to produce such spectra (**Y** block), usually after the implementation of a suitable design of experiments, in order to predict what the concentration of the specimens (analytes) would be when new samples become available, based only on quick spectrometer measurements, and thus avoiding lengthy, complex and expensive laboratory analytical procedures. The fact that spectrometer data often follow Beer's law to a good approximation (the resulting spectrum is a linear combination of the pure component chemical spectra, appropriately weighted by their composition), provides a strong theoretical motivation for the use of model (13.3), and therefore, both PCR and PLS are extensively used in this context (Martens and Næs, 1989; Estienne *et al.*, 2001).

13.2.2 Soft sensors

Soft sensors consist of inferential models that provide on-line estimates for the values of interesting product properties based on readily available measurements, such as temperatures, pressures and flow rates. This is particularly appealing in situations where the equipment required to measure such properties is expensive and/or difficult to implement on-line, but can also be used in parallel, providing a redundant mechanism to monitor measurement devices performance (Kresta *et al.*, 1994; MacGregor and Kourti, 1998).

13.2.3 Experimental design

Experimental design procedures using latent variables instead of the original ones (which in practice amounts to using linear combinations of **X**s) can reduce the number of experiments needed to cover the entire experimental space of interest. This happens because, by moving together groups of variables in the latent variable modelling frameworks, the effective number of independent variables is greatly decreased, while the operational constraints that motivate such groupings are implicitly taken into account (Wold *et al.*, 1986; Gabrielsson *et al.*, 2002).

13.2.4 Quantitative structure activity relationships

The main goal in this area is to relate the structure and physico-chemical properties of compounds (**X** block) to their macroscopic functional, biological or pharmaceutical properties, such as, carcinogenicity, toxicity, degradability, response to treatment, among others (**Y** block). The **X** block variables may be melting points, densities or parameters derived from the underlying molecular structures – the so-called molecular descriptors, such as molecular energy descriptors and carbon atom descriptors (Cholakov *et al.*, 1999). The purpose is therefore to build simple models relating the two group of variables for predicting biological activity or pharmaceutical properties of a wider range of compounds (whose structure and physico-chemical properties are known, that is, the **X** block properties) from the knowledge of a limited number of fully characterised representatives (where both blocks of variables are known), or to optimise their structure in order to improve activity-related indicators. For instance, we may want to predict the performance of a potential new drug, or just to know which properties regulate the response of the **Y** block variables, so that modifications can be introduced in the compounds under study or new ones should be sought that potentially exhibit the desired target properties (Eriksson *et al.*, 2001). Latent variable models, in particular PLS, have had great success in this area, due to the presence of strong relationships either among variables belonging to each block, or between those of the two blocks (see also Burnham *et al.*, 1999, and the references therein).

13.2.5 Process monitoring

Process monitoring is another field where latent variable models have found general acceptance. Both PCA and PLS techniques have been extensively used as estimators of the structure behind normal operation data. Statistical process control (SPC) based on PCA involves using two statistics so as to monitor whether the reference PCA model is still adequate to describe new incoming data (using the Q statistic, representing the distance from the new point to the latent variable subspace, that is, to the space spanned by the loading vectors of the principal components model estimated from reference data) and, if so, whether such data fall inside the normal operation condition (NOC) region defined in the latent variable space (using Hotelling's T^2 statistic, which is the Mahalanobis distance in the space of the latent variables). Basically, Q is a 'lack of model fit' statistic while Hotelling's T^2 is a measure of the variation within the PCA model. These two general and complementary monitoring quantities are normally present in any monitoring procedure based on a latent variable framework (Figure 13.1).

Using these two statistics, processes with tens or hundreds of variables can be easily and effectively monitored (Macgregor and Kourti, 1995; Wise and Gallagher, 1996). Control charts can also be implemented for each individual latent variable (score), whose control limits ($t_{i,\alpha}$) are usually derived under the assumption that they follow normal distributions with zero mean (the variables were originally centred) and standard deviation estimated by the square root of the eigenvalue associated with that latent variable, that is, $t_{i,\alpha} = \pm\sqrt{\lambda_i} \times t_{n-1,\alpha/2}$, where $t_{n-1,\alpha/2}$ is the upper 100 $\alpha/2$ percentage point for the t-distribution with $n - 1$ degrees of freedom (n is the number of samples in reference data) and λ_i is the eigenvalue related with the ith latent variable. Underlying this monitoring scheme is latent variable model (13.2), but we can also monitor processes using the latent variable model structure (13.3), estimated by PLS. In this case, the statistics adopted are usually the Hotelling's T^2 statistic applied to the latent variables, and $SPE_y = \sum_{i=1}^{m_y} (y_{\text{new},i} - \hat{y}_{\text{new},i})^2$, where $\hat{y}_{\text{new},i}$ are the predicted values for the m_y **Y** block variables in the ith observation. When these variables are measured infrequently relatively to the acquisition rate of the **X** block variables (as often happens with quality variables relatively to process variables), then the statistic $SPE_x = \sum_{i=1}^{n} (x_{\text{new},i} - \hat{x}_{\text{new},i})^2$ is used instead (or a modified version of it, which weights the variables according to their modelling power for **Y**,

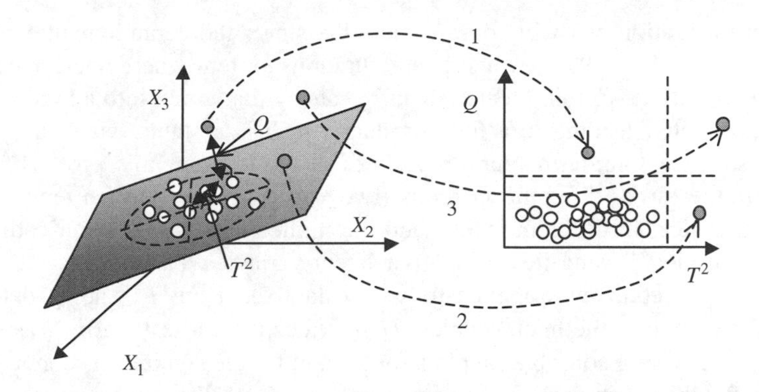

Figure 13.1 Illustration of a multivariate PCA monitoring scheme based on Hotelling's T^2 and Q statistics. Observation 1 falls outside the control limits of the Q statistic (the PCA model is not valid for this observation), despite its projection on the principal components subspace falling inside the NOC region. Observation 2, on the other hand, corresponds to an abnormal event in terms of its Mahalanobis distance to the centre of the reference data, but it still complies with the correlation structure of the variables, that is, with the estimated model. Observation 3 illustrates an abnormal event from the standpoint of both criteria.

$SPE_x = \sum_{i=1}^{n} w_i(x_{\text{new},i} - \hat{x}_{\text{new},i})^2)$, where $\hat{x}_{\text{new},i}$ is the projection of observation i in the latent variable subspace (Kresta *et al.*, 1991). The upper control limit for Hotelling's T^2 statistic is calculated as for the case of SPC based on PCA, while the *SPE* statistic is based on a χ^2 approximation, with parameters determined by matching the moments of this distribution with those for the reference 'in-control' data (Kourti and MacGregor, 1995; Nomikos and MacGregor, 1995; MacGregor *et al.*, 1994).

One important feature of latent variable frameworks in the context of process monitoring is the availability of informative diagnostic tools. The statistics referred to above do detect abnormal situations effectively but do not provide any clue as to what may have caused such behaviour. However, with the assistance of these diagnostic tools, one can move a step further in order to reduce the number of potential root causes or even track down the ultimate source of the abnormal behaviour. This can be done through the use of contribution plots (Kourti and MacGregor, 1996; MacGregor *et al.*, 1994; Westerhuis *et al.*, 2000; Eriksson *et al.*, 2001), which are basically tools that 'interrogate' the underlying latent variable model (estimated through PCA or PLS) about those variables that contribute the most to an unusual value found for the monitoring statistics. For instance, in the case of SPC based on PCA, there are contribution plots available for each individual score, for the overall contribution of each variable to Hotelling's T^2 statistic and for the contribution of each variable to the Q statistic. A hierarchical diagnostic procedure proposed by Kourti and MacGregor (1996) involves following first the behaviour of the T^2 statistic and, if an out-of-limit value is detected, checking the score plots for high values and then the variable contributions for each significant score.

As an illustration, we will consider here the slurry-fed ceramic melter process (Wise and Gallagher, 1996), which is a continuous system where nuclear waste is combined with glass-forming materials into a slurry that is fed into a high temperature glass melter, leading to a final product which is a stable, vitrified material for disposal in a long-term repository. Data acquired from this process consist of temperatures at 20 different locations (two vertical rows with ten temperature-measuring devices at different levels) and the molten glass level. After estimating the reference model using PCA with four latent variables, based upon 450 observations, the procedure was applied to a new data set. Both T^2 and Q detect an abnormal situation at the final samples (Q provides the earliest alarms), as shown in Figure 13.2. Using contribution plots for T^2 and for the significant scores (in this case only the third principal component was significant), the root cause was easily determined as being related to an error in the fifth temperature reading device. The Q contribution plot, not presented here, confirms this diagnosis.

The above-mentioned procedures are especially suited for continuous processes where the assumption of stationarity in the mean and covariance structure holds as a good approximation of reality. Given the presence of dynamics or autocorrelation in the variables, we can still adopt this class of procedures by expanding the

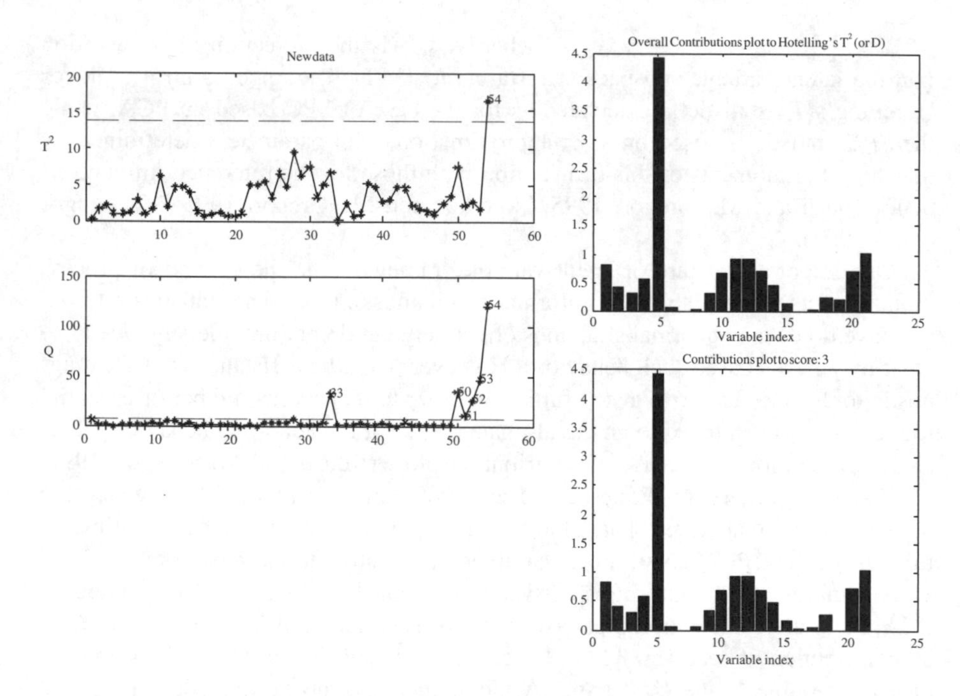

Figure 13.2 Slurry-fed ceramic melter process example: both T^2 and Q detect an abnormal operating condition in the final samples. Contribution plots identify a problem associated with the fifth temperature measurement device at observation 54.

original data matrix with time-lagged variables, in order to model both the cross-covariance and autocorrelation structures. See Ku *et al.* (1995) and Ricker (1988). Lakshminarayanan *et al.* (1997) described an approach where the dynamic modelling is introduced in the inner relationship of PLS using an autoregressive system with exogenous inputs (ARX) or Hammerstein model structures, and Kaspar and Ray (1993) presented an alternative procedure where a dynamic transformation is applied to the **X** block time series, before using PLS.

A special class of intrinsically dynamic processes is concerned with batch processes, which have gained importance over the last decades given their higher operation flexibility. Batch processes typically generate data structures with the following three components: a table with the initial conditions of each batch (batch recipe, charge conditions), **Z**; a three-way table of process operating conditions during the time duration of each batch, for all batches, with dimensions [*batch run* × *variable* × *time*], **X**; and another table of product quality measurements, **Y**. Techniques such as multiway PCA and PLS were developed in order to accommodate these types of structures for process monitoring and prediction purposes (Nomikos and MacGregor, 1995; Westerhuis *et al.*, 1999). They are equivalent to applying PCA or PLS to the unfolded version of the three-way data matrix, **X** [*batch run* × *variable* × *time*], that consists of a two-way matrix with dimensions [*batch run* × (*variable* × *time*)] (Figure 13.3). Intrinsically multiway modelling extensions have also been put forward (such as Smilde *et al.*, 2004).

When the number of variables becomes very large, monitoring and diagnosis procedures can become quite cumbersome and difficult to interpret. Under these conditions and if the variables have some natural grouping, such as belonging to different production sections or product streams, the analysis can still be carried out maintaining this natural blocking in order to make interpretation of results easier, using procedures such as multiblock and hierarchical PLS or PCA (Smilde *et al.*, 2003; Westerhuis *et al.*, 1998; MacGregor *et al.*, 1994).

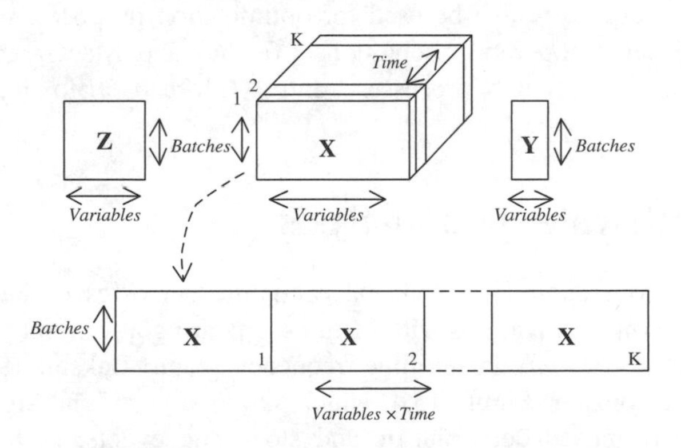

*Figure 13.3 Typical batch data structure, and the unfolding of the three-way **X** data matrix of operating conditions that makes possible the application of PCA and PLS for monitoring or prediction purposes.*

13.2.6 Image analysis

With the growing availability of inexpensive digital imaging systems, new approaches have been developed to take advantage of the information provided by sensors of this type. For instance, monitoring of the performance of an industrial boiler can be adequately carried out by acquiring successive (RGB) digital images of the turbulent flame, and using them to access operation status (Yu and MacGregor, 2004). Even though images change rapidly, their projection onto the latent variable space (using PCA) is quite stable at a given operating condition. However, projections do change significantly if the feed or operating conditions suffer modifications. The success of PCA in the extraction of a stable pattern for a given operating condition is compatible with a view of the variation in the RGB intensities for all the pixels (usually forming a 256×256 array) as a latent variable process of type (13.2), whose number of latent variables is indeed quite low (the monitoring scheme is essentially based on the first two latent variables). This kind of approach is also being used for on-line product quality control in the food and others industries where visual inspection plays a key role.

13.2.7 Product design, model inversion and optimisation

In this context latent variable models, fitted using historical data from a given process, where the process constraints and operating policies are already implicitly incorporated in the data correlation structure, are used to address different tasks. In product design, the model is used to find an operating window where a product can be manufactured with a desired set of properties (MacGregor and Kourti, 1998; Jaeckle and MacGregor, 2000). Such operating windows are derived from the definition of the desired quality specifications for the new product and an inversion, over the latent variable model, from the **Y** to the **X** space. The solution thus found will not only comply with the targeted quality specifications, but also be compatible with past operating policies (Jaeckle and MacGregor, 1998). Such a model can also be used for optimisation purposes, in particular to find the 'best' set of operating conditions (Yacoub and MacGregor, 2004) and the 'best' policies for batch processes control (Flores-Cerrilo and MacGregor, 2004).

13.3 Multiscale data analysis

Data acquired from natural phenomena, economic activities or industrial plants usually present complex patterns with features appearing at different *locations* and with different *localisations* in the time-frequency plane (Bakshi, 1999). To illustrate this point, consider Figure 13.4, which shows an artificial signal composed of several superimposed deterministic and stochastic events, each with its own characteristic time/frequency behaviour. The signal deterministic features consist of a ramp that begins right from the start, a step perturbation at sample 513, a permanent oscillatory component and a spike at observation 256. The stochastic

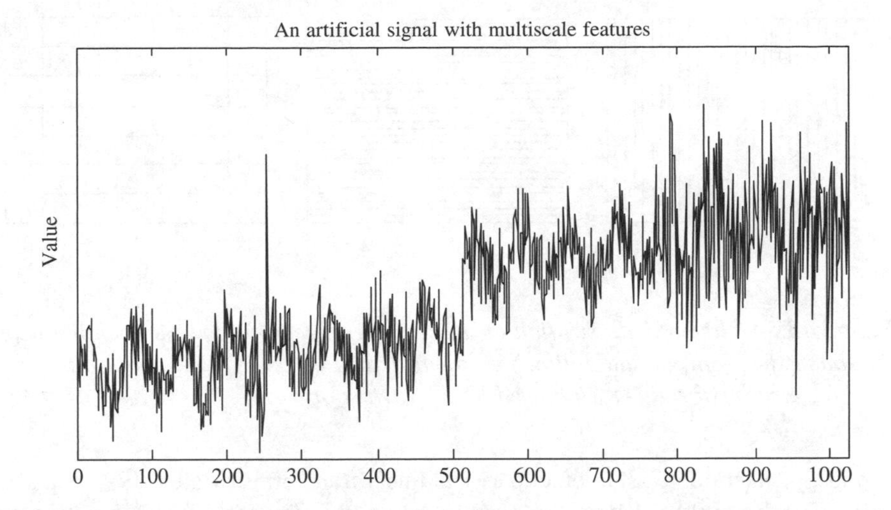

Figure 13.4 An artificial signal containing multiscale features, which results from the sum of a linear trend, a sinusoid, a step perturbation, a spike (deterministic features with different frequency localisation characteristics) and white noise (a stochastic feature whose energy is uniformly distributed in the time/frequency plane).

features consist of Gaussian white noise whose variance increases after sample 768. Clearly these events have different time/frequency locations and localisations: for instance, the spike is completely localised in the time axis, but also fully delocalised in the frequency domain; on the other hand, the sinusoidal component is very well localised in the frequency domain but delocalised in the time axis. White noise contains contributions from all frequencies and its energy is uniformly distributed in the time/frequency plane, but the linear trend is essentially a low-frequency perturbation, and its energy is almost entirely concentrated in the lower-frequency bands. All these patterns appear simultaneously in the signal and, in data analysis, one should be able to deal with them without compromising one kind of feature over the others. This can only be done, however, by adopting a mathematical 'language' suited to efficiently describe data with such multiscale characteristics.

The use of transforms, such as the Fourier transform, often provides an adequate way to conduct data analysis tasks of this kind. Transforms allow for alternative ways of representing raw data, as an expansion of basis functions multiplied by transform coefficients. These coefficients constitute the transform and, if the methodology is properly chosen, data analysis becomes much more efficient and effective when conducted over them, instead of over the original data. For instance, the Fourier transform is the appropriate mathematical language for describing periodical phenomena or smooth signals, since the nature of its basis functions allows for compact representations of such trends, that is, only a few Fourier transform coefficients are needed to provide a good representation of the signal. The same

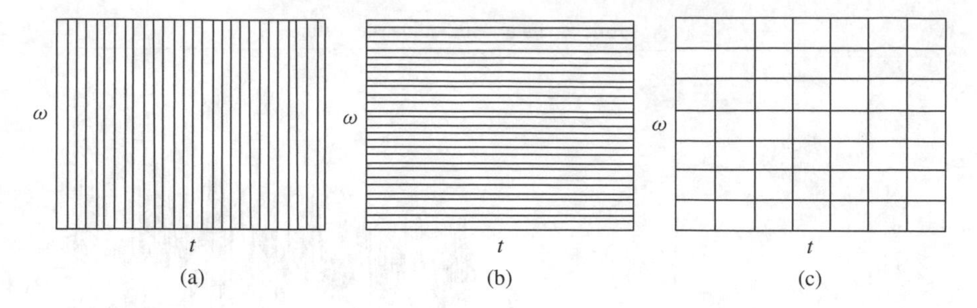

Figure 13.5 Schematic illustration of the time/frequency windows associated with the basis function of the following linear transforms: (a) Dirac δ transform; (b) Fourier transform; (c) windowed Fourier transform.

applies, in other contexts, to other classical linear transforms (Mallat, 1998; Kaiser, 1994; Bakshi, 1999) such as the ones based on the discrete Dirac δ function or the windowed Fourier transform. However, none of these linear transformations are able to cope effectively with the diversity of features present in signals such as that illustrated in Figure 13.4, and they are therefore referred to as single-scale approaches. A proper analysis of this signal with such techniques would require a large number of coefficients, meaning that they are not 'adequate' languages for a compact translation of its key features in the transform domain. This happens because the form of the time/frequency windows (Vetterli and Kovačevič, 1995; Mallat, 1998) associated with their basis functions (Figure 13.5) does not change across the time/frequency plane in such a manner that enables a precise coverage of the localised high-energy zones relative to the several features present in the signal, using only a few 'tiles'.[1]

Therefore, in order to cope with such multiscale features, a more flexible tiling of the time/frequency space is required, which is provided by basis functions called wavelets (Figure 13.6), leading to the so-called wavelet transforms. In practice, it is often the case that signals are composed of short duration events of high frequency and low frequency events of long duration. This is exactly the kind of tiling associated with the wavelet basis sets, whose relative frequency bandwidth is a constant (that is, the ratio between a measure of the size of the frequency band and the mean frequency,[2] $\Delta\omega/\omega$, is constant for each wavelet function), a property also referred to as a 'constant-Q' scheme (Rioul and Vetterli, 1991).

[1]A tile is a region of the time-frequency plane where a basis function concentrates a significant amount of its energy.

[2]The location and localisation of the time and frequency bands, for a given basis function, can be calculated from the first moment (the mean, a measure of location) and the second centred moment (the standard deviation, a measure of localisation), of the basis function and its Fourier transform. The localisation measures define the form of the boxes that tile the time/frequency plane in Figures 13.5 and 13.6. However, the time and frequency widths (that is the localisation) of these boxes do always conform to the lower bound provided by the Heisenberg principle ($\sigma(g) \cdot \sigma(\hat{g}) \geq 1/2$, where $\hat{g}$ represents the Fourier transform of g; Kaiser, 1994; Mallat, 1998). These boxes are often referred to as 'Heisenberg boxes'.

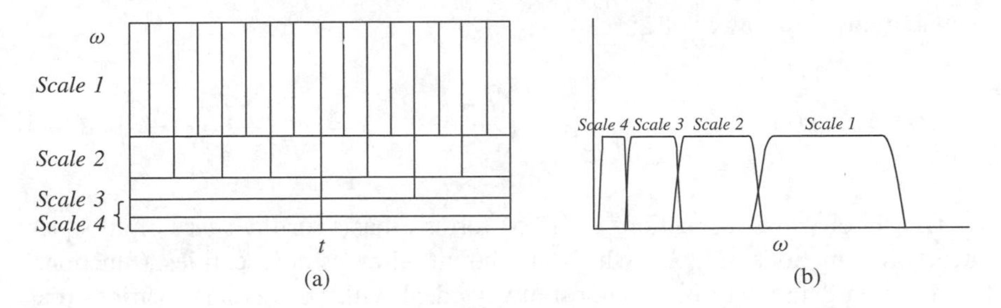

Figure 13.6 (a) Tiling of the time-frequency plane provided by wavelet basis functions, and (b) an illustration of how the wavelet functions divide the frequency domain, where we can see that they work as band-pass filters. The shape of the windows and frequency bands, for a given wavelet function, depends upon the scale index value: for low values of the scale index, the windows have good time localisation and cover a long frequency band; windows with high values for the scale index have large time coverage with good frequency localisation.

Wavelets are a particular type of function whose location and localisation characteristics in the time/frequency are ruled by two parameters: both the localisation in this plane and the location in frequency domain are determined by the scale parameter, s, whereas the location in the time domain is controlled by the time translation parameter, b. Each wavelet, $\psi_{s,b}(t)$, can be obtained from the so-called 'mother wavelet', $\psi(t)$, through a scaling operation (that 'stretches' or 'compresses' the original function, establishing its form) and a translation operation (that controls its positioning in the time axis):

$$\psi_{s,b}(t) = \frac{1}{\sqrt{|s|}}\psi\left(\frac{t-b}{s}\right). \tag{13.4}$$

The shape of the mother wavelet is such that it has an equal area above and below the time axis, which means that, besides having a compact localisation in this axis, they should also oscillate around it, features from which the name 'wavelets' (small waves) is derived. In the *continuous wavelet transform*, scale and translation parameters can vary continuously, leading to a redundant transform[3] (a one-dimensional signal is being mapped onto a two-dimensional function). Therefore, in order to construct a basis set, it is sometimes possible to sample these parameters appropriately, so that the set of wavelet functions parameterized by such new indices (scale index, j, and translation index, k) cover the time-frequency plane in a non-redundant way. This sampling consists of applying a dyadic grid in which b is sampled more frequently for lower values of s, with s growing

[3]The redundancy of the continuous wavelet transform is not necessarily undesirable, as it makes this transform translation-invariant (which does not happen with the orthogonal wavelet transform). Another consequence is that its coefficients do not need to be calculated very precisely in order to still obtain good reconstructions (Daubechies, 1992; Hubbard, 1998).

exponentially at a power of 2:

$$\psi_{j,k}(t) = \psi_{s,b}(t) \Bigg|_{\substack{s = 2^j \\ b = k \cdot 2^j}} = \frac{1}{2^{j/2}} \psi\left(\frac{t - k \cdot 2^j}{2^j}\right) = \frac{1}{2^{j/2}} \psi\left(\frac{t}{2^j} - k\right). \quad (13.5)$$

The set of wavelet functions in (13.5) forms a basis for the space of all square integrable functions, $L^2(\mathbb{R})$, which are infinite-dimensional entities (functions). However, in data analysis we almost always deal with vectors and matrices (data tables, images), which are dimensionally finite entities. Despite this fact, the above concepts can still be adopted for finite-dimensional entities, using the concept described in the following paragraph.

Working in a hierarchical framework for consistently representing images with different levels of resolution, that is, containing different amounts of information regarding what is being portrayed, Mallat (1989, 1998) developed the unifying concept of a *multiresolution approximation*. This is a sequence, $\{V_j\}_{j \in \mathbb{Z}}$, of closed subspaces of $L^2(\mathbb{R})$, with the following six properties:

$$\forall (j, k) \in \mathbb{Z}^2, f(t) \in V_j \Leftrightarrow f(t - 2^j k) \in V_j; \quad (13.6)$$

$$\forall j \in \mathbb{Z}, V_{j+1} \subset V_j; \quad (13.7)$$

$$\forall j \in \mathbb{Z}, f(t) \in V_j \Leftrightarrow f\left(\frac{t}{2}\right) \in V_{j+1}; \quad (13.8)$$

$$\lim_{j \to +\infty} V_j = \bigcap_{j=-\infty}^{+\infty} V_j = \{0\}; \quad (13.9)$$

$$\lim_{j \to -\infty} V_j = Closure \bigcup_{j=-\infty}^{+\infty} V_j = L^2(\mathbb{R}); \quad (13.10)$$

$$\exists \phi : \{\phi(t - k)\}_{k \in \mathbb{Z}} \text{ is a Riesz basis of } V_0. \quad (13.11)$$

Property (13.6) states that any translation applied to a function belonging to the subspace V_j, proportional to its scale (2^j), generates another function belonging to the same subspace. Property (13.7) says that any entity in V_{j+1} also belongs to V_j, that is, $\{V_j\}_{j \in \mathbb{Z}}$ is a sequence of nested subspaces: $\cdots \subset V_{j+1} \subset V_j \subset V_{j-1} \subset \cdots \subset L^2(\mathbb{R})$. In practice this means that projections to approximation functions with higher scale indices should result in coarser versions of the original function (or a lower-resolution, coarser version of the original image), whereas projections to the richer approximation spaces, with lower scale indices, should result in finer versions of the projected function (or a finer version of the original image, with higher resolution). Property (13.8) requires that any dilation ('stretching') by a factor of 2, applied to a function belonging to subspace V_j, results in a function belonging to the next coarser subspace V_{j+1}. However, if we keep 'stretching' it, in the limit as $j \to +\infty$, this function becomes a constant. This means that, in order for such a limiting case to belong to $L^2(\mathbb{R})$, it must coincide with the constant zero function, {0}. This is the only function that belongs to all the approximation spaces, from the finest (lower scale indices) to the coarsest (higher scale indices), as stated by property (13.9). We can also conclude from this property that the projection to coarser approximation spaces successively results in both coarser and residual approximations of the original functions:

$$\lim_{j\to+\infty} \mathrm{Pr}_{V_j} f = 0 \text{ in } L^2(\mathbb{R}) \Leftrightarrow \lim_{j\to+\infty} \|\mathrm{Pr}_{V_j} f\| = 0. \tag{13.12}$$

On the other hand, following from property (13.10), any function in $L^2(\mathbb{R})$ can be successively better approximated by a sequence of increasingly finer subspaces:

$$\lim_{j\to-\infty} \mathrm{Pr}_{V_j} f = f \text{ in } L^2(\mathbb{R}) \Leftrightarrow \lim_{j\to-\infty} \|f - \mathrm{Pr}_{V_j} f\| = 0. \tag{13.13}$$

Property (13.11) concerns the existence of a Riesz basis for the space V_0, which consists of the so called *scaling function*, $\phi(t)$, along with its integer translations. In what follows this basis is an orthonormal one, which, according to (13.6) and (13.8), means that the set $\{\phi_{j,k}\} = \{2^{-j/2}\phi(2^{-j}t - k)\}$ is also an orthonormal basis for the space V_j.

Therefore, at this point, we have a well-characterised sequence of nested subspaces, each equipped with a basis set resulting from translation and scaling operations of the scaling function. Let us now introduce the complementary concept of approximation subspaces: the *detail subspaces*, $\{W_j\}_{j\in\mathbb{Z}}$. As V_{j+1} is a proper subspace of V_j ($V_{j+1} \subset V_j$, $V_{j+1} \neq V_j$), we will call the orthogonal complement of V_{j+1} in V_j, W_{j+1}. Therefore, we can write $V_j = V_{j+1} \oplus^{\perp} W_{j+1}$, which means that any function in V_j can uniquely be written as a sum of elements, one belonging to the approximation space V_{j+1} and another to the detail space W_{j+1}. These elements are just the projections onto these subspaces. As $V_j \subset V_{j-1}$, we can also state that $V_{j-1} = V_j \oplus^{\perp} W_j = V_{j+1} \oplus^{\perp} W_{j+1} \oplus^{\perp} W_j$. This means that if we have a function (or a signal or an image), f_0, belonging to V_0, we can represent it as a projection into the approximation level at scale j, f_j, plus all the details relative to the scales in between ($\{w_i\}_{i=1,\ldots,j}$), since

$$V_0 = V_j \overset{\perp}{\oplus} W_j \overset{\perp}{\oplus} \cdots \overset{\perp}{\oplus} W_2 \overset{\perp}{\oplus} W_1 \tag{13.14}$$

In terms of projection operations:

$$f_0 = f_j + \sum_{i=1}^{j} w_i \Leftrightarrow f_0 = \mathrm{Pr}_{V_j} f_0 + \sum_{i=j}^{1} \mathrm{Pr}_{W_i} f_0 \tag{13.15}$$

It can be shown that an orthonormal basis for the detail space W_j can be given by a set of wavelet functions, $\{\psi_{j,k}\}_{k\in\mathbb{Z}}$. These sets are mutually orthogonal for different scale indices, j, as they span orthogonal subspaces of $L^2(\mathbb{R})$. By extending decomposition (13.14) in order to incorporate all scales, and considering properties (13.9) and (13.10), we can also conclude that

$$L^2(\mathbb{R}) = \overset{\perp}{\oplus}{}_{i=-\infty}^{+\infty} W_i,$$

which means that the set of all wavelet functions with the discrete parameterization adopted do indeed form a basis set of this space. The projections, f_j and $\{w_i\}_{i=1,\ldots,j}$ in (13.15), can be written in terms of the linear combination of basis functions multiplied by the expansion coefficients (calculated as inner products of the signal and the basis functions): the approximation coefficients, a_k^j $(k \in \mathbb{Z})$, and the detail coefficients, d_k^i $(i = 1, \ldots, j;\ \ k \in \mathbb{Z})$. These are usually referred to as the (discrete) wavelet transform or wavelet coefficients:

$$f_0 = \sum_k a_k^j \phi_{j,k} + \sum_{i=1}^{j} \sum_k d_k^i \psi_{i,k} \tag{13.16}$$

where

$$a_k^j = \langle f_j, \phi_{j,k} \rangle, \ \ d_k^i = \langle f_j, \psi_{i,k} \rangle \tag{13.17}$$

Still within the scope of the multiresolution approximation framework, Mallat (1989) proposed a very efficient recursive scheme for computing the wavelet coefficients as well as for signal reconstruction (starting from the wavelet coefficients), which basically consists of a pyramidal algorithm, based upon convolutions with quadrature mirror filters, a well-known technique in the engineering (discrete) signal processing community. In fact, this framework unifies several approaches used in a number of fields under different names, such as *wavelet series expansions* from harmonic analysis, *multiresolution signal processing* from computer vision and *subband coding* or *filter banks* from speech/image processing (Rioul and Vetterli, 1991; Bruce *et al.*, 1996; Burrus *et al.*, 1998), adding important mathematical insight to all of them, while providing, at the same time, strong and intuitive concepts, such as the notion of approximations and details as projections to particular subspaces of $L^2(\mathbb{R})$.

In practice, for finite-dimensional elements (data arrays), it is usually assumed that data is already the projection to space V_0, f_0, and therefore the wavelet coefficients appearing in the decomposition formula (13.16) can be easily computed using Mallat's efficient algorithm. Thus, one essentially applies the analysis and reconstruction quadrature mirror filters associated with a given wavelet, without ever using any wavelet function explicitly. In fact, wavelets very often do not even have a closed formula in the time domain, although they can be plotted as accurately as required, by iterating over such filters (Strang and Nguyen, 1997). By way of illustration, we can decompose the signal in Figure 13.4, which contains 2^{10} points at scale $j = 0$, into a coarser, lower-resolution version at scale $j = 5$ with 2^5 approximation coefficients appearing in the expansion ($f_5 = \sum_{k=0}^{2^5-1} a_k^5 \phi_{5,k}$) plus all the detail signal from scale $j = 1$ (with 2^9 detail coefficients, $w_1 = \sum_{k=0}^{2^9-1} d_k^1 \psi_{1,k}$) up to scale $j = 5$ (with 2^5 detail coefficients, $w_5 = \sum_{k=0}^{2^5-1} d_k^1 \psi_{5,k}$). The total number of wavelet coefficients is equal to the cardinality of the original signal, meaning that no information is being 'created' or 'disregarded', but simply transformed ($2^{10} = 2^5 + 2^5 + 2^6 + \cdots + 2^9$). The projections onto the approximation and detail spaces are presented in Figure 13.7, where it is possible to see that

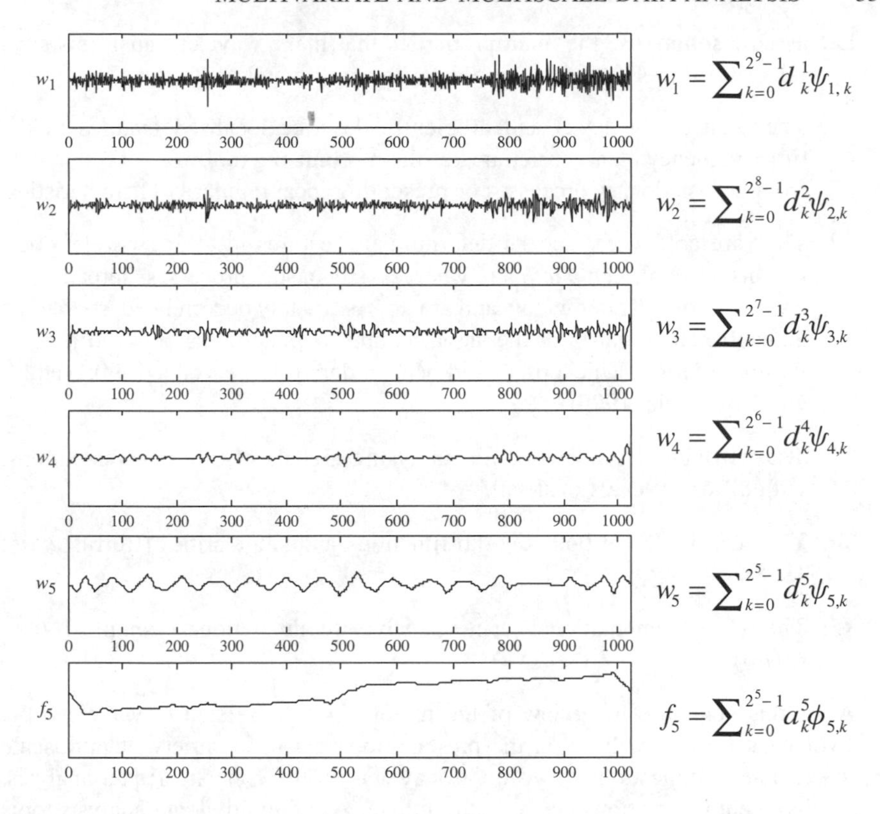

Figure 13.7 The signal in Figure 13.4 decomposed into its coarser version at scale $j = 5$, f_5, plus all the detail signals lost across the scales ranging from $j = 1$ to $j = 5$, $\{w_i\}_{i=1,\ldots,5}$. The filter used here is the Daubechies compactly supported filter with three vanishing moments.[4]

the deterministic and stochastic features appear quite clearly separated according to their time/frequency localisation characteristics: coarser deterministic features (ramp and step perturbation) appear in the coarser versions of the signal (containing the lowest-frequency contributions), the sinusoid is well captured in the detail at scale $j = 5$, noise features appear more clearly at the high-frequency bands (details for $j = 1, 2$), where the increase in the variance is also quite visible, as well as the spike at observation 256 (another high-frequency perturbation). This illustrates the ability of wavelet transforms to separate the deterministic and stochastic contributions present in a signal, according to their time/frequency location.

[4]A wavelet has p vanishing moments if $\int_{-\infty}^{+\infty} t^k \psi(t)dt = 0$ for $0 \leq k < p$. This is an important property in the fields of signal and image compression – it can induce a higher number of low-magnitude detail coefficients, if the signal does have local regularity characteristics.

Let us now summarise the main properties that make wavelet transforms such a potentially useful tool for addressing data processing and analysis tasks:

1. They can easily detect and efficiently describe localised features in the time/frequency plane, which makes them promising tools for analysing data from non-stationary processes or presenting local regularity characteristics.

2. They are able to extract the deterministic features in a few wavelet coefficients (*energy compaction*), whereas stochastic processes spread their energy across all coefficients and are approximately decorrelated, so that the autocorrelation matrix of the signal is approximately diagonalized (*decorrelation ability*) (Dijekerman and Mazumdar, 1994; Bakshi, 1999; Vetterli and Kovačevič, 1995).

3. They provide a framework for analysing signals at different resolutions, with different levels of detail.

4. They can represent both smooth functions and singularities (Burrus *et al.*, 1998).

5. They are computationally inexpensive (computational complexity of $O(N)$).

After this very brief overview of the major concepts related to wavelets, the focus in what follows will be on the presentation of a wide variety of multiscale approaches and applications based upon wavelets in the context of data analysis. These also include situations where both multiscale and multivariate analysis tools appear integrated, in order to explore their complementary roles in the extraction of structure and knowledge from data sets.

13.3.1 Signal and image denoising

Denoising is concerned with uncovering the true signal from noisy data in which it is immersed (for a more formal interpretation of the term, see Donoho, 1995). It is one of the fields in which wavelets have been most usefully applied, leading to significant improvements over more classical approaches. This is due to their ability to concentrate deterministic features in a few high-magnitude coefficients while stochastic phenomena are spread over all coefficients, which lays the ground for the implementation of wavelet-domain coefficient thresholding schemes. Donoho and Johnstone (1992, 1994) pioneered the field and proposed a simple but effective denoising scheme for estimating a signal with additive, independent and identically distributed zero-mean Gaussian noise,

$$y_i = x_i + \sigma \cdot \varepsilon_i, \ \ \varepsilon_i \overset{\text{iid}}{\sim} N(0, 1) \quad (i = 1, \dots, N), \tag{13.18}$$

which consists of the following three steps:

1. Compute the wavelet transform of the sequence $\{y_i\}_{i=1:N}$ (the authors suggested the boundary-corrected or interval-adapted filters developed by Cohen *et al.*, 1993).

2. Apply a thresholding operation to the detail wavelet coefficients (the authors suggested soft-thresholding), using the threshold $T = \hat{\sigma}\sqrt{2\ln(N)}$ ($\hat{\sigma}$ is an estimator of the noise standard deviation):[5]

   - 'hard-thresholding': $$HT(x) = \begin{cases} x \Leftarrow |x| > T \\ 0 \Leftarrow |x| \leq T \end{cases} \quad (13.19)$$

   - 'soft-thresholding': $$ST(x) = \begin{cases} \text{sign}(x) \cdot (|x| - T) \Leftarrow |x| > T \\ 0 \Leftarrow |x| \leq T \end{cases} \quad (13.20)$$

3. Compute the inverse wavelet transform to obtain the denoised signal.

This simple scheme is called VisuShrink, since it provides better visual quality than other procedures based on the mean squared error alone. By way of illustration, Figure 13.8 depicts the denoising of a nuclear magnetic resonance (NMR) spectrum using a Symmlet-8 filter with a decomposition depth of $j = 5$.

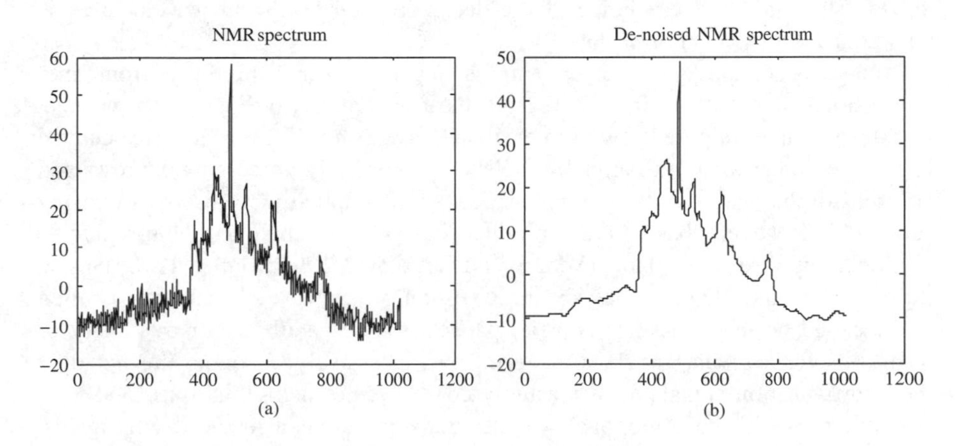

Figure 13.8 Denoising of an NMR spectrum: (a) original NMR spectrum; (b) denoised NMR spectrum. Computations conducted with the package WaveLab 8.02 (developed at the Department of Statistics, Stanford University, and available at http://www-stat.stanford.edu/~wavelab/) in the MATLAB environment.

[5]Usually a robust approach is applied to the wavelet coefficients at the finest scale, $\hat{\sigma} = \text{Med}\left(\left|\{d_k^1\}_{k=1,\ldots,N/2}\right|\right) / 0.6745$.

This procedure corresponds to a non-linear estimation approach, since the wavelet thresholding scheme is adaptive and signal-dependent, in opposition for instance to the optimal linear Wiener filter (Mallat, 1998) or to thresholding policies that tacitly eliminate all the high-frequency coefficients (for certain scales selected a priori), sometimes also referred to as smoothing techniques (Depczynsky *et al.*, 1999b).

Since the first results published by Donoho and Johnstone, there have been numerous contributions concerning modifications and extensions to the above procedure, to improve its performance in certain application scenarios. For instance, orthogonal wavelet transforms lack the translation-invariant property and this often causes the appearance of artefacts (also known as pseudo-Gibbs phenomena) in the neighbourhood of discontinuities. Coifman and Donoho (1995) proposed a translation-invariant procedure that essentially consists of averaging out several denoised versions of the signal (using orthogonal wavelets), for several shifts, after unshifting. In simple terms, the procedure consists of 'Average [Shift – Denoise – Unshift]', a scheme which Coifman called 'cycle spinning'. With this procedure, not only are pseudo-Gibbs phenomena near discontinuities greatly reduced, but also the results are often so good that lower sampling rates may be used. The choice of a proper thresholding criterion has also been the subject of various contributions, with several alternative approaches being proposed (Nason, 1995; Jansen, 2001). The simultaneous choice of the decomposition level and wavelet filter is also addressed by Pasti *et al.* (1999).

Image denoising does not encompass any significant difference from one-dimensional signal denoising, apart from the fact that a two-dimensional wavelet transform is now required. Two-dimensional wavelet transforms can be calculated using one-dimensional orthogonal wavelets, by alternately processing the rows and columns of the matrix with pixel intensities, which implicitly corresponds to using separable 2D wavelet basis (tensor products of 1D basis functions). Non-separable 2D functions are also available (Vetterli and Kovačevič, 1995; Mallat, 1998; Jansen, 2001). In 1D and 2D data analysis, an extension of the wavelet transform is often used, called wavelet packets. Wavelet packets provide a library or dictionary of basis sets for a given wavelet transform, by successively decomposing not only the approximation signals at increasingly coarser scales using high-pass and low-pass filters (as in the orthogonal wavelet transform), but also the detail signals. As a result, the frequency bands in Figure 13.6(b) are successively halved at each stage, and in the end one has $2^{N/2}$ different basis sets for a signal of length N (Mallat, 1998). In order to choose a basis set for a given application, efficient algorithms have been developed to find the one that optimises a chosen quality criteria, such as entropy (Coifman and Wickerhauser, 1992). Therefore, wavelet packets with basis selection provide an adaptive multiscale representation framework that matches time/frequency features of each signal (1D or 2D), in a more flexible and effective way than orthogonal wavelets, because now we not only have more freedom in covering the time/frequency plane using different arrangements of tiles, but also can select the best tiling for a specific application, such as the one that maximises energy compaction, something very useful in compression applications. This added flexibility does not come, however, without a cost, since the computational complexity is no longer $O(N)$, as happens with the orthogonal wavelet transform, but $O(N \log(N))$ (Coifman and Wickerhauser, 1992; Burrus *et al.*, 1998; Mallat, 1998).

The approaches referred to so far involve implementing denoising schemes by off-line data processing. Within the scope of *on-line data rectification*, where the goal is also to remove errors from measurements in order to make them useful for other tasks, such as process control, monitoring and faults diagnosis, Nounou and Bakshi (1999) proposed an approach for cases where no underlying process model is available. It consists of implementing the classical denoising algorithm with a boundary-corrected filter, in a sliding window of dyadic length, retaining only the last point of the reconstructed signal for on-line use (*on-line multiscale rectification*). If there is some redundancy between the different variables acquired, Bakhsi *et al.* (1997) present a methodology where PCA is used to build up an empirical model for exploring such redundancies.

13.3.2 Signal, image and video compression

The main goal of compression is to represent the original signal (audio, image or even a sequence of images) with as few bits as possible without compromising its final use. The usual steps involved in a ('lossy') compression scheme are *signal transformation* $\rightarrow$ *quantisation* $\rightarrow$ *entropy coding*. First the signal is expanded into a suitable basis set (that is transformed); then, the expansion coefficients (that is the transform) are mapped into a countable discrete alphabet; finally, another mapping is applied, where a new arrangement of coefficients is built up such that the average number of bits per symbol is minimised. The first and last steps are reversible, so that we can move forward and backward without losing any information, but the second one involves some approximation, and therefore once we have performed quantisation we can no longer fully recover the original coefficients during decompression. It is in this step that many of the small wavelet coefficients are set to zero, and a high percentage of the compression arises from dropping many wavelet coefficients in this way (Strang and Nguyen, 1997; Vetterli and Kovačevič, 1995). At present, the approach just described is being used for instance by the FBI services to store fingerprints with compression ratios of the order of 15:1, using *wavelet transformation* (linear phase 9/7 filter) together with *scalar quantisation* (Strang and Nguyen, 1997). A wavelet-based compression scheme is also used in the JPEG 2000 compression spec, with compression levels of up to 200:1 being obtained for images in the 'lossy' mode (where the potential of using wavelets can be fully used; a 9/7 wavelet filter is used) while the typical 2:1 compression ratio is obtained in the 'lossless' mode (that is without the quantisation step, using a 5/3 wavelet filter). To get some practical insight into the reasons for the success of wavelet-based compression we present in Figure 13.9 a fingerprint digitised image along with another version in which only 5 % of the original wavelet packet coefficients were retained in order to reconstruct the image, with all the remaining ones set to zero (a

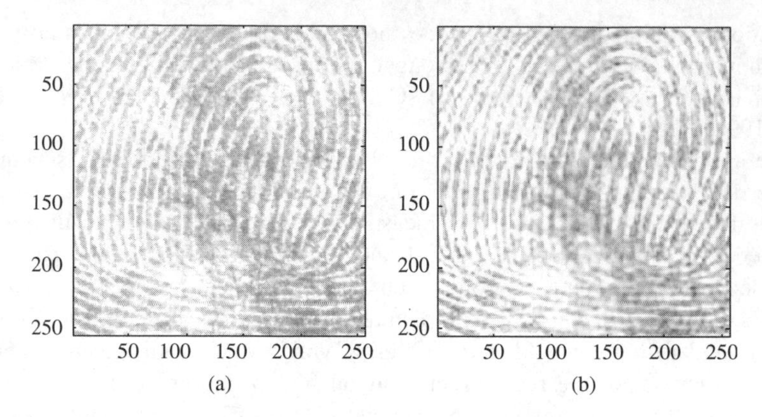

Figure 13.9 (a) Original digitised fingerprint image and (b) a compressed version in which 95 % of the wavelet packet coefficients were set equal to zero.

basis was selected using the best-basis algorithm of Coifman and Wickerhauser, as implemented in the WaveLab toolbox). Despite the high percentage of disregarded coefficients, the quality of the reconstructed image remains quite close to the original one.

13.3.3 Process monitoring

Several process monitoring approaches based on a multiscale representation of data have been developed to detect various time/frequency disturbances that affect normal operation. Top and Bakshi (1998) suggested the idea of following the trends of wavelet coefficients at different scales using separate control charts. The state of the process is confirmed by converting the signal back to the time domain, using only coefficients from the scales where they have exceeded the control limits, and checking them against a detection limit based on such scales where significant events were detected. The approximate decorrelation ability of the wavelet transform makes this approach suitable even for autocorrelated processes, and the use of scale-dependent limits effectively accommodates the signal power spectrum. Furthermore, with the energy compaction property of wavelet transforms, deterministic events are quickly detected. The multiscale nature of the framework led the authors to point out that it unifies the Shewhart, cumulative sum and exponentially weighted moving average procedures, as these control charts differ in the scale at which they represent data (Bakshi, 1999).

Regarding multivariate applications, Bakshi (1998) developed multiscale principle component analysis (MSPCA), which combines the wavelet transform's ability to approximately decorrelate autocorrelated processes and enable the detection of deterministic features present in the signal, together with PCA's ability to extract the variables' correlation structure. MSPCA can then be applied to process monitoring, by computing independently principal components models and control limits for SPC-PCA control charts at each scale. Once again, the scales at which significant activity is detected are those that will be used to reconstruct the covariance matrix at the finest scale, in a scale-recursive procedure, and a final test is performed at this scale to confirm or refute the occurrence of abnormal perturbations.

Several other works report improvements and changes to this original procedure. In Kano *et al.* (2002) the process monitoring procedure based on MSPCA is integrated with monitoring methods designed to detect changes over data distribution features. Misra *et al.* (2002) used MSPCA with variable grouping and the analysis of contribution plots when an event is detected in the control charts at any scale, in order to monitor the process and simultaneously perform early fault diagnosis. Reis and Saraiva (2006a) presented an extension of multiscale statistical process control for integrating multiresolution data, as the original procedure was designed to handle signals represented at a single resolution or scale. A non-linear process monitoring approach, similar to MSPCA, but with a non-linear modelling step based upon an input-training neural network, was presented by Fourie and de Vaal (2000).

MSPCA is an example where tools from multivariate and multiscale data analysis are combined in order to deal with the multivariate and multiscale nature of data, by taking the 'best of both worlds'. But there are other monitoring approaches where this combination is also explored. Trygg *et al.* (2001) applied a 2D wavelet transformation to compress data from near-infrared spectra collected over time and estimated a PCA model for this 2D compressed matrix. Such a model was then used to check whether new incoming spectra deviate from those collected during normal operation using PCA diagnostic tools. In the context of a PLS-based monitoring scheme, Teppola and Minkkinen (2000) used a multiscale decomposition in order to remove seasonal and low-frequency contributions to variation. A scale (or frequency band) selection methodology is also used by Luo *et al.* (1999) to choose the frequency bands on which one should focus when trying to detect sensor faults. Bakhtazad *et al.* (2000) developed a framework for the detection and classification of abnormal situations using the so-called multidimensional wavelet domain hidden Markov trees (a multidimensional hidden Markov tree built over the wavelet coefficients calculated from the scores of a PCA model, estimated from pre-processed raw data; see also Crouse *et al.*, 1998; Sun *et al.*, 2003).

13.3.4 System identification

System identification, according to L.A. Zadeh, is 'the determination, on the basis of input and output, of a system within a specified class of systems, to which the system under test is equivalent' (Åstrom and Eykhoff, 1971). It plays a central role in any application that requires adequate representations for input/output relationships (Ljung, 1999). There are many ways in which wavelets can be used in this context and their application scenarios range from time-invariant systems (Kosanovich *et al.*, 1995) to non-linear black-box modelling.

Observing that all standard linear-in-parameter system identification methods can be understood as projections onto a given basis set, Carrier and Stephanopoulos (1998) applied wavelet bases in order to improve the performance of those methods in estimating reduced-order models and non-linear systems, as well as systems corrupted with noise and disturbances. Tsatsanis and Giannakis (1993) also use wavelets as basis functions in the identification of time-varying systems (see also Doroslovački and Fan, 1996). Nikolaou and Vuthandam (1998) presented a methodology for estimating finite-impulse response models by compressing the kernel (sequence of coefficients that characterise the model) using a wavelet expansion. Applications in non-linear black-box modelling include the use of wavelets in the identification of Hammerstein model structures (Hasiewicz, 1999) and neural networks with wavelets as activation functions (Zhang and Benveniste, 1992; Bakshi and Stephanopoulos, 1993).

The wavelet decomposition may also suggest alternative model structures. Basseville *et al.* (1992) introduced the theory of multiscale autoregressive processes within the scope of dynamic systems built upon trees, something that is also suggested by the wavelet decomposition scheme and provides a natural and simple way to model the relationships between data from different sensors at different resolutions, with applications to sensor fusion (see also Willsky, 2002). By 'translating' discrete state-space models in a discrete time grid into multiscale models in homogeneous trees whose nodes and branches define a new 2D time-scale grid where the wavelet coefficients of the inputs and outputs signals evolve and interact, Stephanopoulos *et al.* (1997) laid the ground for a framework that can efficiently be used to simulate linear systems, for multiscale optimal control and for multiscale state estimation and optimal data fusion.

Examples of applications in the field of time series analysis include the following: estimation of parameters that define long-range dependency (Abry *et al.*, 1998; Veitch and Abry, 1999; Whitcher, 2004; Percival and Walden, 2000) and the analysis of $1/f$ processes (Wornell, 1990; Wornell and Oppenheim, 1992; Percival and Walden, 2000); prediction in time series using either a multiscale model structure (Renaud *et al.*, 2003) or a multiscale version of a state-space model structure (Chui and Chen, 1999); and scale-wise decomposition of the sample variance of a time series – wavelet-based analysis of variance (Percival and Walden, 2000).

13.3.5 Regression, classification and clustering

Wavelets have been applied in regression analysis, with the purpose of compressing the predictor space, when it consists of a spectrum (as in multivariate calibration), in order to reduce the variance of predictions by eliminating components with low predictive power (Depczynsky *et al.*, 1999a; Jouan-Rimbaud *et al.*, 1997; Alsberg *et al.*, 1998; Cocchi *et al.*, 2003). With analogous purposes, in classification (or supervised machine learning), where a preliminary dimensional reduction step is usually involved (Pal and Mitra, 2004; Walczak *et al.*, 1996), the wavelet transform has also been used to generate those elements in the reduced dimension space that can be more effectively classified into one among several classes (Walczak *et al.*, 1996; Alsberg, 2000; Cocchi *et al.*, 2001; Theodoridis and Koutroumbas, 2003; Bakshi and Stephanopoulos, 1995). Applications to clustering (unsupervised machine learning) explore the interaction of patterns with the scale in which they are presented, to enhance the clustering analysis (Alsberg, 1999; Han and Kamber, 2001; Wang, 1999; Murtagh and Starck, 1998).

13.3.6 Numerical analysis

Within the scope of numerical analysis, wavelets have been used in the solution of systems of equations (Tewfik, 1994) and differential equations (Nikolaou and You, 1994; Louis *et al.*, 1997). Beylkin *et al.* (1991) address the issue of fast application of dense matrices to vectors, using a class of numerical algorithms based upon wavelets.

13.4 Conclusions and some current trends

In this chapter we have addressed two important issues in data analysis, which are especially relevant when dealing with large amounts of data. The first concerns the effective extraction of inner relationships among variables, and adequate frameworks were presented that are built around the concept of latent variable modelling. The second issue concerns the extraction of data regularities due to the presence of autocorrelation or multiscale phenomena, for which we have presented several approaches able to deal with such features, based upon wavelet transform theory. We have also illustrated the practical usefulness underlying these two central concepts with examples from different fields. Now, we will conclude this chapter with some observations on current topics and trends in multivariate and multiscale statistical approaches.

The specification and use of measurement uncertainties (ISO, 1993; Lira, 2002) in addition to measured values, is a practice that is increasingly being adopted, due to the development of more reliable and 'informative' sensors, on the one hand, and of the necessary metrological procedures, on the other. Data analysis methods should therefore also take advantage of these developments, and in fact several efforts have been made recently to explicitly incorporate measurement uncertainties into several data analysis tasks. Wentzell *et al.* (1997a) developed the so-called maximum likelihood principal components analysis, which estimates optimal PCA models in a maximum likelihood sense, when data are affected by measurement errors exhibiting an uncertainty structure known a priori. The reasoning underlying MLPCA was also applied by Wentzell *et al.* (1997b) to input/output modelling approaches closely related to PCA, such as principal components regression and latent root regression, giving rise to their maximum likelihood versions: maximum likelihood principal components regression (MLPCR) and maximum likelihood latent root regression. The MLPCR approach was further modified by Martínez *et al.* (2002), who replaced ordinary least squares in the regression step of MLPCR by multivariate least squares, which explicitly takes data uncertainty into account.

Reis and Saraiva (2005) proposed new modifications to some current linear regression methodologies in order to incorporate data uncertainty information in their formulations, and conducted a Monte Carlo comparison study encompassing several methods. Faber and Kowalski (1997) considered the influence of measurement errors in the calculation of confidence intervals for parameters and predictions made using PCR and partial least squares.

As far as multiscale approaches are concerned, new fields of application are emerging, such as the monitoring of profiles, where wavelet-based approaches have already proved to be promising, namely in effectively extracting the several phenomena going on at different scales in order to improve their simultaneous supervision (Reis and Saraiva, 2006b). However, we would like to stress that the interest in these classes of approaches does not always arise from the multiscale nature of the underlying phenomena. For instance, the fact that data is acquired and/or presented at multiple resolutions[6] can motivate the use of such approaches, regardless of whether or not the phenomena themselves have relevant multiscale features (Willsky, 2002).

References

Abry, P., Veitch, D. and Flandrin, P. (1998) Long-range dependence: Revisiting aggregation with wavelets. *Journal of Time Series Analysis*, **19**(3), 253–266.

Alsberg, B.K. (1999) Multiscale cluster analysis. *Analytical Chemistry*, **71**, 3092–3100.

Alsberg, B.K. (2000) Parsimonious multiscale classification models. *Journal of Chemometrics*, **14**, 529–539.

Alsberg, B.K., Woodward, A.M., Winson, M.K., Rowland, J.J. and Kell, D.B. (1998) Variable selection in wavelet regression models. *Analytica Chimica Acta*, **368**, 29–44.

Åström, K.J. and Eykhoff, P. (1971) System identification – a survey. *Automatica*, **7**, 123–162.

Bakhtazad, A., Palazoğlu, A. and Romagnoli, J.A. (2000) Detection and classification of abnormal process situations using multidimensional wavelet domain hidden Markov trees. *Computers and Chemical Engineering*, **24**, 769–775.

Bakshi, B.R. (1998) Multiscale PCA with application to multivariate statistical process control. *AIChE Journal*, **44**(7), 1596–1610.

Bakshi, B.R. (1999) Multiscale analysis and modelling using wavelets. *Journal of Chemometrics*, **13**, 415–434.

Bakshi, B.R. and Stephanopoulos, G. (1993) Wave-Net: A multiresolution, hierarchical neural network with localised learning. *AIChE Journal*, **39**(1), 57–81.

Bakshi, B.R. and Stephanopoulos, G. (1995) Reasoning in time: Modelling, analysis, and pattern recognition of temporal process trends. In J. Anderson, G. Stephanopoulos and C. Han (eds), *Intelligent Systems in Process Engineering, Part II: Paradigms from Process Operations* (Advances in Chemical Engineering 22), pp. 485–548. Academic Press, San Diego, CA.

[6] One example can be found in the data fusion problem of satellite measurements with measurements taken from stations located at the surface (Willsky, 2002).

Bakshi, B.R., Bansal, P. and Nounou, M.N. (1997) Multiscale rectification of random errors without process models. *Computers and Chemical Engineering*, **21**(Suppl.), S1167–S1172.

Basseville, M., Benveniste, A., Chou, K.C., Golden, S.A., Nikoukhah, R. and Willsky, A.S. (1992) Modelling and estimation of multiresolution stochastic processes. *IEEE Transactions on Information Theory*, **38**(2), 766–784.

Beylkin, G., Coifman, R. and Rokhlin, V. (1991) Fast wavelet transforms and numerical algorithms I. *Communications on Pure and Applied Mathematics*, **XLIV**, 141–183.

Bruce, A., Donoho, D. and Gao, H.-Y. (1996) Wavelet analysis. *IEEE Spectrum* (October), 26–35.

Burnham, A.J.,; MacGregor, J.F. and Viveros, R. (1999) Latent variable multivariate regression modelling. *Chemometrics and Intelligent Laboratory Systems*, **48**, 167–180.

Burrus, C.S., Gopinath, R.A. and Guo, H. (1998) *Introduction to Wavelets and Wavelet Transforms – A Primer*. Prentice Hall, Upper Saddle River, NJ.

Carrier, J.F. and Stephanopoulos, G. (1998) Wavelet-based modulation in control-relevant process identification. *AIChE Journal*, **44**(2), 341–360.

Cholakov, G.S., Wakeham, W.A. and Stateva, R.P. (1999) Estimation of normal boiling point of hydrocarbons from descriptors of molecular structure. *Fluid Phase Equilibria*, **163**, 21–42.

Chui, C. and Chen, G. (1999) *Kalman Filtering – with Real-Time Applications*, 3rd edition. Springer-Verlag, Berlin.

Cocchi, M., Seeber, R. and Ulrici, A. (2001) WPTER: Wavelet packet transform for efficient pattern recognition of signals. *Chemometrics and Intelligent Laboratory Systems*, **57**, 97–119.

Cocchi, M., Seeber, R. and Ulrici, A. (2003) Multivariate calibration of analytical signals by WILMA (Wavelet Interface to Linear Modelling Analysis). *Journal of Chemometrics*, **17**, 512–527.

Cohen, A., Daubechies, I. and Vial, P. (1993) Wavelets on the interval and fast wavelet transforms. *Applied and Computational Harmonic Analysis*, **1**, 54–81.

Coifman, R.R. and Donoho, D.L. (1995) Translation-invariant de-noising. Technical report, Department of Statistics, Stanford University.

Coifman, R.R. and Wickerhauser, M.V. (1992) Entropy-based algorithms for best basis selection. *IEEE Transaction on Information Theory*, **38**(2), 713–718.

Crouse, M.S., Nowak, R.D. and Baraniuk, R.G. (1998) Wavelet-based statistical signal processing using hidden Markov models. *IEEE Transactions on Signal Processing*, **46**(4), 886–902.

Daubechies, I. (1992) *Ten Lectures on Wavelets*. SIAM, Philadelphia.

Depczynsky, U., Jetter, K., Molt, K. and Niemöller, A. (1999a) Quantitative analysis of near infrared spectra by wavelet coefficient regression using a genetic algorithm. *Chemometrics and Intelligent Laboratory Systems*, **47**, 179–187.

Depczynsky, U., Jetter, K., Molt, K. and Niemöller, A. (1999b) The fast wavelet transform on compact intervals as a tool in chemometrics – II. Boundary effects, denoising and compression. *Chemometrics and Intelligent Laboratory Systems*, **49**, 151–161.

Dijekerman, R.W. and Mazumdar, R.R. (1994) Wavelet representations of stochastic processes and multiresolution stochastic models. *IEEE Transactions on Signal Processing*, **42**(7), 1640–1652.

Donoho, D.L. (1995) De-noising by soft-thresholding. *IEEE Transactions on Information Theory*, **41**(3), 613–627.

Donoho, D.L. and Johnstone, I.M. (1992) Ideal spatial adaptation by wavelet shrinkage. Technical report, Department of Statistics, Stanford University

Donoho, D.L. and Johnstone, I.M. (1994) Ideal spatial adaptation by wavelet shrinkage. *Biometrika*, **81**(3), 425–455.

Doroslovački, M.I. and Fan, H. (1996) Wavelet-based linear system modelling and adaptive filtering. *IEEE Transactions on Signal Processing*, **44**(5), 1156–1167.

Eriksson, L., Johansson, E., Kettaneh-Wold, N. and Wold, S. (2001) *Multi- and Megavariate Data Analysis – Principles and Applications*. Umetrics AB.

Estienne, F., Pasti, L., Centner, V., Walczak, B., Despagne, F., Rimbaud, D.J., de Noord, O.E. and Massart, D.L. (2001) A comparison of multivariate calibration techniques applied to experimental NIR data sets. Part II. Predictive ability under extrapolation conditions. *Chemometrics and Intelligent Laboratory Systems*, **58**, 195–211.

Faber K. and Kowlaski B.R. (1997) Propagation of measurement errors for the validation of predictions obtained by principal component regression and partial least squares. *Journal of Chemometrics*, **11**, 181–238.

Flores-Cerrilo, J. and MacGregor, J.F. (2004) Control of batch product quality by trajectory manipulation using latent variable models. *Journal of Process Control*, **14**, 539–553.

Fourie, S.H. and de Vaal, P. (2000) Advanced process monitoring using an on-line nonlinear multiscale principal component analysis methodology. *Computers and Chemical Engineering*, **24**, 755–760.

Gabrielsson, J., Lindberg, N.-O. and Lundstedt, T. (2002) Multivariate methods in pharmaceutical applications. *Journal of Chemometrics*, **16**, 141–160.

Geladi, P. and Kowalski, B.R. (1986) Partial least-squares regression: A tutorial. *Analytica Chimica Acta*, **185**, 1–17.

Han, J. and Kamber, M. (2001) *Data Mining – Concepts and Techniques*. Morgan Kaufmann, San Francisco.

Hasiewicz, Z. (1999) Hammerstein system identification by the Haar multiresolution approximation. *International Journal of Adaptive Control and Signal Processing*, **13**, 691–717.

Helland, I.S. (1988) On the structure of partial least squares regression. *Communications in Statistics – Simulation and Computation*, **17**(2), 581–607.

Höskuldsson, A. (1988) PLS regression methods. *Journal of Chemometrics*, **2**, 211–228.

Hubbard, B.B. (1998) *The World According to Wavelets*. A.K. Peters, Natick, MA.

ISO (1993) *Guide to the Expression of Uncertainty*. ISO, Geneva.

Jackson, J.E. (1991) *A User's Guide to Principal Components*. John Wiley & Sons, Inc., New York.

Jaeckle, C. and MacGregor, J.F. (1998) Product design through multivariate statistical analysis of process data. *AIChE Journal*, **44**(5), 1105–1118.

Jaeckle, C. and MacGregor, J.F. (2000) Product transfer between plants using historical process data. *AIChE Journal*, **46**(10), 1989–1997.

Jansen, M. (2001) *Noise Reduction by Wavelet Thresholding*. Springer-Verlag, New York.

Jouan-Rimbaud, D., Walczak, B., Poppi, R.J., de Noord, O.E. and Massart, D.L. (1997) Application of wavelet transform to extract the relevant component from spectral data for multivariate calibration. *Analytical Chemistry*, **69**, 4317–4323.

Kaiser, G. (1994) *A Friendly Guide to Wavelets*. Birkhäuser, Boston.

Kano, M., Nagao, K., Hasebe, S., Hashimoto, I., Ohno, H., Strauss, R. and Bakshi, B.R. (2002) Comparison of multivariate statistical process monitoring methods with applications to the Eastman challenge problem. *Computers and Chemical Engineering*, **26**, 161–174.

Kaspar, M.H. and Ray, W.H. (1993) Dynamic PLS modelling for process control. *Chemical Engineering Science*, **48**, 3447–3461.

Kosanovich, K.A., Moser, A.R. and Piovoso, M.J. (1995) Poisson wavelets applied to model identification. *Journal of Process Control*, **4**, 225–234.

Kourti, T. and MacGregor, J.F. (1995) Process analysis, monitoring and diagnosis, using multivariate projection methods. *Chemometrics and Intelligent Laboratory Systems*, **28**, 3–21.

Kourti, T. and MacGregor, J.F. (1996) Multivariate SPC methods for process and product monitoring. *Journal of Quality Technology*, **28**(4), 409–428.

Kresta, J.V., MacGregor, J.F. and Marlin, T.E. (1991) Multivariate statistical monitoring of process operating performance. *Canadian Journal of Chemical Engineering*, **69**(1991): 35–47.

Kresta, J.V., Marlin, T.E. and MacGregor, J.F. (1994) Development of inferential process models using PLS. *Computers and Chemical Engineering*, **18**(7), 597–611.

Ku, W., Storer, R.H. and Gergakis, C. (1995) Disturbance detection and isolation by dynamic principal component analysis. *Chemometrics and Intelligent Laboratory Systems*, **30**, 179–196.

Lakshminarayanan, S., Shah, S.L. and Nandakumar, K. (1997) Modelling and control of multivariable processes: Dynamic PLS approach. *AIChE Journal*, **43**(9), 2307–2322.

Lira, I. (2002) *Evaluating the Measurement Uncertainty – Fundamentals and Practical Guidance*. Institute of Physics Publishing, Bristol.

Ljung, L. (1999) *System Identification: Theory for the User*, 2nd edition. Prentice Hall, Upper Saddle River, NJ.

Louis, A.K., Maass, P. and Rieder A. (1997) *Wavelets – Theory and Applications*. John Wiley & Sons, Ltd, Chichester.

Luo, R., Misra, M. and Himmelblau, D.M. (1999) Sensor Fault detection via multiscale analysis and dynamic PCA. *Industrial and Engineering Chemistry Research*, **38**, 1489–1495.

MacGregor, J.F. and Kourti, T. (1995) Statistical process control of multivariate processes. *Control Engineering Practice*, **3**(3), 403–414.

MacGregor, J.F. and Kourti, T. (1998) Multivariate statistical treatment of historical data for productivity and quality improvements. In J. Pekny, G. Blau and B. Carnahan, (eds), *Foundations of Computer-Aided Process Operations* (AIChE Symposium Series 94), pp. 31–41. American Institute of Chemical Engineers, New York.

MacGregor, J.F., Jaeckle, C., Kiparissides, C. and Koutoudi, M. (1994) Process monitoring and diagnosis by multiblock PLS methods. *AIChE Journal*, **40**(5), 826–838.

Mallat, S. (1989) A theory for multiresolution signal decomposition: The wavelet representation. *IEEE Transaction on Pattern Analysis and Machine Intelligence*, **11**(7), 674–693.

Mallat, S. (1998) *A Wavelet Tour of Signal Processing*. Academic Press, San Diego, CA.

Martens, H. and Næs, T. (1989) *Multivariate Calibration*. John Wiley & Sons, Ltd, Chichester.

Martínez, A., Riu, J. and Rius, F.X. (2002) Application of the multivariate least squares regression to PCR and maximum likelihood PCR techniques. *Journal of Chemometrics*, **16**, 189–197.

Misra, M., Yue, H.H., Qin, S.J. and Ling, C. (2002) Multivariate process monitoring and fault diagnosis by multi-scale PCA. *Computers and Chemical Engineering*, **26**, 1281–1293.

Murtagh, F. and Stark, J.-L. (1998) Pattern clustering based on noise modelling in wavelet space. *Pattern Recognition*, **31**(7), 845–855.

Nason, G.P. (1995) Choice of the threshold parameter in wavelet function estimation. In A. Antoniadis and G. Oppenheim (eds), *Wavelets and Statistics*, pp. 261–299. Springer-Verlag, New York.

Nelson, P.R.C., Taylor, P.A. and MacGregor, J.F. (1996) Missing data methods in PCA and PLS: Score calculations with incomplete observations. *Chemometrics and Intelligent Laboratory Systems*, **35**, 45–65.

Nikolaou, M. and Vuthandam, P. (1998) FIR model identification: Parsimony through kernel compression with wavelets. *AIChE Journal*, **44**(1), 141–150.

Nikolaou, M. and You, Y. (1994) Use of wavelets for numerical solution of differential equations. In R.L. Motard and B. Joseph (eds), *Wavelet Applications in Chemical Engineering*, pp. 209–274. Kluwer Academic, Boston.

Nomikos, P. and MacGregor, J.F. (1995) Multivariate SPC charts for monitoring batch processes. *Technometrics*, **37**(1), 41–59.

Nounou, M.N. and Bakshi, B.R. (1999) On-line multiscale filtering of random and gross errors without process models. *AIChE Journal*, **45**(5), 1041–1058.

Pal, S.K. and Mitra, P. (2004) *Pattern Recognition Algorithms for Data Mining*. Chapman & Hall/CRC, Boca Raton, FL.

Pasti, L., Walczak, B., Massart, D.L. and Reschiglian, P. (1999) Optimisation of signal denoising in discrete wavelet transform. *Chemometrics and Intelligent Laboratory Systems*, **48**, 21–34.

Percival, D.B. and Walden, A.T. (2000) *Wavelets Methods for Time Series Analysis*. Cambridge University Press, Cambridge.

Reis, M.S. and Saraiva, P.M. (2005) Integration of data uncertainty in linear regression and process optimisation. *AIChE Journal*, **51**(11), 3007–3019.

Reis, M.S. and Saraiva, P.M. (2006a) Multiscale statistical process control with multiresolution data. *AIChE Journal*, **52**(6), 2107–2119.

Reis, M.S. and Saraiva, P.M. (2006b) Multiscale statistical process control of paper surface profiles. *Quality Technology and Quantitative Management*, **3**(3), 263–282.

Renaud, O., Starck, J.-L. and Murtagh, F. (2003) Prediction based on a multiscale decomposition. *International Journal of Wavelets, Multiresolution and Information Processing*, **1**(2), 217–232.

Ricker, N.L. (1988) The use of biased least-squares estimators for parameters in discrete-time pulse-response models. *Industrial and Engineering Chemistry Research*, **27**, 343–350.

Rioul, O. and Vetterli, M. (1991) Wavelets and signal processing. *IEEE Signal Processing Magazine*, **8**(4), 14–38.

Shi, R. and MacGregor, J.F. (2000) Modelling of dynamic systems using latent variable and subspace methods. *Journal of Chemometrics*, **14**, 423–439.

Smilde, A.K., Westerhuis, J.A. and de Jong, S. (2003) A framework for sequential multiblock component methods. *Journal of Chemometrics*, **17**, 323–337.

Smilde, A.K., Bro, R. and Geladi, P. (2004) *Multi-way Analysis with Applications in the Chemical Sciences*. John Wiley & Sons, Ltd, Chichester.

Stephanopoulos, G., Karsligil, O. and Dyer, M. (1997) A multi-scale systems theory for process estimation and control. Basis report of a paper presented at the NATO-ASI conference on 'Nonlinear Model Based Process Control', Antalya, Turkey, 10–20 August.

Strang, G. and Nguyen, T. (1997) *Wavelets and Filter Banks*. Wellesley-Cambridge Press, Wellesley, MA.

Sun, W., Palazoğlu, A. and Romagnoli, J.A. (2003) Detecting abnormal process trends by wavelet domain hidden Markov models. *AIChE Journal*, **49**(1), 140–150.

Teppola, P. and Minkkinen, P. (2000) Wavelet-PLS regression models for both exploratory data analysis and process monitoring. *Journal of Chemometrics*, **14**, 383–399.

Tewfik, A.H. (1994) Fast positive definite linear system solvers. *IEEE Transactions on Signal Processing*, **42**(3), 572–585.

Theodoridis, S. and Koutroumbas, K. (2003) *Pattern Recognition*, 2nd edition. Elsevier, Amsterdam.

Top, S. and Bakshi, B.R. (1998) Improved statistical process control using wavelets. In J. Pekny, G. Blau and B. Carnahan, (eds), *Foundations of Computer-Aided Process Operations* (AIChE Symposium Series 94), pp. 332–337. American Institute of Chemical Engineers, New York.

Trygg, J., Kettaneh-Wold, N. and Wallbäcks, L. (2001) 2D wavelet analysis and compression of on-line process data. *Journal of Chemometrics*, **15**, 299–319.

Tsatsanis, M.K. and Giannakis, G.B. (1993) Time-varying identification and model validation using wavelets. *IEEE Transactions on Signal Processing*, **41**(12), 3512-3274.

Veitch, D. and Abry, P. (1999) A wavelet-based joint estimator of the parameters of long-range dependence. *IEEE Transactions on Information Theory*, **45**(3), 878-897.

Vetterli, M. and Kovačevič, J. (1995) *Wavelets and Subband Coding*. Prentice Hall, Englewood Cliffs, NJ.

Walczak, B., van den Bogaert, V. and Massart, D.L. (1996) Application of wavelet packet transform in pattern recognition of near-IR data. *Analytical Chemistry*, **68**, 1742–1747.

Walczak, B. and Massart, D.L. (2001) Dealing with missing data. *Chemometrics and Intelligent Laboratory Systems*, **58**, 15–27, 29–42.

Wang, X.Z. (1999) *Data Mining and Knowledge Discovery for Process Monitoring and Control*. Springer-Verlag, London.

Wentzell, P.D., Andrews, D.T., Hamilton, D.C., Faber, K. and Kowalski, B.R. (1997a) Maximum likelihood principal component analysis. *Journal of Chemometrics*, **11**, 339–366.

Wentzell, P.D., Andrews, D.T. and Kowalski, B.R. (1997b) Maximum likelihood multivariate calibration. *Analytical Chemistry*, **69**, 2299–2311.

Westerhuis, J.A., Kourti, T. and MacGregor, J.F. (1998) Analysis of multiblock and hierarchical PCA and PLS models. *Journal of Chemometrics*, **12**, 301–321.

Westerhuis, J.A., Hourti, T. and MacGregor, J.F. (1999) Comparing alternative approaches for multivariate statistical analysis of batch process data. *Journal of Chemometrics*, **13**, 397–413.

Westerhuis, J.A., Gurden, S.P. and Smilde, A.K. (2000) Generalised contribution plots in multivariate statistical process monitoring. *Chemometrics and Intelligent Laboratory Systems*, **51**, 95–114.

Whitcher, B. (2004) Wavelet-based estimation for seasonal long-memory processes. *Technometrics*, **46**(2), 225–238.

Willsky, A.S. (2002) Multiresolution Markov models for signal and image processing. *Proceedings of the IEEE*, **90**(8), 1396–1458.

Wise, B.M. and Gallagher, N.B. (1996) The process chemometrics approach to process monitoring and fault detection. *Journal of Process Control*, **6**(6), 329–348.

Wold, S., Sjöström, M., Carlson, R., Lundstedt, T., Hellberg, S., Skagerberg, B., Wikström, C. and Öhman, J. (1986) Multivariate design. *Analytica Chimica Acta*, **191**, 17–32.

Wold, S., Berglung, A. and Kettaneh, N. (2002) New and old trends in chemometrics. How to deal with the increasing data volumes in R&D&P (research development and production) – with examples from pharmaceutical research and process modelling. *Journal of Chemometrics*, **16**, 377–386.

Wornell, G.W. (1990) A Karhunen-Loève-like expansion for $1/f$ processes via wavelets. *IEEE Transactions on Information Theory*, **36**(4), 859–861.

Wornell, G.W. and Oppenheim, A.V. (1992) Estimation of fractal signals from noisy measurements using wavelets. *IEEE Transactions on Signal Processing*, **40**(3), 611–623.

Yacoub, F. and MacGregor, J.F. (2004) Product optimisation and control in the latent variable space of nonlinear PLS models. *Chemometrics and Intelligent Laboratory Systems*, **70**, 63–74.

Yu, H. and MacGregor, J.F. (2004) Digital imaging for process monitoring and control with industrial applications. In *7th International Symposium on Advanced Control of Chemical Processes ADCHEM*.

Zhang, Q. and Benveniste, A. (1992) Wavelet networks. *IEEE Transactions on Neural Networks*, **3**(6), 889–897.

Appendix 13A A short introduction to PCR and PLS

Principal components regression is a method that handles the presence of collinearity in the predictor variables. PCR uses those uncorrelated linear combinations of the input variables (**X**) that most explain the input space variability, provided by principal components analysis (PCA), as the new set of predictors onto which the single response variable y is to be regressed. These predictors are orthogonal, and therefore the collinearity problem is solved by disregarding the linear combinations with little power to explain variability. This procedure also leads to an effective stabilization of the matrix inversion operation involved in the parameter estimation task. Thus, if $\mathbf{X}_c(n \times m)$ and $y_c(n \times 1)$ are the mean-centred predictors and response variables, respectively (usually they are also scaled), **T** is the PCA $(n \times p)$ matrix of scores, **P** the $(m \times p)$ matrix of PCA loadings and **E** the $(n \times m)$ residual PCA matrix (obtained for a given number of latent dimensions, m), then we have, in the first place, the following matrix equality from PCA:

$$\mathbf{X}_c = \mathbf{T}\mathbf{P}^T + \mathbf{E}$$

PCR consists of regressing y_c on the PCA scores (**T**), using ordinary least squares:

$$\hat{b}^*_{\text{PCR}} = (\mathbf{T}^T\mathbf{T})^{-1}\mathbf{T}^T y_c,$$

where $\hat{b}^*_{\text{PCR}}$ is the vector of coefficients to be applied to the scores.

Partial least squares is an algorithm widely used in the chemometrics community that can handle correlated predictors with several responses, where moderate amounts of noise and missing data may be present, in the task of estimating a linear multivariate model linking a block of predictor variables (**X**) to a block of response variables (**Y**). Both the **X** and **Y** blocks should be appropriately scaled, as the objective function of PLS is concerned with the calculation of linear combinations of the two blocks that produce scores with maximum *covariance* (Höskuldsson, 1988; Shi and Macgregor, 2000). Basically, as in PCR, PLS finds a set of uncorrelated linear combinations of the predictors, belonging to some lower-dimensional subspace of the **X** block variables space. However, in PLS, this subspace is not the one that, for a given dimensionality, has the highest capacity to explain the overall variability in the **X** block (as happens in PCR), but the one that, while still covering the **X**-variability well, provides a good description of the variability exhibited by the **Y** block variable(s). This is accomplished by finding the linear combinations of the **X** block, $t_i = \mathbf{X}_i w_i$, and **Y** block variables, $u_i = \mathbf{Y}_i q_i$, with maximal covariance (Hoskuldssön, 1988; Kourti and MacGregor, 1995). There are several alternative ways of calculating the PLS model, most of them leading to very similar results. In fact, Helland (1988) proved the equivalence between two such algorithms. One procedure is presented in Table 13.A1, for the case of a single response variable,

Table 13.A1 NIPALS algorithm for PLS with one response variable (Geladi and Kowalsky, 1986).

Step 1. Pre-treatment: centre **X** and y; Scale **X** and y.

Begin **FOR** cycle

$a = 1$: # latent variables

Step 2. Calculate the ath **X**-weight vector (w):

$w = \mathbf{X}^T u/(u^T u)$; $w_{new} = w_{\text{old}}/\|w_{\text{old}}\|$ (normalization)

Note: for $a = 1$, the **Y**-scores, u, are equal to y.

Step 3. Calculate the ath **X**-scores vector (t): $t = \mathbf{X}w$

Step 4. Calculate the ath **X**-loadings vector (p): $p = \mathbf{X}^T t/(t^T t)$

Step 5. Rescale **X**-loadings, **X**-scores and **X**-weights:

$p_{\text{new}} \leftarrow p_{\text{old}}/\|p_{\text{old}}\|$; $t_{\text{new}} \leftarrow t_{\text{old}} \times \|p_{\text{old}}\|$; $w_{\text{new}} \leftarrow w_{\text{old}} \times \|p_{\text{old}}\|$

Step 6. Regression of u on t (calculation of b): $b = u^T t/(t^T t)$

Step 7.Calculation of **X** and **Y** residuals:

$\mathbf{E}_a = \mathbf{E}_{a-1} - t_a \mathbf{p}_a^T (\mathbf{X} = \mathbf{E}_0)$
$\mathbf{F}_a = \mathbf{F}_{a-1} - \mathbf{b}_a t_a (y = \mathbf{F}_0)$

Note: Continue calculations with $\mathbf{E}_a$ playing the role of **X** and $\mathbf{F}_a$ the role of $y(\mathbf{u})$.

End **FOR** cycle

y (in this case the algorithm is not iterative, but convergence of the iterative algorithm for the case of more than one response is usually very fast – see Martens and Næs, 1989; Jackson, 1991; Geladi and Kowalsky, 1986).
Therefore, PLS provides a way of relating the **X** and **Y** matrices through a latent variable model structure as follows:

$$\mathbf{X} = \mathbf{T}\mathbf{P}^T + \mathbf{E}$$

$$\mathbf{Y} = \mathbf{T} \cdot \mathrm{diag}(B) \cdot \mathbf{Q} + \mathbf{F}$$

where **T** is the matrix whose columns are the **X**-scores, **P** is the matrix whose columns are the **X**-loadings, **E** is the final residual matrix for the **X** block, B is the vector with the regression coefficients between the **X** block and the **Y** block scores (t and u, respectively) for each dimension (b), diag is the operator that transforms a vector into a matrix with its elements along the main diagonal, **Q** is the matrix with the **Y** loadings in their columns (for the single **Y** case, it is a row vector with dimension given by the number of latent variables, and with ones in its entries), and **F** is the final residual matrix for the **Y** block. When there is only one response variable, it is also possible to relate the two blocks directly in the standard multivariate linear format: $y = \mathbf{X}\mathbf{B}^* + \mathbf{F}$, with $B^* = \mathbf{W}(\mathbf{P}\mathbf{W})^{-1}B$.

14

Simulation in industrial statistics

David Ríos Insua, Jorge Muruzábal,* Jesus Palomo, Fabrizio Ruggeri, Julio Holgado and Raúl Moreno

14.1 Introduction

Typically, once an organisation has realised that a system is not operating as desired, it will look for ways to improve its performance. Sometimes it will be possible to experiment with the real system and, through observation and the aid of statistics, reach valid conclusions to improve the system. Many chapters in this book illustrate this point. However, experiments with a real system may entail ethical and/or economic problems, which may be avoided by dealing with a prototype, that is, a physical model of the system. Sometimes, it is not feasible or possible to build a prototype, yet we may develop a mathematical model, consisting of equations and constraints, that captures the essential behaviour of the system. This analysis may sometimes be done using analytical or numerical methods. In extreme cases, however, the model may be too complex and we must use simulation instead.

Large complex system simulation has become common practice in many industrial areas such as predicting the performance of integrated circuits, the behaviour of controlled nuclear fusion devices, the properties of thermal energy storage devices or the stresses in prosthetic devices. Essentially, simulation consists of building a computer model that describes the behaviour of a system and experimenting with

*Jorge Muruzábal passed away on 5 August 2006, this being one of his last contributions in a life devoted to the search for truth.

Statistical Practice in Business and Industry Edited by S.Y. Coleman, T. Greenfield, D.J. Stewardson and D.C. Montgomery

this computer model to reach conclusions that support decisions to be made with the real system. The system might be an existing one that needs to be improved or a future one to be built in an optimal way.

In this chapter, we introduce key concepts, methods and tools from simulation with the practitioner of industrial statistics in mind. In order to give a flavour of various simulation applications and distinguish between discrete event (DES) and Monte Carlo (MC) simulation, we describe several real cases. We give an example in gas pipeline reliability to illustrate simulation methods in Bayesian statistics, an example in vehicle crash test modelling, to illustrate large-scale modelling, and a detailed description of a DES model to predict the behaviour of a complex workflow line, which we shall later use throughout to illustrate various sections in this chapter.

14.1.1 Designing a workflow line

The study of moderately complex workflow lines, so as to optimise their performance, quickly leads to mathematical models which may only be dealt with through simulation. The specific context we consider here is that of modelling massive data-capturing lines for financial institutions. This task has been outsourced by banks to companies working in a very competitive environment, which, therefore, look for gains in efficiency.

The process specifically considered here is as follows. Every day several bags arrive at the line containing bank documents to be digitised and recorded. Such documents must be stored on CDs for later use in the bank, with several financial checks in between. The line is meant to deliver the completed task in a given time window. Otherwise, the company is obliged to pay a fine. From input (bags with financial documents) to output (CDs with financial information) the process goes through several stages. Our aim here is to forecast whether the company will deliver the work before the deadline and, if not, detect bottlenecks and suggest line reconfigurations.

We now briefly describe a typical configuration and the whole process within this massive data-capturing line, with the various stages and the operations, resources, objectives and intervening agents involved:

- Paper preparation. The documents received are sorted into the different types by several persons for later processing.

- Document scanning, to produce digitised documents. Two scanners are used (one for back-up). Some of the documents may be illegible and would need to be marked for later special treatment.

- Optical recognition (OCR) of relevant features, such as dates or imports, so as to reduce later workload. Some of these may be wrongly recognised. Several workstations are involved at this stage.

- Recording. Documents are classified as legible or illegible. These are filled manually. Up to 40 people may take part in this process.

- Resolution of illegible documents. In this phase, a person looks for the original document, to digitise it correctly. An operator controls a workstation dedicated to this task.
- Coding. A second image recording takes place with all images that the recorder could not deal with. An operator controls a workstation dedicated to this task.
- Date verification. If necessary, dates are corrected. An operator controls a workstation dedicated to this task.
- Field verification. If necessary, fields are corrected. An operator controls a workstation dedicated to this task.
- Final verification. Every now and then, a document is inspected for correct recording. If necessary, corrections are made. An operator controls a workstation dedicated to this task.
- Balancing the totals. All documents within a batch are checked to verify their imports. A person with a workstation and a balancing program takes care of this.
- Modification. Those documents which did not pass the balancing phase are corrected by contrast with the original documents.
- Export. A file is generated and recorded on CD.

At each of the stages of the scenario described above, the difficulties that need to be dealt with are that the resources are shared by processes, several random processes are involved, several parameters are under control, and others need to be estimated. The basic problem we want to solve is whether the company will be able to finish the work before a limit time h, given the current system configuration, defined by the parameters under control z (number of OCRs, number of workers at each work stage, ...). Formally, if T denotes the actual completion time window, we want to find out whether $\Pr(T \leq h|z)$ will be large enough to guarantee the completion of work. Recall that, otherwise, the company will have to pay a large penalty. If this probability is not big enough, we must reconfigure the system, modifying z. Because of the complexity of the model, we are unable to compute such probability analytically and, consequently, simulation is used.

14.1.2 Reliability of a gas pipeline system

Gas distribution networks can be considered repairable systems whose failures (gas escapes) may be described by non-homogeneous Poisson processes; see the review paper by Ruggeri (2005) and references therein.

In a consulting case, both homogeneous and non-homogeneous Poisson processes were used for 'old' cast iron and steel pipes, respectively, since the former pipes are not ageing, whereas the latter are subject to corrosion. Both maximum likelihood and Bayesian estimation were used on data collected from a gas company. The final objective was to fit a model suitable for the description of the network so as to predict which pipes and operating conditions were more likely to lead to gas escapes, identifying influential factors on escapes and related costs, due, for example, to preventive maintenance and repair. An important issue to note was that the data were only fairly complete for recent years. We also had available experts whom we interviewed to obtain their prior beliefs on the parameters of interest, tested through sensitivity analysis.

Simulation techniques were fundamental in this study. We merely quote several applications:

- Modelling. Typical repairable systems have a 'bathtub'-shaped intensity function, although more complex behaviour is possible. It is quite common to consider various models at different time intervals; this is a typical change-point problem. The estimation of change points and their number and changes in the parameters have been addressed in a Bayesian framework using a simulation technique called reversible jump Markov chain Monte Carlo, due to Green (1995), in which samples from the posterior distributions of the parameters are obtained along with the distribution of the number and the location of change points.

- Parameter estimation. Standard optimisation techniques are used to get maximum likelihood estimates, whereas Markov chain Monte Carlo (MCMC) techniques (mostly Gibbs and Metropolis–Hastings) are now routine in Bayesian analysis and they were used in our study.

- Missing data. Installation dates of failed pipes were sometimes missing and were drawn from a probability distribution.

- Comparison between estimates. Maximum likelihood and Bayes estimates were compared through extensive simulations, drawing data from known processes and trying to estimate their parameters.

14.1.3 Reducing vehicle crash tests

Traditionally, the automobile and aerospace industries have used physical experiments for design and security test purposes, such as car crash tests and wind tunnels. However, as the market becomes more competitive, the time to delivery gets shorter and budgets get tighter. This implies that the number of prototypes built on the scale required to gather sufficient information for decision or research purposes is reduced to the point that it is not sufficient to produce good estimates. In these cases, simulation becomes essential to compensate for the lack of information. A fully Bayesian approach would account for the uncertainty in simulation input parameters with MCMC methods. A problem in this case is that to ensure convergence we have to run the simulation over a large number of iterations, but the simulation model is very complex, so that the whole research and development process is computationally intensive and slow. Therefore, due to time constraints, a proper experimental design should be performed in order to cover the input parameter space with as few iterations as possible (Atkinson and Donev, 1992).

By way of example, we considered the simulation of a model of a vehicle being driven over a road with two major potholes; see Bayarri *et al*. (2007) for full details. The model included 30 parameters but, to simplify, only nine were considered uncertain, the rest being assumed known. The uncertainty on these parameters comes from deviations of manufacturing properties of the materials from their nominal values (measured parameters but with uncertainty), and from non-measured inputs (calibration parameters). The outputs of the simulation were the vertical tensions on the suspension system of a particular vehicle while hitting two consecutive potholes. Each simulation model run takes around four days to produce one output for a particular experimental design vector. Moreover, due to inconsistencies within it, the run may sometimes fail to converge. As a result, we will obtain curves for each of the experimental design vectors, for the vertical tension at different points on the vehicle along the distance that contains both potholes.

The advantages of using a simulation approach in this problem are that the physical experiment is expensive, and that it provides insights into testing a potential vehicle with specifications that are currently infeasible. For example, it could be of interest for the car maker to test the behaviour of a current vehicle model with particular springs that have not yet been developed, but could be ordered from a vendor.

The simplifications usually assumed on these large simulation models cause big differences from the physical experiments. To overcome this drawback, we need to develop methods to estimate the bias. This information can be used in various useful ways – for example, by using it as a target (bias equal to zero) to improve the simulation model, say by variable and model selection, or to produce bias-corrected prediction curves of reality, without performing the physical experiment, for the same physically tested vehicle, another vehicle of the same type, or another vehicle of different type stemming from the simulation curves obtained; see Berger *et al*. (2004) for more details.

14.1.4 Outline of chapter

We have sketched several real case studies concerning applications of simulation in industrial statistics. We shall use the first one to illustrate the basic four-step process in any simulation experiment, once we have estimated the corresponding simulation model:

1. Obtain a source of random numbers.
2. Transform them into inputs to the simulation model.
3. Obtain outputs of the simulation model, stemming from the inputs.
4. Analyse the outputs to reach conclusions.

We devote a section to each of these four steps. Pointers to recent and/or relevant literature and URLs are given.

14.2 Random number generation

We start by introducing key concepts in random number generation, the basic ingredient in any simulation analysis. The issue is how to choose a source that provides a sufficient quantity of (pseudo)random numbers to carry out a simulation experiment.

Sometimes we may appeal to historical data from the relevant process. For example, in a reservoir management problem (see Ríos Insua and Salewicz, 1995), we used the available historical time series of inflows for simulation of various management policies, as this series was long enough and we had no possibility of controlling the source of randomness. However, very frequently, we either lack a sufficiently long series or we want somehow to control the source of randomness. For example, if we are simulating a computer centre, one of the typical sources of uncertainty will be the arrivals process, modelled, say, through a Poisson process. Typically, we will be interested in studying the performance of the system for a long enough period and under different conditions. The first issue suggests that we might need to store an excessive amount of historical data. The second may imply costly transformations, when we modify the parameters. One possibility would be to use some physical mechanism to generate random numbers and store them for later access in some storage device. However, this procedure tends to be too slow in many application areas.

As an alternative, we may use algorithmic procedures for random number generation. The idea, due to Von Neumann (1951), is to produce numbers which appear to be random using arithmetic computer operations: starting with an initial *seed* $(u_0, u_{-1}, \ldots, u_{-p+1})$, generate a sequence from $u_i = d(u_{i-1}, \ldots, u_{i-p})$, for a certain function d. Clearly, once the seed is chosen, the sequence is determined. For that reason, we adopt the following criterion:

Definition 14.1 *A sequence* (u_i) *is of random numbers if non-overlapping h-tuples of subsequent numbers are approximately uniformly distributed on* $(0, 1)^h$*, for* $h = 1, 2, \ldots, n$*, and n sufficiently large for the application of interest.*

Formally, any test applied to a finite part of (u_i) which tried to detect relevant deviations from randomness would not reject the null hypothesis that the numbers are random.

For simulation purposes, these properties of uniformity on (0,1) and independence are complemented with others referring to computational efficiency: speed, little memory consumed, portability, implementation simplicity, reproducibility, mutability, sufficiently long time period, and so on. The last of these is becoming more and more important as we aim to undertake detailed simulations of more complex systems, which demand longer and longer random number series.

The most popular random number generators are the *congruential* family, due to Lehmer (1951). They follow the recursion

$$x_{n+1} = (ax_n + b)\text{mod } m,$$
$$u_n = x_n/m,$$

for a multiplier a, bias b, modulus m and seed x_0. When $b = 0$, they are called *multiplicative*. In spite of their apparent simplicity and predictability, a careful choice of (a, b, m) allows us to provide sufficiently large and random series for many purposes. In this sense, the generator

$$x_{n+1} = (16\,807x_n)\text{mod } (2^{31} - 1),$$
$$u_n = x_n/(2^{31} - 1),$$

originally proposed by Lewis *et al.* (1969), has long been a minimal standard generator, implemented, for example, in Press *et al.* (1992).

However, a number of powerful tests have been designed to reject the randomness of congruential generators; see L'Ecuyer (1998) or L'Ecuyer and Simard (2001). Such tests explode the reticular structure of the generators. For that reason, and because of the need to have longer-period generators (for example, the minimal standard has period $2^{31} - 2$, which is insufficient for many purposes), many other generators have been designed, including perturbations of random number generators, such as shuffling (Bays and Durham, 1976), shift register generators, Fibonacci lagged generators, non-linear generators and mixtures of generators. For example, the S-Plus statistical package implements, among others, Marsaglia's Super-Duper algorithm, which combines a multiplicative and a Tausworthe generator modified to skip outcomes of zero.

In the basic generators introduced so far, the theoretical properties are easy to analyse because of the highly regular structure. However, this is not desirable in terms of randomness. Recently, methods that combine generators from different families have been proposed. They aim at providing better uniformity properties by reducing the regularity of their structure. Examples of these methods are combinations of linear congruential/multiple recursive or Tausworthe generators with any linear or non-linear generator type through the functions $u_n = (u_{Gen_1} + u_{Gen_2})$ mod 1 or $u_n = Bin(u_{Gen_1})$ XOR $Bin(u_{Gen_2})$, respectively. See L'Ecuyer and Granger-Piche (2003) for a full description and their statistical properties and L'Ecuyer *et al.* (2002) for a description of an implementation. For example, one of the most powerful generators is the *Mersenne twister*, proposed by Matsumoto and Nishimura (1998), with period length $2^{19\,937} - 1$.

14.2.1 Pointers to the literature

In general, the practitioner should be somewhat careful with the random number generators provided in commercial software. These may give incorrect answers if the period length is not long enough for a particular application. For example, Sun's

Java standard library, available at http://java.sun.com/j2se/1.3/docs/api/java/util/Random.html, has period length 2^{48}, and Visual Basic's is 2^{24}. See L'Ecuyer (2001) and McCullough (1999) for a set of tests performed on these popular random number generators.

Good sources of random number generators are at StatLib at http://www.stat.cmu.edu/. Other important sites in relation to random number generation are L'Ecuyer's page at http://www.iro.umontreal.ca/~lecuyer and http://random.mat.sbg.ac.at. A set of statistical tests for random number generators are available at http://csrc.nist.gov/rng/rng5.html. Implementations of the minimal standard and other variants can be found in Press *et al*. (1992).

14.2.2 Case study

In our case study, we used the simulation tool Extend (http://www.imaginethatinc.com) as a discrete event simulation environment. The version we used includes as default random number generator the minimum standard and Schrage's (1979) generator as an alternative.

14.3 Random variate generation

The next step in a simulation experiment is to convert the random numbers into inputs appropriate for the model at hand. For example, in our case study we had to generate variates from a binomial distribution to simulate the number of illegible documents within a batch, variates from a gamma distribution to simulate times in modifying a document, and so on. Random variate generation is essentially based on the combination of six general principles: the inversion method, the composition method, the rejection method, the ratio-of-uniforms method, the use of pretests, the use of transformations and MCMC methods.

The most popular traditional method is based on *inversion*. Assuming we are interested in generating from distribution X with distribution function F, we have available a source U of random numbers and we have an efficient way to compute F^{-1}, the inversion method goes through

Generate $U \sim \mathcal{U}(0, 1)$

Output $X = F^{-1}(U)$

Within Bayesian statistics (see French and Ríos Insua, 2000), the most intensely used techniques are MCMC methods, which assume we have found a Markov chain $\{\theta^n\}$ with state θ and with its stationary distribution being the (posterior) distribution of interest. The strategy is then to start with arbitrary values of θ, let the Markov chain run until practical convergence, say after t iterations, and use the next m observed values from the chain as an approximate sample from the distribution of interest.

The key question is how to find Markov chains with the desired stationary distribution. There are several generic strategies to design such chains. One of these is the popular Gibbs sampler. Suppose that $\theta = (\theta_1, \ldots, \theta_k)$. The simplest version of the Gibbs sampler requires efficient sampling from the conditional distributions $(\theta_1 \mid \theta_2, \ldots, \theta_k)$, $(\theta_2 \mid \theta_1, \theta_3, \ldots, \theta_k)$, $\ldots$, $(\theta_k \mid \theta_1, \theta_2, \ldots, \theta_{k-1})$. Starting from arbitrary values, the Gibbs sampler iterates through the conditionals until convergence:

1. Choose initial values $(\theta_2^0, \ldots, \theta_k^0)$. $i = 1$
2. Until convergence is detected, iterate through
   - Generate $\theta_1^i \sim \theta_1|\theta_2^{i-1}, \ldots, \theta_k^{i-1}$
   - Generate $\theta_2^i \sim \theta_2|\theta_1^i, \theta_3^{i-1}, \ldots, \theta_k^{i-1}$
   - ...
   - Generate $\theta_k^i \sim \theta_k|\theta_1^i, \ldots, \theta_{k-1}^i$
   - $i = i + 1$

This sampler is particularly attractive in many scenarios, because the conditional posterior density of one parameter given the others is often relatively simple (perhaps after the introduction of some auxiliary variables). Given its importance, we provide a simple example from Berger and Ríos Insua (1998).

Example Suppose the posterior density in a given problem is

$$p_\theta(\theta_1, \theta_2 \mid x) = \frac{1}{\pi} \exp\{-\theta_1(1 + \theta_2^2)\}$$

over the set $\theta_1 > 0$, $-\infty < \theta_2 < \infty$. The posterior conditional distribution of θ_2, given θ_1, is normal with mean zero and variance $1/2\theta_1$, since

$$p_{\theta_2}(\theta_2 \mid \theta_1, x) \propto p(\theta_1, \theta_2 \mid x) \propto \exp(-\theta_1\theta_2^2).$$

Similarly, given θ_2, θ_1 has exponential distribution with mean $1/(1 + \theta_2^2)$. Then a Gibbs sampler for this problem iterates through:

1. Choose initial value for θ_2; such as, the posterior mode, $\theta_2^0 = 0$. $i = 1$
2. Until convergence, iterate through
   - Generate $\theta_1^i = \mathcal{E}/(1 + [\theta_2^{i-1}]^2)$, ($\mathcal{E}$, standard exponential).
   - Generate $\theta_2^i = Z/\sqrt{2\theta_1^i}$, ($Z$, standard normal).

Other strategies to design Markov chains with a desired stationary distribution are the *Metropolis–Hastings algorithm*, the *perfect sampler*, the *slice sampler*, and *random direction interior point samplers*. Complex problems will typically require a mixture of various MC algorithms, known as hybrid methods. As an example, Müller (1991) suggests using Gibbs sampler steps when conditionals are available for efficient sampling and Metropolis steps otherwise. For variable dimension problems, reversible jump (Green, 1995) and birth–death (Stephens, 2000) samplers are highly relevant strategies. Adaptive strategies for sequential methods as in Liu (2001) are also relevant.

14.3.1 Pointers to the literature

The literature on modern MCMC methods is vast. Good introductions to Bayesian computation methods may be found in Gelfand and Smith (1990) and Johnson and Albert (1999). Extensive developments are discussed in French and Ríos Insua (2000), Tanner (1996) and Gamerman and Lopes (2006). Cheng (1998) and Devroye (1986) give information about standard methods.

As far as software is concerned, Press *et al.* (1992) provide code to generate from the exponential, normal, gamma, Poisson and binomial distributions, from which many other distributions may be sampled based on the principles outlined above (see also http://www.nr.com). Many generators are available at http://www.netlib.org/random/index.html. WINBUGS (Spiegelhalter *et al.*, 1994) and OpenBUGS are downloadable from http://www.mrc-bsu.cam.ac.uk/bugs, facilitating MCMC sampling in many applied settings. Another useful library is GSL, which is available at http://www.gnu.org/software/gsl.

14.3.2 Case study

As mentioned, we have used Extend for our case study, which includes samplers for the beta, binomial, constant, Erlang, exponential, gamma, geometric, hyperexponential, discrete uniform, lognormal, normal, Pearson, Poisson, continuous uniform, triangular and Weibull distributions. This allows us to deal with most of the distributions in our model, except for cases such as mixtures.

In general, a mixture f is expressed as

$$f(x) = \sum_{i=1}^{n} p_i \cdot g(x|y = i) = \sum_{i=1}^{n} p_i \cdot g_i(x)$$

with $p_i = P(Y = i) > 0$, for all $i = 1, \ldots, n$, $\sum_i p_i = 1$ and density functions g_i. The procedure to generate from such distribution is:

Generate $i \sim \begin{pmatrix} p_1 & p_2 & \cdots & p_n \\ 1 & 2 & \cdots & n \end{pmatrix}$

Output $X \sim g_i$

14.4 Obtaining model outputs

The third step in a simulation process consists of passing the inputs through the simulation model to obtain outputs to be analysed later. We shall consider the two main application areas in industrial statistics: Monte Carlo simulation and discrete event simulation.

14.4.1 Monte Carlo simulation models

A key group of simulation applications in Industrial Statistics use Monte Carlo simulation. These are standard deterministic problems whose analytic solution is too complex, but such that by introducing some stochastic ingredient we are able to obtain a simulation-based solution with reasonable computational effort. Within statistics, we may use MC methods for optimisation purposes (say, to a obtain a maximum likelihood estimate or a posterior mode); for resampling purposes, as in the bootstrap; within MC hypothesis tests and confidence intervals; for computations in probabilistic expert systems, and so on, the key application being Monte Carlo integration, specially within Bayesian statistics. We therefore illustrate it in some detail.

Suppose we are interested in computing

$$I_S = \int_{[0,1]^s} f(u)du$$

where $[0, 1]^s$ is the s-dimensional unit hypercube. We have many numerical methods for such purpose, but they tend to be inefficient as the dimension s grows. As an alternative, we may use simulation-based integration methods, or Monte Carlo integration methods, whose (probabilistic) error bound is dimension-independent, therefore making them competitive as the integral dimension grows. Note first that

$$I_S = E(f),$$

where the expectation is taken with respect to the uniform distribution. We, therefore, suggest the following strategy, based on the strong law of large numbers:

Sample $u_1, \ldots, u_N \sim U[0, 1]^s$

Approximate $\hat{I}_S = \frac{1}{N} \sum_{i=1}^{N} f(u_i)$

Within Bayesian analysis, we are frequently interested in computing posterior moments as in

$$E_{\theta|x}(g(\theta)) = \int g(\theta)\pi(\theta|x)d\theta,$$

where

$$\pi(\theta|x) = \frac{p(x|\theta)\pi(\theta)}{\int p(x|\theta)\pi(\theta)d\theta}$$

is the posterior distribution for an observation x, $\pi(\theta)$ is the prior, and $p(x|\theta)$ is the model. For example, when $g(\theta) = \theta$, we have the posterior mean, whereas when $g(\theta) = I_A(\theta)$, we have the posterior probability of A. To compute these, we may use an MC approximation as in

Sample $\theta_1, \ldots, \theta_N \sim \pi(\theta|x)$

Do $\widehat{E_{\theta|x}(g(\theta))} = \frac{1}{N} \sum_{i=1}^{N} g(\theta_i)$

Sampling will generally be done through a Markov chain algorithm.

To sum up, assume we want to estimate $\theta = E(X)$ through Monte Carlo integration. If it is simple to sample from X, our output process consists of the values X_i sampled from the distribution of X. We have that $F_X = F_{X_i}$, so that $F_X = \lim_{i\to\infty} F_{X_i}$ and $\theta = E_{F_X}(X)$. If, on the other hand, it is not easy to sample from X, we may define a Markov chain $X_i \xrightarrow{d} X$, so that $F_X = \lim_{i\to\infty} F_{X_i}$, our output process again being X_i.

Example Following on from our Gibbs sampler example, we shall typically be interested in approximating posterior expectations of functions $g(\theta_1, \theta_2)$ through

$$E_\theta[g(\theta_1, \theta_2) \mid x] = \int_{-\infty}^{\infty} \int_0^{\infty} g(\theta_1, \theta_2) p(\theta_1, \theta_2 \mid x) d\theta_1 d\theta_2$$
$$\cong \frac{1}{m} \sum_{i=1}^{m} g(\theta_1^i, \theta_2^i).$$

For example, under squared error loss, we estimate θ_1 by its posterior mean, approximated by

$$\hat{\theta}_1 = \frac{1}{m} \sum_{i=1}^{m} \theta_1^i.$$

The output process in this case will be (θ_1^i, θ_2^i).

14.4.2 Discrete event simulation models

The other big application area refers to discrete event simulation, which deals with systems whose state changes at discrete times, not continuously. These methods were initiated in the late 1950s. For example, the first DES-specific language, GSP, was developed at General Electric by Tocher and Owen to study manufacturing problems. To study such systems, we build a discrete event model. Its evolution in time implies changes in the attributes of one of its entities, or model components, and it takes place in a given instant. Such a change is called an event. The time between two instants is an interval. A process describes the sequence of states of an entity throughout its life in the system.

There are several strategies to describe such evolution, which depend on the mechanism that regulates time evolution within the system. When such evolution is based on time increments of the same duration, we talk about *synchronous simulation*. When the evolution is based on intervals, we talk about *asynchronous simulation*.

We illustrate both strategies describing how to sample from a Markov chain with state space S and transition matrix $P = (p_{ij})$, with $p_{ij} = P(X_{n+1} = j | X_n = i)$. The obvious way to simulate the $(n+1)$th transition, given X_n, is

Generate $X_{n+1} \sim \{p_{x_n j} : j \in S\}$

This synchronous approach has the potential shortcoming that $X_n = X_{n+1}$, with the resulting waste of computational effort. Alternatively, we may simulate T_n, the time until the next change of state, and then sample the new state X_{n+T_n}. If $X_n = s$, T_n follows a geometric distribution with parameter p_{ss} and X_{n+T_n} will have a discrete distribution with mass function $\{p_{sj}/(1 - p_{ss}) : j \in S \setminus \{s\}\}$. Should we wish to sample N transitions of the chain, assuming $X_0 = i_0$, we proceed as follows:

Do $t = 0, \ X_0 = i_0$
While $t < N$
  Sample $h \sim \mathcal{G}e(p_{x_t x_t})$
  Sample $X_{t+h} \sim \{p_{x_t j}/(1 - p_{x_t x_t}) : j \in S \setminus \{x_t\}\}$
  Do $t = t + h$

There are two key strategies for asynchronous simulation. One is that of event scheduling. The simulation time advances until the next event and the corresponding activities are executed. If we have k types of events $(1, 2, \ldots, k)$, we maintain a list of events, ordered according to their execution times $(t_1, t_2, \ldots, t_k)$. A routine R_i associated with the ith type of event is started at time $\tau_i = \min(t_1, t_2, \ldots, t_k)$. An alternative strategy is that of *process interaction*; a process represents an entity and the set of actions that experiments throughout its life within the model. The system behaviour may be described as a set of processes that interact, for example, competing for limited resources. A list of processes is maintained, ordered according to the occurrence of the next event. A process may be interrupted, having its routine multiple entry points designated reactivation points.

Each execution of the program will correspond to a replication, which corresponds to simulating the system behaviour for a long enough period of time, providing average performance measures, say X_n, after n customers have been processed. If the system is stable, $X_n \xrightarrow{w} X$. If, for example, processing 1000 jobs is considered long enough, we associate with each replication j of the experiment the output X_{1000}^j. After several replications, we would analyse the results as described in the next section.

14.4.2.1 Discrete event simulation software

The implementation of complex simulation models with several types of events and processes in standard programming languages may be very involved. This would explain the emergence of many simulation environments, as shown in the recent software simulation review in *OR/MS Today* (Swain, 2005), periodically adapted at http://www.lionhrtpub.com/orms/ORMS-archive.html. Banks (1998) provides a good overview of these tools, which include simulation languages and simulators.

Simulation languages are general purpose languages which include tools and utilities designed for simulation, such as:

- a general framework to create and describe a model in terms of processes, events, entities, attributes, resources and interactions between model components;

- a mechanism to control the evolution of time;
- methods to schedule event occurrences;
- random number and variate generators;
- tools to collect and analyse data;
- tools to describe and display graphically the model and its simulation;
- tools to validate and verify the model.

Simulators are software packages which allow for the simulation of complex models in specific application areas, such as manufacturing, supply chain management, material handling and workflow management, of interest in industrial statistics.

14.4.2.2 Case study

We now illustrate the implementation of our reference example in a DES environment, specifically in Extend. In the DES terminology, the processes that we shall consider are the batches of documents and the batch arrival generator, and the resources will be the scanners, the program, the operators and their workstations. To facilitate implementation we associate seven numerical attributes, which establish the different types of documents, that a batch may include within its life in the system:

- Batch headers (R).
- Bills and receipts (L).
- Promissory notes (P).
- Bill and receipts wrongly scanned or illegible (LC).
- Bills and receipts wrongly recognised (LD).
- Promissory notes wrongly scanned or illegible (PC).
- Promissory notes wrongly recognised (PD).

For each stage in the line, we must model the process occurring at that stage, the random processes involved, the parameters to be estimated and the controllable parameters. As an example, we provide that of the scanning process:

- *Description.* Scanning is done in batches. Once scanned, there is a waiting time of around 10 seconds before we can feed in the next batch. The technical specifications suggest a scanning speed of 4.17 documents per second.

- *Random processes*. We need to consider the failure distribution for the scanner. We use a binomial model for each type of document, $x \sim Bin(L, P_s)$, $y \sim Bin(P, P_s)$, where P_s, is the probability of a document being incorrectly scanned, x is the number of illegible bills and y is the number of illegible promissory notes. The estimated scanning time per batch is, therefore: $(L + P + R)/4.17 + 10$ seconds.

- *Parameter estimation*. We use a beta-binomial to estimate P_s. With a uniform prior, the posterior is beta (84, 9916), the posterior mode being $\hat{P}_s = 0.0085$.

- *Controllable parameters*. One person and two scanners are used in this phase. One of the scanners is for back-up.

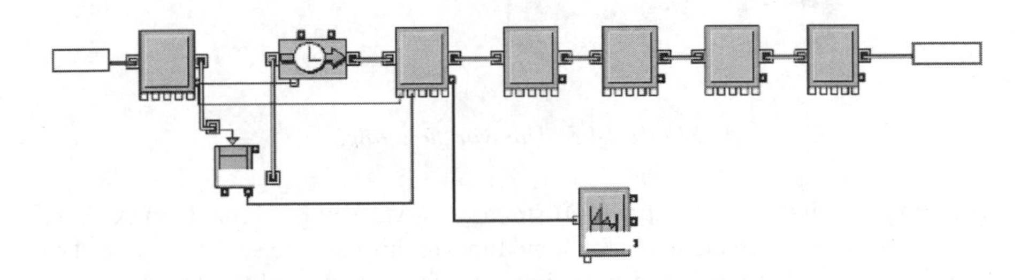

Figure 14.1 Block-modelling the scanning stage.

The above stage may be described in Extend as shown in Figure 14.1. The first block in the figure is called (in the Extend language) Dequation, which computes an expression given the input values; in this case, it will compute the scanning time for the batch, as indicated above. Then the items go through an element called Queue-Fifo, which represents a queue in which we store and leave elements

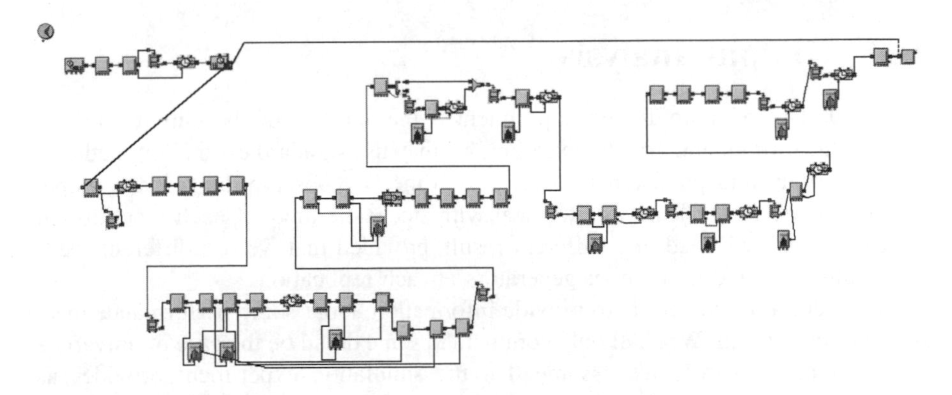

Figure 14.2 The entire workflow line in extend.

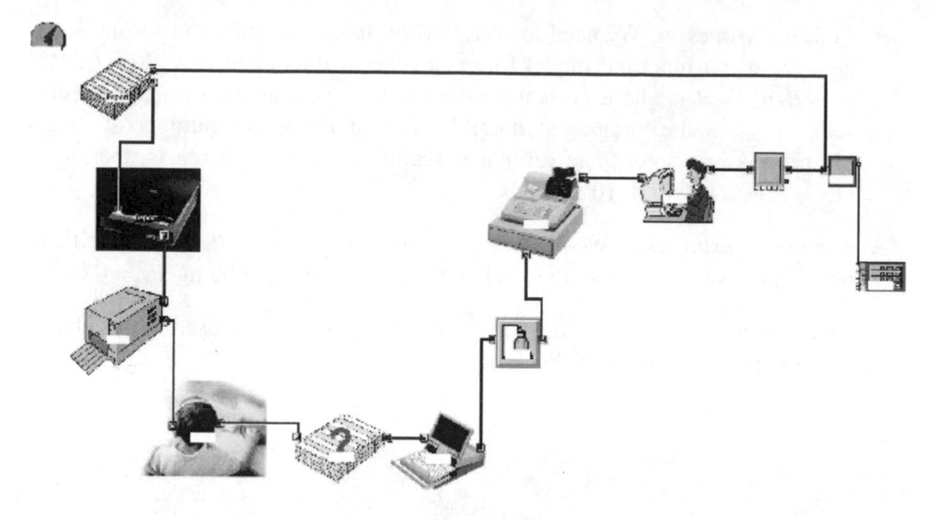

Figure 14.3 The workflow line.

according to a first in first out (FIFO) strategy. Next, they go through an Activity Delay, which will keep elements for some time, in this case the scanning time. The item then leaves that block and goes through a Dequation used to obtain the time that the element remained in queue before leaving the Activity Delay. Items next go through a Dequation which computes the attribute LC, via a binomial model. The next Dequation modifies the attribute L. The next two Dequation blocks affect PC and P.

The whole workflow line is described in Figure 14.2. Clearly, this is too complicated to describe to a practitioner, but we may use appropriate icons to summarise the model as presented in Figure 14.3.

Each run of this simulation model will provide outputs in relation to a day of operation.

14.5 Output analysis

The final stage of a simulation experiment is the analysis of the output obtained through the experiment. To a large extent, we may use standard estimation methods, point estimates and precision measures, with the key observation that the output might be correlated. Clearly, as we deal with stochastic models, each repetition of the experiment will lead to a different result, provided that we use different seeds to initialise the random number generators at each replication.

The general issue here is to provide information about some performance measure θ of our system. We shall only comment in some detail on the case of univariate performance measures. We assume that the simulation experiment provides us with an output process $X = \{X_i\}$, so that θ is a property of the limit distribution $F_X = \lim_{i\to\infty} F_{X_i}$. In fact, most of the time, we shall be able to redefine the output process so that θ is the expected value of X (that is $\int_{-\infty}^{\infty} x dF_X(x)$) or the pth quantile (that is the value x_p such that $F_X(x_p) = p$, for $p \in (0, 1)$). For example, if we are interested in estimating the covariance between two variables $X, Y, \theta = \sigma_{XY} = \int\int (X - \mu_X)(Y - \mu_Y) dF_{XY}(x, y)$, we may define the bivariate output process $\{(X_i - \overline{X}), (Y_i - \overline{Y})\}$, where $\overline{X}, \overline{Y}$ are the sample means of X, Y. Sometimes we are interested in estimating the entire distribution of a system performance measure, which may be done by estimating the parameters of a (parametric) distribution. Alternatively, we may estimate a sufficient number of quantiles so as to approximate the distribution function.

Another important issue is the distinction between stationary and transition behaviour. In fact, this determines the way to carry out the simulation experiment and analyse the data. Transition behaviour refers to short-term system performance. In our example, we are interested in whether we shall be able to complete the workload before the deadline. Stationary behaviour refers to long-term performance. In our example, we are interested in the fraction of batches not processed on time due to the system being busy, or we may be interested in determining the long-term fraction of lost messages due to the system being saturated; see, for example, Conti et al. (2004).

14.5.1 Point estimation

As far as point estimation is concerned, standard methods and concepts such as unbiasedness and consistency apply. For obvious reasons, the concept of asymptotic unbiasedness is especially relevant when dealing with stationary performance.

In that sense, in order to estimate the mean $\mu_X = \lim_{i\to\infty} \mu_{X_i}$, we shall normally use the sample mean

$$\overline{X} = \frac{\sum_{i=1}^{n} X_i}{n}.$$

To estimate proportions or probabilities, we use the sample mean of the indicators of the event of interest. Specifically, if A is such an event, we define $Z_i = I_{\{X_i \in A\}}$ and use

$$\overline{Z} = \frac{\sum_{i=1}^{n} Z_i}{n} = \frac{\#\{X_i \in A\}}{n}$$

to estimate $P(X \in A)$. In order to estimate the variance

$$\sigma_X^2 = \int_{-\infty}^{\infty} (x - \mu_X)^2 dF_X(x),$$

we shall compute $\overline{X}$, $\{(X_i - \overline{X})^2\}$ and use

$$S_X^2 = \frac{1}{n-1} \sum_{i=1}^{n} (X_i - \overline{X})^2$$

which is unbiased for independent and identically distributed (i.i.d.) observations, and asymptotically unbiased in general; for small samples and correlated output we may use various methods to correct the bias.

As far as quantiles are concerned, if $\{X_{(i)}\}$ is the order statistic associated with the output $\{X_i\}$, a simple estimator of $F_X^{-1}(p)$ is

$$(1-\alpha)X_{(r)} + \alpha X_{(r+1)},$$

with $\alpha = p(n+1) - \text{int}(p(n+1))$ and $r = \text{int}(p(n+1))$. An alternative for simultaneous estimation of several quantiles is to use the histogram. Although we lose some information while computing the counts in the histogram cells, we usually obtain good results with small cells, in spite of the inconsistency of histograms of fixed width; see Hartigan (1996) for references on histograms.

To conclude this subsection, we would like to point out three possible sources of bias in these estimators: the initial transition, which is linked with the problem of convergence detection (see Cowles and Carlin, 1996, for a review); non-linearity of transformations; and random sample sizes.

14.5.2 Precision estimation

We also need to estimate the precision of the estimator. We shall use the mean square error which, when the bias is negligible, will coincide with the variance. As basic measure, we shall use the standard deviation of $\hat{\theta}$, $EE(\hat{\theta}) = (\text{Var}(\hat{\theta}))^{1/2}$ and we aim to estimate $\widehat{EE}(\hat{\theta})$ or, equivalently, $\widehat{\text{Var}}(\hat{\theta})$. Thus, when we say that $\hat{\theta} = 16.3289$ with $\widehat{EE}(\hat{\theta}) = 0.1624$, we may find meaningful 16, give some validity to .3 (for example, 16.3 is more meaningful than 15.8) and consider 0.0289 as random digits.

In the i.i.d. case, we use the standard variance estimation theory. For example, when $\hat{\theta} = \overline{X}$, with fixed n, $\text{Var}(\overline{X}) = n^{-1}\text{Var}(X_i)$ and an estimator is S^2/n, where

$$S^2 = \frac{\sum_{i=1}^{n} X_i^2 - n\overline{X}^2}{n-1} \tag{14.1}$$

is the sample variance. If N is random and independent of the observations, $\text{Var}(\overline{X}) = \text{Var}(X_i)/E(N)$ and an unbiased estimator is $\hat{V}_0 = S^2/N$, with S^2 as before and N in place of n, with $N \geq 2$.

If we are estimating $p = P(A)$, we use

$$\hat{\theta} = \hat{p} = \frac{\sum_{i=1}^{N} I_{\{X_i \in A\}}}{N}$$

with N possibly random, but independent of the observations. We have that $\text{Var}(\hat{p}) = (p(1-p))/E(N)$ and

$$\widehat{\text{Var}}(\hat{p}) = \frac{\hat{p}(1-\hat{p})}{N-1}.$$

is an unbiased estimator.

For other estimators we would proceed similarly. A shortcoming of this approach is that it is somewhat ad hoc, in the sense that we need to develop methods for each estimator. An alternative general, and popular, variance estimation method in simulation is that of macro-micro replications. Given a set of n replications, we actually assume that it consists of k independent macroreplications with m microreplications $(X_{1j}, \ldots, X_{mj})$, $j = 1, \ldots, k$, and $km = n$. Each microreplication provides an observation of the output process; each macroreplication provides an estimator $\hat{\theta}_j$, $j = 1, \ldots, k$, based on the m observations of such replication, with the same expression as $\hat{\theta}$. The mean of the k macroreplications,

$$\overline{\theta} = \frac{1}{k} \sum_{j=1}^{k} \hat{\theta}_j,$$

is an alternative to the estimator $\hat{\theta}$. Clearly, when $m = n$, $k = 1$, $\hat{\theta} = \overline{\theta}$. As $\overline{\theta}$ is a sample mean, its variance will be estimated via

$$\hat{V}_1 = \frac{1}{k} \frac{\sum_{j=1}^{k} \hat{\theta}_j^2 - k\overline{\theta}^2}{k-1}.$$

For a discussion on how to choose m and k, see Schmeiser (1990). Recall also that the bootstrap and the jackknife (see Efron, 1982), provide methods to estimate the variance.

14.5.3 Dependent output

We need to be especially careful when simulation output is correlated, as happens in an MCMC sampler. To illustrate the issues involved, assume that $(X_1, \ldots, X_n)$ are observations from a stationary process and we estimate θ through $\overline{X}$. If $\mathrm{Var}(X) = \sigma_X^2$ and $\rho_j = \mathrm{Corr}(X_i, X_{i+j})$ we have

$$\mathrm{Var}(\overline{X}) = \frac{d\sigma_X^2}{n}$$

with

$$d = 1 + 2\sum_{j=1}^{n-1} \left(1 - \frac{j}{n}\right) \rho_j.$$

In the i.i.d. case, $d = 1$. When the process is positively correlated, $\mathrm{Var}(\overline{X}) > \sigma^2/n$. Moreover,

$$E\left(\frac{S^2}{n}\right) = \frac{e\sigma_X^2}{n}$$

with

$$e = 1 - \frac{2}{n-1} \sum_{j=1}^{n-1} \left(1 - \frac{j}{n}\right) \rho_j,$$

so that we underestimate the variability of $\overline{X}$. Similarly, if the process in negatively correlated, we overestimate it.

To mitigate the problem, several methods have been devised. The most popular is that of macro-micro replications, described above, which is known as the batch method for dependent data; correlation substitution; time series methods; regenerative simulation; and thinning. Further details may be seen in Balci (1994) and Ríos Insua *et al*. (1997).

14.5.4 Confidence intervals

We shall usually employ $\mathrm{Var}(\hat{\theta})$ to obtain a confidence interval for θ: we aim to use the output process $(X_1, \ldots, X_n)$ to obtain a random interval $[L, U]$ so that $P(\{L \leq \theta \leq U\}) = 1 - \alpha$ for some α. The standard confidence interval theory applies, with convenient modifications to take into account the issue of dependent data and the various methods used to estimate the precision.

14.5.4.1 Pointers to the literature

We have concentrated on analysing only one performance measure. Multiple performance measures require special care when defining multiple confidence intervals, which is done through Bonferroni-type inequalities; see Ríos Insua *et al*. (1997) for further details.

Special care must be taken when using simulation software, since it often provides precision measures under the hypothesis of independence. Beware. Alternatively, you may use your favourite statistical environment to undertake this task, if it allows for time series analysis data, after preparation of simple programs. Convergence detection is an important issue, for which CODA (http://www.mrc-bsu.cam.ac.uk/bugs/classic/coda04/readme.shtml) is useful. Fishman (1996) provides ample information about his LABATCH.2 package, available at http://www.unc.edu/~gfish/labatch.2.html. MCMCpack also contains some useful utilities and functions, and is available at http://www.r-project.org.

14.5.4.2 The example

Assume we want to approximate $I = \int_{-\infty}^{\infty} (x + x^2) f(x) dx$, where f is a normal density with mean 1 and standard deviation 2. We use Monte Carlo integration and repeat the experiment 50 times with sample size 50 at each iteration. The values obtained are:

6.769	5.603	5.614	9.229	7.189	3.277	4.312	7.070	5.195	4.496
5.775	4.646	5.670	7.134	4.931	4.403	6.783	7.152	5.834	4.958
7.159	7.270	8.379	5.037	5.143	5.757	7.399	5.236	4.749	5.729
7.015	6.156	3.985	5.643	5.720	6.878	6.367	7.520	7.093	6.605
6.356	6.567	7.784	5.256	6.302	5.460	4.808	5.880	3.846	5.962

Note that if we used just one replication, we would obtain highly disparate values, from 3.277 to 9.229. We use the sample mean as estimator, which is 5.983, which gives us $I = 6$. The mean square error is 1.475.

14.5.4.3 Case study

As indicated, our interest is in estimating the probability that the process is completed before the deadline, in our case 28 800 seconds, that is, we need to estimate $Pr(T \leq 28\,800)$ for each configuration line. For this purpose, we perform 1000 replications, equivalent to 1000 days of operation, and record whether the process finishes before 28 800 seconds. The output process is therefore N_i, where $N_i = 1$ if $T_i \leq 28\,800$ and 0 otherwise, T_i being the final time of the ith iteration. Therefore, we estimate the probability as

$$\frac{\sum_{i=1}^{1000} N_i}{1000}.$$

To complete the study, we record other output values, specifically average waiting times at each queue and average times at each process, 22 outputs in total.

We show below the statistics generated by Extend after simulating 1000 days under the current configuration:

- Probability of finishing work on time: 0.441
- Average finishing time: 31 221.1 sec.

Device	Batch av. que. time	Batch av. resp. time
Scanner	5.93	41.03
OCR	3779.69	3861.32
Recording	0	1888.72
Illegible resolution	0.75	11.21
Coding	961.33	1038.89
Date verification	59.83	87.23
Field verification	6183.34	6325.18
Quality verification	44.79	1836.54
Balancing	1.75	1.77
Modification	5.64	123.23

Note that the probability of timely completion is too low (0.44) and the average finishing time (31 221) is much longer than that allowed (the respective mean square errors were 0.12 and 925). Note that the longest waiting times hold for the OCR and field verification, the main bottlenecks.

We therefore reconfigure the system, doubling the resources for the OCR phase and field verification. This leads to the following results:

- Probability of finishing process on time: 0.941
- Average finishing process time: 21 881.1 sec.

Device	Batch av. que. time	Batch av. resp. time
Scanner	6.05	41.15
OCR	454.17	541.34
Recording	369.27	2465.18
Illegible resolution	1.79	12.18
Coding	2608.79	2685.28
Date verification	83.29	110.64
Field verification	1351.67	1478.59
Quality verification	2179.06	2280.67
Balancing	0.154	1.73
Modification	68.93	131.21

The modifications induce a reduction in the modified stages and, more importantly, have been sufficient to guarantee completion of the process (probability 0.94). Note also that the average queueing times have been balanced, therefore attaining a more balanced load. Although some processes, such as recording, require more, the design is much better on the whole.

As indicated, the above results allow us to answer the initial question. To complete the study, we could investigate the performance under more extreme work conditions. For example, we could study the possibility of expanding our business, looking for new customers.

For example, suppose that five vanloads of documents arrive every day (the current workload is randomly distributed between one and four). The resulting statistics would be:

- Probability of finishing on time: 0.264
- Average finishing time: 30 569

Device	Batch av. que. time	Batch av. resp. time
Scanner	5.81	40.91
OCR	638.89	723.92
Recording	629.56	2748.63
Illegible resolution	1.91	12.51
Coding	3766.47	3842.59
Date verification	87.56	114.95
Field verification	1823.13	1952.27
Quality verification	3228.38	3330.5
Balancing	0.13	1.84
Modification	72.99	135.37

Under such extreme conditions, the average processing time is much worse and there is little guarantee of completing the process on time. Hence, it does not seem advisable to expand the business so much.

14.6 Tactical questions in simulation

In this section we turn to issues concerning tactical questions in simulation: how we design simulation experiments, how we combine simulation with optimisation, and the issue of variance reduction.

14.6.1 How many iterations?

The number n of replications chosen will affect the precision of estimators and the cost of computation. As variance estimators are normally $O(n^{-1})$, the basic idea will be to take n sufficiently large to guarantee a certain precision. Note that an initial idea about n somehow limits our computational effort, whereas if we view it as excessive, we may try to apply variance reduction techniques. We briefly illustrate the ideas in the simplest case.

Suppose we make n observations of the output process X_i, and $X_1, \ldots, X_n$ are i.i.d. with $E(X) = \theta$, $\text{Var}(X) = \sigma^2$. We use $\overline{X} = n^{-1}\sum_{i=1}^{n} X_i$ and $S/\sqrt{n}$, with $S^2 = (n-1)^{-1}\sum_{i=1}^{n}(X_i - \overline{X})^2$, respectively, as the estimator and the precision of the estimator. To determine n, we fix $1-\alpha$ as an acceptable confidence level. For n sufficiently large,

$$\left[\overline{X} - z_{1-\alpha/2}\frac{S}{\sqrt{n}},\ \overline{X} + z_{1-\alpha/2}\frac{S}{\sqrt{n}}\right]$$

is a confidence interval of level $1-\alpha$ and width $2z_{1-\alpha/2}S/\sqrt{n}$. If we fix the maximum allowed width as d, then

$$2z_{1-\alpha/2}\frac{S}{\sqrt{n}} \leq d$$

or

$$\left(2z_{1-\alpha/2}\frac{S}{d}\right)^2 \leq n.$$

Sometimes we have an initial idea about σ or S, which we may use to suggest an initial size. If this is not the case, we may use a pilot sample to estimate S and apply the previous method; typically we shall need to iterate with an argument such as the following, where S_n is the sample variance when the size is n:

Do $n_0 = 30$
Generate $\{X_j\}_{j=1}^{30}$
Compute $S_{30},\ i = 0$
While $2z_{1-\alpha/2}S_{n_i}/\sqrt{n_i} > d$
  $i = i + 1$
  Compute min n_i : $\left(2z_{1-\alpha/2}\frac{S_{n_{i-1}}}{d}\right)^2 \leq n_i$
  Generate $\left\{X_j\right\}_{j=n_{i-1}+1}^{n_i}$
  Compute S_{n_i}

Let us apply the previous procedure to estimate

$$\int_{-\infty}^{\infty} (x + x^2) f(x) dx,$$

with f the normal density $N(1, 4)$. The following table includes the results when $\alpha = 0.005$ and $d = 0.1$.

n	$\hat{I}$	S	Width
30	5.06	5.22	5.216
85849	5.996	8.2	0.158
212521	6.005	8.27	0.101
216225	6.003	8.26	0.099

For the general case, we shall use confidence intervals

$$\left[\hat{\theta} - t_{\nu,1-\alpha/2}\sqrt{\hat{V}},\ \hat{\theta} + t_{\nu,1-\alpha/2}\sqrt{\hat{V}}\right]$$

where $\hat{\theta}$ is the estimator of interest and $\hat{V}$ is a variance estimator distributed as χ^2 with ν degrees of freedom and independent of the distribution of $\hat{\theta}$. Typically, $\hat{V}$ depends on n, that is, $\hat{V} = \hat{V}(n)$. Moreover, often $\nu = \nu(n)$, so that

$$2t_{\nu(n),1-\alpha/2}\hat{V}(n) \leq d,$$

which we solve for n. The former sequential procedure may be easily extended, especially if the distribution of $t_{\nu(n)}$ can be approximated by a normal distribution with $\nu(n) \geq 30$.

14.6.2 Regression metamodels

In most simulation-based industrial statistical applications, we are interested in either understanding the behaviour of a system (that is how changes in operation conditions affect performance) or improving its functioning. Algebraically, we describe the relation of the output z_0 of the real system to the inputs $y_1, y_2, \ldots$ through a function

$$z_0 = f_0(y_1, y_2, \ldots; R_0)$$

where R_0 designates the sources of randomness in a generic form. We identify those inputs that we consider relevant $y_1, y_2, \ldots, y_k$ and describe the relation between the model output z_1 and the inputs through the function

$$z_1 = f_1(y_1, y_2, \ldots, y_k; R_1)$$

where R_1 designates the randomness sources of the model, our objective being to estimate

$$\theta(y_1, \ldots, y_k) = E_{R_1}(f_1(y_1, \ldots, y_k; R_1)).$$

When computing such expectation is difficult, we may appeal to simulation to estimate $\theta(y_1, \ldots, y_k)$. For this purpose, the relation between the output $(X_i)_{i=1}^n$ and the inputs through a simulation program is given by a function

$$X(y_1, \ldots, y_k) = f_2(y_1, \ldots, y_k; R_2)$$

where R_2 designates the random seeds used to initialise the random number generators.

The tools described above allow us to estimate the performance $\theta(y_1, \ldots, y_k)$, given the inputs, together with a precision measure. Analogously, should we be interested in determining optimal inputs we should solve

$$\min\ \theta(y_1, \ldots, y_k)$$

under appropriate constraints. One possibility would be to use an optimisation algorithm that requires only function evaluations, such as that of Nelder and Mead, and then estimate the function at each new input value, through the simulation algorithm; see Ríos Insua *et al*. (1997) for an example.

This approximation may be extremely costly from a computational point of view. In this case it may be more interesting to associate with the problem a new model called the *regression metamodel*. For this purpose, we just need to represent the estimation problem in such a way that if $Z_3(y_1, \ldots, y_k) = \hat{\theta}(y_1, \ldots, y_k)$ we introduce the representation

$$Z_3 = f_3(y_1, \ldots, y_k, \delta) + \epsilon \tag{14.2}$$

where f_3 represents a parametric function with parameters δ to be estimated and an error term ϵ. An example based on neural networks may be seen in Müller and Ríos Insua (1998). We may then use the regression metamodel for prediction and optimisation purposes as required.

The optimisation problem is described as

$$\begin{aligned} \min\ \theta_0 &= E_R(f_0(y_1, \ldots, y_k; R)) \\ \text{s.t.}\ \ \theta_i &= E_R(f_i(y_1, \ldots, y_k; R)) \leq b_i,\ i = 1, \ldots, r \\ &(y_1, \ldots, y_k) \in S \end{aligned}$$

where R designates the randomness sources, r the number of output constraints, and the last constraint refers to inputs. We have a simulation model with m responses of interest $\hat{\theta}_i = f_i(y_1, \ldots, y_k; \underline{R})$, $i = 0, 1, \ldots, r$, where, as before, $\underline{R}$ designates the random numbers employed.

We conclude this subsection by mentioning that a number of specific simulation-optimisation methods have been developed. In the case of finite sets of alternatives, we should mention methods, on the one hand, based on ranking and selection and, on the other, based on multiple comparisons. Among methods for continuous problems, we should mention response surface methods and stochastic approximation methods such as Robbins and Monro (1951) and Kiefer and Wolfowitz (1952). Other methods include algorithms based on perturbation analysis (Glasserman, 1991) and on likelihood ratios (Kleijnen and Rubinstein, 1996).

14.6.3 Experimental design and simulation

We have described simulation as a (computer-based) experimental methodology. As such, all principles of good experimentation, as reviewed in Chapter 7 of this book, are relevant. Further details may be found in Box *et al.* (1978) and Chaloner and Verdinelli (1995). An excellent review focusing on simulation applications is given in Kleijnen *et al*. (2005).

An important difference with respect to other types of experiments refers to the key point that we control the source of randomness, and we may take advantage of this. For example, in simulation we have the common random number technique, which uses the same random numbers for simulations under different input conditions.

14.6.4 Variance reduction

We have emphasised the need for quality measures of simulation estimators through precision estimates. In this respect, it seems natural to improve the quality of estimators, typically by looking for estimators with similar bias but smaller variance. The techniques addressed towards this purpose are called variance reduction techniques.

Given a basic simulation experiment, the idea is to introduce another experiment in which sampling is done differently or in which we observe a different output variable, which leads to a better-quality estimator. Trivially, we may reduce the variance by augmenting the sample size n, as we have seen that variance estimators are $O(n^{-1})$. But this entails an increase in computational effort, which will often be unacceptable. The objective would be, on the other hand, to reduce the mean squared error, keeping the computational effort fixed, or reduce the computational effort, keeping the mean squared error fixed.

For that purpose several techniques have been developed, including antithetic variates, control variates, conditioning, importance sampling, common random numbers and stratified sampling. Computational savings may be tremendous, however application of these techniques is far from simple, frequently demanding ingenuity and small pilot samples to ascertain whether we really can achieve a variance reduction. Fishman (2003) provides ample information.

We conclude this section by remarking that, again, variance reduction is a topic on the border between simulation and experimental design.

14.7 Discussion and conclusions

We have provided an illustrated introduction to key concepts and methods in simulation from the standpoint of industrial statistics. Further details are given in various texts, including Fishman (1996, 2001, 2003), Ríos Insua *et al.* (1997), Law and Kelton (2001), Banks (1998) and Schmeiser (1990). We hope that this chapter provides the reader with a broad view of simulation methods and applications.

Discrete event simulation is one of the most frequently used techniques in business and industrial statistics in such problems as manufacturing systems and local area network modelling where the performance of a system whose state evolves discretely in time may not be computed analytically. The standard approach proceeds by building a simulation model, estimating the model parameters, plugging the estimates into the model, running the model to forecast performance evaluation, and analysing the output. However, although this has not been acknowledged in the DES literature, this approach typically greatly underestimates uncertainty in predictions, since the uncertainty in the model parameters is not taken into account, by assuming parameters fixed at estimated values. In other fields, this issue of input uncertainty influencing model uncertainty has generated an important body of literature; see, for example, Draper (1995) or Chick (2001). Applying Bayesian methods in discrete event simulation seems a fruitful area of research.

Note that we have scarcely mentioned continuous time simulations, typically based on stochastic differential equations. Normally, a synchronous approach will be adopted; see Neelamkavil (1987) for further information.

Finally, the reader may find the following websites concerned with on-line executable simulations of interest: http://www.cis.ufl.edu/~fishwick/websim.html and http://www.national.com/appinfo/power/webench/websim/.

Acknowledgments

This work was supported by funds under the European Commission's Fifth Framework 'Growth Programme' via Thematic Network 'Pro-ENBIS', contract reference G6RT-CT-2001-05059, and by grants from MEC, the Government of Madrid and the Everis Foundation.

References

Atkinson, A.C. and Donev, A.N. (1992) *Optimum Experimental Designs*. Oxford University Press, Oxford.

Balci, O. (ed.) (1994) Simulation and modelling. *Annals of Operations Research*, **53**.

Banks, J. (1998) *Handbook of Simulation*. John Wiley & Sons, Inc., New York.

Bayarri, M.J., Berger, J.O., Garcia-Donato, G., Liu, F., Palomo, J., Paulo, R., Sacks, J., Cafeo, J. and Parthasarathy, R. (2007) Computer model validation with functional output. *Annals of Statistics*. To appear.

Bays, C. and Durham, S. (1976) Improving a poor random number generator. *ACM Transactions on Mathematical Software*, **2**, 59–64.

Berger, J.O., Garcia-Donato, G. and Palomo, J. (2004) Validation of complex computer models with multivariate functional outputs. Technical Report, SAMSI, Durham, NC.

Berger, J.O. and Ríos Insua, D. (1998) Recent developments in Bayesian inference, with applications in hydrology. In E. Parent *et al.* (eds), *Statistical and Bayesian Methods in Hydrological Sciences*, pp. 43–62. Unesco, Paris.

Box, G.E.P., Hunter, W.G. and Hunter, J.S. (1978) *Statistics for Experimenters*. John Wiley & Sons, Inc., New York.

Chaloner, K. and Verdinelli, I. (1995) Bayesian experimental design: A review. *Statistical Science*, **10**, 273–304.

Chick, S.E. (2001) Input Distribution selection for simulation experiments: Accounting for input uncertainty. *Operations Research*, **49**, 744–758.

Conti, P.L., Lijoi, A. and Ruggeri, F. (2004) A Bayesian approach to the analysis of telecommunication systems performance. *Applied Stochastic Models in Business and Industry*, **20**, 305–321.

Cowles, K. and Carlin B. (1996) Markov chain Monte Carlo convergence diagnostics. *Journal of the American Statistical Association*, **91**, 883–904.

Devroye, L. (1986) *Non-Uniform Random Variate Generation*. Springer-Verlag, New York.

Draper (1995) Assessment and propagation of model uncertainty (with discussion). *Journal of the Royal Statistical Society, Series B*, **57**, 45–97.

Efron, B. (1982) *The Jackknife, the Bootstrap and Other Resampling Plans*. SIAM, Philadelphia.

Fishman, G.S. (1996) *Monte Carlo: Concepts, Algorithms and Applications*. Springer-Verlag, New York.

Fishman, G.S. (2001) *Discrete Event Simulation*. Springer-Verlag, New York.

Fishman, G.S. (2003) *Monte Carlo*. Springer-Verlag, New York.

French, S. and Ríos Insua, D. (2000) *Statistical Decision Theory*. Arnold, London.

Gamerman, D. and Lopes, H. (2006) *Markov Chain Monte Carlo: Stochastic Simulation for Bayesian Inference*, 2nd edition. Taylor & Francis, Boca Raton, FL.

Gelfand, A. and Smith, A.F.M. (1990) Sampling based approaches to calculating marginal densities. *Journal of the American Statistical Association*, **85**, 398–409.

Glasserman, P.(1991) *Gradient Estimation via Perturbation Analysis*. Kluwer, Boston.

Green, P. (1995) Reversible jump Markov chain Monte Carlo computation and Bayesian model determination. *Biometrika*, **82**, 711–732.

Hartigan, J. (1996) Bayesian histograms. In J. Bernardo, J. Berger, A. Dawid, and A. Smith (eds), *Bayesian Statistics 5*, pp. 211–222. Oxford University Press, Oxford.

Johnson, V.E. and Albert, J.H. (1999) *Ordinal Data Modelling*. Springer-Verlag, New York.

Kiefer, J. and Wolfowitz, J. (1952) Stochastic estimation of the maximum of a regression function. *Annals of Mathematical Statistics*, **23**, 462–466.

Kleijnen, J.P.C. and Rubinstein, R.Y. (1996) Sensitivity analysis by the score function method. *European Journal of Operations Research*, **88**, 413–427.

Kleijnen, J., Sanchez, S., Lucas, T. and Cioppa, T. (2005) State-of-the-art review: A user's guide to the brave new world of designing simulation experiments. *INFORMS Journal on Computing*, **17**, 263–289.

Law, A.M. and Kelton, W.D. (2001) *Simulation Modelling and Analysis*. McGraw-Hill, New York.

L'Ecuyer (1998) Random number generators and empirical tests. In H. Niederreiter, P. Hellekalek, G. Larcher and P. Zinterhof (eds), *Monte Carlo and Quasi-Monte Carlo Methods 1996*, Lecture Notes in Statistics 127, pp. 124–138. Springer-Verlag, New York.

L'Ecuyer, P. (2001) Software for uniform random number generation: Distinguishing the good and the bad. In *Proceedings of the 2001 Winter Simulation Conference*, pp. 95–105. IEEE Press, Piscataway, NJ.

L'Ecuyer, P. and Granger-Piche, J. (2003) Combined generators with components from different families. *Mathematics and Computers in Simulation*, **62**, 395–404.

L'Ecuyer, P. and Simard, R. (2001) On the performance of birthday spacings tests for certain families of random number generators. *Mathematics and Computers in Simulation*, **55**, 131–137.

L'Ecuyer, P., Simard, R., Chen, E.J. and Kelton, W.D. (2002) An object-oriented random number package with many long streams and substreams. *Operations Research*, **50**, 1073–1075.

Lehmer, D.H. (1951) Mathematical methods in large-scale computing units. In *Proceedings of the Second Symposium on Large Scale Digital Computing Machinery*, pp. 141–146, Harvard University Press, Cambridge, MA.

Lewis, P.A., Goodman, A.S. and Miller, J.M. (1969) A pseudo-random number generator for the system/360. *IBM Systems Journal*, **8**, 136–143.

Liu, J. (2001) *Monte Carlo Strategies for Scientific Computing*, Springer-Verlag, New York.

Matsumoto, M. and Nishimura, T. (1998) Mersenne twister: A 623-dimensionally equidistributed uniform pseudo-random number generator. *ACM Transactions on Modelling and Computer Simulation*, **8**, 3–30.

McCullough, B.D. (1999) Assessing the reliability of statistical software: Part II. *American Statistician*, **53**, 149–159.

Müller, P. (1991) A generic approach to posterior integration and Bayesian sampling. Technical Report 91-09, Statistics Department, Purdue University.

Müller, P. and Ríos Insua, D. (1998) Issues in Bayesian analysis of neural network models. *Neural Computation*, **10**, 571–592.

Neelamkavil, F. (1987) *Computer Simulation and Modelling*. John Wiley & Sons, Inc., New York.

Press, W.H., Teukolsky, S.A., Vetterling, W.T. and Flannery, B.P. (1992) *Numerical Recipes in C*. Cambridge University Press, Cambridge.

Ríos Insua, D. and Salewicz, K. (1995) The operation of Kariba Lake: A multiobjective decision analysis. *Journal of Multicriteria Decision Analysis*, **4**, 203–222.

Ríos Insua, D., Ríos Insua, S. and Martin, J. (1997) *Simulación, Métodos y Aplicaciones*. RA-MA, Madrid.

Robbins, H. and Monro, S. (1951) A stochastic approximation method, *Annals of Mathematical Statistics*. **22**, 400–407.

Ruggeri, F. (2005) On the reliability of repairable systems: Methods and applications. In A. Di Bucchianico, R.M.M. Mattheij and M.A. Peletier (eds), *Progress in Industrial Mathematics at ECMI 2004*, pp. 535–553. Springer-Verlag, Berlin.

Schmeiser, B. (1990) Simulation methods. In D.P. Heyman and M.J. Sobel (eds), *Stochastic Models*. North Holland, Amsterdam.

Schrage, L. (1979) A more portable FORTRAN random number generator. *ACM Transactions on Mathematical Software*, **5**, 132–138.

Spiegelhalter, D., Thomas, A., Best, N. and Gilks, W. (1994) BUGS: Bayesian inference using Gibbs sampling, version 0.30. MRC Biostatistics Unit, Cambridge.

Stephens, M. (2000) Bayesian analysis of mixture models with an unknown number of components: An alternative to reversible jump methods. *Annals of Statistics*, **28**, 40–74.

Swain, J. (2005) Gaming reality. *OR/MS Today*, December. http://www.lionhrtpub.com/orms/orms-12-05/frsurvey.html (accessed October 2007).

Tanner, M.A. (1996) *Tools for Statistical Inference: Methods for the Exploration of Posterior Distributions and Likelihood Functions*, 3nd edition. Springer-Verlag, New York.

Von Neumann, J. (1951) Various techniques in connection with random digits. *NBS Applied Mathematics Series*, **12**, 36–38.

15

Communication

Tony Greenfield and John Logsdon

15.1 Introduction

There are many styles of writing: the simple children's story, popular journalism, the quality press, popular magazines, technical magazines, scientific journals, and many more. The choice is wide and it depends on the target readership. You may vary vocabulary, sentence length, grammar, but the rules of good English usually apply. The rules may be broken, for effect, so long as the meaning is clear.

We shall discuss style later in the chapter. We mention it here so that when you study the table of communication media (Table 15.1) you will appreciate that your choice of media relates to your choice of style and the need to develop style appropriately.

You have just finished a project. You modelled the process, analysed the information, and reported implications informally to the client. You applied the continuous improvement iterative Deming approach to your work to make sure that it was fit for purpose.

So why did you start it? It may seem a bit late to think of this now, but the reason will affect how you write.

Was it just a topic that grabbed your interest, with no other interested people in mind? Did you come across a problem, either published or in conversation with colleagues, on which you believed you could cast more light? Perhaps there are applications that haven't yet been realised such as a new software product that needs marketing, for example, but nobody has thought of it before. Your role is as an originator.

Or did some colleagues offer you a problem, on which they had been working, in their belief that you could take the solution further forward? Your role is as a collaborator.

Statistical Practice in Business and Industry Edited by S.Y. Coleman, T. Greenfield, D.J. Stewardson and D.C. Montgomery

Or was it a problem that arose in a business or industrial context and for which a manager or client sought your help? Here, you are a sole specialist, a consultant.

Who might be interested in what you have done: who needs to know? Will the solution benefit a small special interest group, or will it be of wider interest: to the whole industry, to the public, to financiers, to government?

The reason for your project's existence and its origin will determine the style, the length, the medium for its onward communication. Perhaps your purpose will specify several styles, several lengths and several media for the promotion of your work.

Your first thought on completing a project may be of a peer-reviewed paper in an academic journal: a publication to augment your CV, to tell other workers in your field that you are with them, that you are as clever. But, in this book, we are not so concerned with personal career enhancement as with promoting the vision of ENBIS:

> 'To promote the widespread use of sound, science driven, applied statistical methods in European business and industry.'

There are several different reasons for communicating your research. Although you will use the same core of information in all communications, you will treat it in different ways for different readerships. Your understanding of your readers is imperative for the way in which you present the technical content. Your work will be useless if the reader cannot understand it.

15.2 Classification of output

A simple approach is to consider a table of communication media as in Table 15.1. The columns are audience, which gives a guide to the complexity, and the rows are the length of article. Into the cells of the table we enter descriptions of various media. These are not cast in stone. A conference presentation may be given to a highly specialised audience rather than to a technically aware audience.

We shall discuss most of these in terms of the written and spoken word. We omit discussion of some, such as training books, easy technical books, academic books and obscure technical books, because they would demand far too much detail and discussion, far beyond our scope.

It's a long list, but we believe you should consider every item as a possible output for your work if you are to fulfil the vision of ENBIS.

Throughout, we shall keep to the fore the following questions:

- Why was this work done?
- For whom was it done?

Table 15.1 Communication table

	Public	Technical	Specialist
Short	• News items for popular newspapers • Letters to newspapers and magazines • Promotional material for companies	• News items for technical magazines • Posters for conferences	• Company internal memoranda
Medium	• Feature articles for popular newspapers • Public lectures • Documentary radio scripts	• Feature articles for technical magazines • Articles for company house journals • Platform presentations for conferences • Company training • Training courses for wider industrial audiences	• Short technical seminars • Internal technical reports for companies • Academic papers • Advanced training courses
Long	• Populist book • Television documentary	• Training books • Easy technical books	• Academic books • Obscure technical books

- To whom do you want to communicate information about the work?
- Why would they be interested?
- What information for what audiences?
- Who may benefit from the work?

We shall also discuss communication from the start of a project:

- Who originated it?
- What exchange, style and content of memoranda were needed to clarify the purpose of the project?

- What communication measures were needed to establish high-quality and timely data collection?
- What support was needed from colleagues or specialists?
- What progress memoranda and reports were written and for whom?

15.3 Figures and tables

Before we discuss writing, it is appropriate to consider the figures and tables that you may wish to use in your work. These are the space between the words. It is as well to think carefully about their presentation and form.

Figures (including graphics) and tables may be used in almost every one of the media in Table 15.1. They are essential in reports of statistical results. Graphics reveal data: they show the shape, the relationships, the unexpected or erroneous values; they draw the reader's attention to the sense and substance of the data. So here is some brief guidance before we discuss the media in more detail.

- Use tables and graphics to present complex ideas with clarity, precision and efficiency.
- If the data are simple or changes are not significant you may not need a figure: keep to the text.
- Be sure of the purpose of the figure: description of the data; exploration of the data; tabulation of the data, for reference and analysis.
- Use tables or graphs for essential repetitive data.
- Use tables for precision when exact results are critical.

15.3.1 Figures (including graphics)

- Use graphs when the shape of the data, such as trends or groups, is more important than exact values.
- Be sure that the graphic shows the data, so that you persuade the reader to think about the substance rather than the methodology or graphic design.
- Design the graphic so that it encourages the reader's eye to compare different pieces of data.
- Reveal the data at several levels of detail, from a broad overview to the fine structure.
- Give every graph a clear, self-explanatory title.

- State all measurement units.
- Choose scales on graphs carefully.
- Label axes clearly.
- Avoid chart junk.
- Improve by trial and error since you rarely get the graphic right first time.
- Beware of the graphic artist who aims to beautify the image but fails to elucidate the data. So insist on checking the figures after the artist has done the work.
- Beware of those journalists, politicians and businessmen who deliberately use misleading scales to spin their messages.

15.3.2 Tables

- Right-justify numbers in tables.
- Line up decimal points in columns.
- Round numbers so that the two most effective digits are visible.
- Avoid distortion of the information in the data.
- Add row and column averages or totals where these are appropriate and may help.
- Consider reordering rows and/or columns to make the table clearer.
- Consider transposing the table.
- Give attention to the spacing and layout of the table.

Several textbooks offer guidance about the presentation of data and of data analysis. In *A Primer in Data Reduction*, Andrew Ehrenberg (1982) wrote four chapters on communicating data, called 'Rounding', 'Tables', 'Graphs' and 'Words'. These are worth reading before you write any papers or your report.

Rarely are measurements made to more than two or three digits of precision. Yet results of analysis are often shown to many more digits. Finney (1995) gives an example: $2.39758632 \pm 0.03245019$ 'computed with great numerical accuracy from data at best correct to the nearest 0.1 %'. Such numbers are crass and meaningless, but computers will automatically produce them if the programmer has forgotten to use the format statement. Would you then report them, pretending scientific precision, or would you round them to an understandable level that means something?

In his discussion of tables, Ehrenberg says that:

- rows and columns should be ordered by size;
- numbers are easier to read downwards than across;
- table lay-out should make it easier to compare relevant figures;
- a brief verbal summary should be given for every table.

The briefest and (in our view) best of guides about the presentation of results is reprinted as an article from the *British Medical Journal* (Altman *et al.*, 1983). This has good advice for all research workers, not just those in the medical world, and we suggest that you obtain a copy. Here are a few of its points:

- Mean values should not be quoted without some measure of variability or precision. The standard deviation (SD) should be used to show the variability among individuals and the standard error of the mean (SE or SEM) to show the precision of the sample mean. You must make clear which is presented.
- The use of the symbol ± to attach the standard error or standard deviation to the mean (as in 14.2 ± 1.9) causes confusion and should be avoided. The presentation of means as, for example, 14.2 (SE 1.9) or 14.2 (SD 7.4) is preferable.
- Confidence intervals are a good way to present means together with reasonable limits of uncertainty and are more clearly presented when the limits are given – 95 % confidence interval (10.4, 18.0) – than with the ± symbol.
- Spurious precision adds no value to a paper and even detracts from its readability and credibility. It is sufficient to quote values of t, χ^2, and r to two decimal places.
- A statistically significant association does not itself provide direct evidence of a causal relationship between the variables concerned.

15.4 Public audiences and readers

We turn back to the issue of the audience and refer again to Table 15.1.

We consider the public to be people who do not have any particular technical skill or understanding. As soon as they start to read an item you must seize and develop their interest.

15.4.1 Short items

We include here:

- news items for popular newspapers,

- letters to newspapers and magazines,
- promotional material for companies,

Your head is full of so much technical detail that you may find it easy to write several thousand words about what you have done, developed and discovered. It is hard to encapsulate all that detail into 100 words. So why should you?

We suggest that you bear in mind the following:

- You will reach and influence several types of reader through short news items.
- Letters to editors should explain the issues in a short space so they are less likely to be cut and the reader will be better able to understand your point.
- Promotional material has to be quickly understood by many people with differing understanding before they move on to the next glossy handout.

Consider the British tabloid press. Disregard their comments and their selection of news; think about the way in which they write the news. They do it very well: short sentences, key points well highlighted, one point per paragraph made in only two or three sentences. Their standard of writing is of the highest.

Think about your own reading habits. If you are like other professionals you probably rely heavily on the popular media to keep yourself informed about disciplines other than your own. Imagine an audience:

Readers	Reason
Your peers	If they have read a brief account of your work in a newspaper, they may be persuaded to read your academic paper or listen to you speak at a conference. They may be encouraged to learn more because the subject interests them; perhaps they are already working in the field.
Academic managers (faculty deans, university vice chancellors)	They love to see their faculties and universities named in the press.
Funding agencies	You may improve the chance of future funding if your name and your work are widely known.
Company managers	You may improve the chances of industrial collaboration for future work and for invitations to present courses.
General readers	They just like to know what is happening, perhaps even to broaden their own horizons.
Journalists	They are always seeking ideas for news and features that they can write about for their own outlets.

15.4.1.1 News items for popular newspapers

The difficulty is to write the short news item and to persuade a news editor to use it. Here are a few pointers:

All paragraphs	Short, direct statements; avoid qualifications, adjectives and adverbs. Do not use words or phrases that may be obscure to some readers. Define any necessary technical words.
Opening paragraph	Describe your most important finding: a short crisp direct statement. This is called the lead and it is the most important paragraph in the story. Do not start with any of the background detail.
Second paragraph	Who did the work, where was it done, when was it done, why was it done and how was it done?
Third paragraph	What impact will the main result have on people, the community, the economy, industry, science. Include a direct quotation from some celebrity (local, political, scientific, industrial).
Fourth paragraph	Any other results.
Fifth paragraph	What will you do next?
Sixth paragraph	Some detail about you and your colleagues.

This order will appeal to the news editor:

- Facts are arranged in decreasing order of importance.
- The news editor has the right to cut the story and to rewrite it to any length, and he can do this most easily by cutting back from the end. If you write in this order, you make it easy for him to do this and avoid the danger of misrepresentation through rewriting.
- Do not expect to see proofs. Newspaper publishing works on short and urgent deadlines. Today's news is tomorrow's history.
- Include pictures, preferably with people in them. Any graphs or tables must be totally clear to all readers.
- Keep the total length to less than 200 words. Type the number of words at the end of the item. Number every page, top right, using a key word so that the editor can instantly see who's article it is – such as *ship 1, ship 2,* .... Add your name and contact details at the end. Do not assume that the enclosed letter has stayed with the story.
- A news item is not exclusive to a single newspaper. Send the story to all newspapers, local as well as national, weekly as well as daily, by the quickest route possible and address it to the news editor.

- But ensure that the newspaper readership could be interested in your fascinating news. There is little point in submitting a news item on a new statistical process to a paper that specialises in sport, unless the process has a sporting connection such as estimating scores in football matches.

15.4.1.2 Letters to newspapers and magazines

Letters to an editor (often a different person than the actual editor) should follow the same format. Consider a letter as an example of a news item or a point of view that you wish to air.

While you can open a new topic, a good correspondence column will also reflect a number of views from different writers about items published earlier. It is not the editor's role to censor your views, but realise all the same that the editor is responsible for what appears in his newspaper.

Letters to the editor need to be brief:

- Do not exceed 100 words to a newspaper, perhaps 250 words to a magazine, depending on its type.
- If you make it easy for the letters editor to cut the letter from the end you are more likely to get most of your point over.
- As with a news article, make the letter concentrate on one point; many letters ramble far too much. The more it rambles, the less likely a letters editor will be tempted to accept it or to spend time rewriting it.
- Do not expect the letters editor to get back to you with suggested modifications. Although some magazine letters editors will do this, newspaper letters editors do not have time. If the letter is that bad, they will not print it.
- Remember that there are copy dates for letters as well as for news items and this may well be earlier. Note: the word *copy* refers to text that is submitted for publication.

15.4.1.3 Promotional material for companies

Promotional material comprises more than just the words. It is the total package that is generally self-contained and glossy. It is about generating a 'wow' response in the reader or the viewer. People read words only when they have to, so think of visuals.

Promotional copy works to more relaxed time-scales but, if it is for inclusion in a magazine or journal, copy dates still apply. Remember:

- Artwork needs creativity time. The difference between an amateur attempt, one man and his Mac, and professional design needs to be appreciated.
- Unless you are completely confident with graphics tools and have very clear visual communication skills, employ a creative designer.

- He will see aspects that you won't appreciate, partly because you are too close to the issue and also because you may not understand the psychology of visual communication.
- Brevity is the key. Advertising copy is just the same as a news item except that either you are trying to persuade someone to buy your product instead of another product or you are introducing a new concept.
- Remember you are in the business of promotion, not information.
- You, or your designer, are in charge of the text, so it won't get edited or corrected.
- Don't forget to include adequate contact information: a URL or email address at least.
- If the publicity is for a magazine, ensure that it is there well before the copy date. This will mean that you are more likely to get a good place on the page or page in the document. Latecomers may be put into odd positions that are not so eye-catching.
- Have a brochure professionally printed. There are low-cost, short-run printing facilities available. A professional designer will know them and will probably get a better deal than you would.

15.4.2 Medium-length items

We include here, in order of difficulty:

- feature articles for popular newspapers;
- public lectures;
- documentary radio scripts.

Clear writing or presentation are just as important for these as for short articles. The overall length for all of these is about 1000 words.

Unlike short news items, medium-length and longer articles need a clear overall structure that has a beginning, which introduces the issue, a middle where the analysis or argument is made, and an end where conclusions are drawn. This is also true of medium-length and longer articles for more specialist audiences. The major conclusions can also be included in an introductory paragraph as a taster for the rest of the article.

15.4.2.1 Feature articles for popular newspapers

The reasons for writing features articles for popular newspapers are similar to those for news items. Editors will welcome articles provided they recognise the subject to be of interest to at least a large minority of their readers.

Write to the editor before you write the article and ask him if he might welcome your contribution. He will then be more likely to read it when it arrives than if it were to arrive out of the blue.

The length of a newspaper feature is generally between 500 and 1500 words. It is exclusive to one newspaper. If you want full-length articles to appear in several newspapers, you must rewrite the article for each, directing it towards the readers of each.

Newspapers, local and regional as well as national, often have special reasons for dealing with a subject. There may be new legislation or new public works that impinge on the readership; there may be a review of local industry, public services or education.

A newspaper may run a special topic-advertising supplement for which they sell several pages of advertising space and must balance this with a proportionate amount of editorial; this is where your feature article may be welcomed. Write to the editor and ask him whether any special topics are going to be covered in the next year and whether he would welcome your contribution. If he answers positively you will be committed to a contribution on time.

When you do submit your article, address it to the features editor, preferably by name, with a short covering letter.

15.4.2.2 Public lectures

Too many public lectures miss the audience and hence the point. Before you start, write the script or prepare the PowerPoint™ presentation, consider the type of audience you are likely to have and the context:

- Is it genuinely open to the public or are there subtle controls?
- Will it be televised, broadcast via radio or television, or webcast on the internet? If so, you have to consider this wider audience as well; they may actually be more important than the studio audience.
- Is it part of a wider event, such as a conference or public meeting, where there are other speakers?
- How much time do you have? A public lecture of 30 minutes is rather short, but if it is more than an hour you will lose attention unless you are a particularly good speaker and the topic is fascinating.
- Are you expected to supply notes or a synopsis of your lecture before or after the event? Even a printout of the PowerPoint slides is helpful.
- Check the acoustics and sound system; can you be heard clearly over the whole auditorium? There are still too many public lectures held in nineteenth-century halls where people at the back can't hear. Sometimes the public address system is to blame but, whatever the cause, poor sound leads to muttering and general loss of focus by the audience and hence of the talk.

- Are questions to the speaker to be allowed? If there are questions, is there a radio microphone?

The best lectures are given by people who are confident and calm and prepared to talk *with* rather than *at* the audience.

Now to the content:

- Usually, a lecture with more than one slide per minute will be very disturbing; the fewer slides the better. We suggest a 50-minute talk should have no more than 30 slides.

- If you script the lecture, consider one word per second too fast. A talk designed to last 50 minutes should be no more than 2000 words in length.

- Remember to define your terms and use a minimum of technical language.

- Structure the talk clearly, remembering that, unlike a written article, the listener cannot easily jump back to remind himself of a point you made ten slides ago.

- You may use a slide projector, overhead or other visual aid. Use these intelligently, without hopping backwards and forwards. If you need to refer to a slide twice, copy the slide so it appears in the correct place without fumbling. The more polished a presentation, the more likely it is that your audience will understand what you are trying to say.

- Inexperience often shows in people who read from the slides. Use the slides as a prompt rather than as text to follow.

- Do not expect the audience to read a lot of text.

- Extravagant graphics can be very entertaining but do they really add to the talk?

- Humour can be used to relax the audience, and yourself, but don't go over the top and remember that you are not a stand-up comic in a night club.

- If there are questions, don't ridicule the questioner who asks a silly question. If they have not understood a point, regard it as your failure.

- Do not use sarcasm, even if a questioner is rude to you. It is better to avoid the temptation and be ultra-polite which will get much more audience sympathy.

15.4.2.3 Documentary radio scripts

Documentary radio scripts are more difficult than public lectures, except that you may be more relaxed in the absence of a live audience. A script may involve more than one person talking as a dialogue rather than the monologue of a public lecture, or it may be a scripted interview. If the programme is more of a discussion event, this becomes largely unscripted, and we don't consider this here.

Most of the issues have already been covered in the discussion of content in public lectures. Additional points are:

- Is there any audience or is the presentation in a closed studio? If there is no live audience, there is no immediate feedback. The lack of feedback and real presentation environment is what makes radio presenting difficult.

- Are there any other people involved in the documentary?

- Is the broadcast live or will you have the chance to record it again if you make a mess of it?

- How long is it? Very few radio programmes have monologues lasting more than ten minutes although two presenters are more acceptable for the listener for perhaps 30 minutes.

- Don't speak too quickly. A ten-minute talk is only about 500 words.

- Remember that the listeners can't go back in the talk, unless they are recording it.

- You have to convey your message in words without any pictures at all. That's the greatest difficulty. You can't say 'In Figure 2, we see that the probability peaks at . . . '. The audience also cannot see you waving your arms about or pointing to a picture.

- Can you refer to some other resource, such as a book or an internet site, where pictures can be found? Remember to mention these during the talk but also to say that the resource will be repeated at the end so that people can have time to write it down. And don't forget to do this.

- Again, use humour sparingly and never be sarcastic.

15.4.3 Longer items

We include here, in order of difficulty:

- populist books

- television documentaries.

A book is easier to write but more difficult to understand than a television documentary, but with a documentary the audience is generally more relaxed. What the reader needs from a book is to become substantially informed; a TV programme is lighter but at the same time can be a very powerful information source.

15.4.3.1 Populist books

A book is a collection of chapters on a common theme. Each chapter is a collection of thoughts on a common sub-theme. A typical chapter may be between 5000 and 10 000 words long, so a book may be 50 000 words long. Each chapter must have a beginning, a middle and an end.

Bear in mind that we are considering populist works here so there is little room for formulae or complex pictures. Novels and historical books as well as academic books can be much longer: many hundreds of thousands of words. But we are considering books designed to popularise a subject, much as Hawking's *A Brief History of Time* has done for cosmology.

Clear phrasing, short sentences and a few sentences per paragraph are essential for all writing. In a populist book in particular, do not be afraid to repeat yourself. It is better to say the same thing two or three times in different contexts as this enables the reader to understand the concept from a wider perspective. It is like learning anything: the more sources you have about a topic, the more you can ensure that you have understood the fundamentals.

So what in particular is needed for a populist book on statistics, and what is definitely not needed? Here are some general features:

- Ensure that the publisher is experienced in the area and understands the target audience. Will the publisher be able to give advanced publicity to the book, to market it effectively? Have you considered publishing it on the internet with print-on-demand, or is it better to try to get it into the bookshops? What about the book launch?
- You need a suitable title so that the bookshop will put it in the right section and so that bookshop browsers will not confuse it with another book or dismiss it before taking it off the shelf. Keep the title short and to the point.
- When is publication due?
- When are the initial and final proof copies due?
- Will the book be hardback or paperback?
- The jacket cover, overall print design and layout must relate to the promotional material.
- Will you include software? It may be better to put this on a website so that any changes can be made before or during publication.
- It may be possible to link the book with a television documentary.

Turning now to the business of writing and editing:

- The writing style needs to be decided within the overall guidelines in this chapter. This must be agreed at the outset with the publisher, or you may find yourself in disagreement with the publisher and you will lose the argument.

- Is this to be a single authorship, a joint authorship or an edited book where individual chapters are written by single or joint authors?
- Don't fill the book with unnecessary equations.
- Ensure all graphs are clearly labelled and can be reproduced properly.
- An edited book will have several contributors, each with an individual style. It is not easy to edit several contributions with very different styles into stylistic conformity without acrimony. All the contributors should agree in advance that they will cooperate in this with the editor.
- A book with more than three authors should be described as edited and the author of each chapter should be named. Otherwise, readers may think that not all the authors have contributed and one or more names are included for other reasons.
- You need to decide what pictures (photographs and drawings) are needed and where will they be placed in the book. It is cheaper to keep them all together, but this may destroy the continuity of the book.

15.4.3.2 Television documentaries

Suppose you write a script for a 90-minute feature film. The number of words spoken in that time will rarely be more than 5000, yet the script could be as long as 40 000 words. The remaining 35 000 would be about the location, the characters and other technical aspects. Here are some tips:

- Find a commissioning editor who understands the statistical issues you are trying to communicate.
- Ask the editor for an estimate of the cost of production and for help in finding funds. Television programmes are not cheap to make.
- Few people can speak naturally to camera. Leave this to the professionals except, perhaps, for an interview. Restrict yourself to advising about presentation.
- Avoid too much 'talking heads'. A documentary should contain a mixture of presentation techniques but should not be just a televised lecture.
- Use visual images to their full, especially moving images if you can, for communicating your ideas.

15.5 Technical readers and audiences

Technical people are those with some technical knowledge. They may not be aware of statistical techniques, other than the most rudimentary, but they may have a scientific or technical background and wish to understand the statistical aspects in more detail. In many ways such people are a joy to meet and to try to educate, but they may not have an intuitive understanding of statistical concepts. It is very difficult for some people to make that jump.

15.5.1 Short items

We include here:

- news items for technical magazines;
- posters for conferences.

15.5.1.1 News items for technical magazines

You may be surprised by the number of technical magazines. Some may have only a few readers, others have tens of thousands. Do not ignore the small-circulation magazines; their readers are specialists and if any are interested in your work, most of them will be. There are directories of technical magazines and your librarian should be able to help you to compile a list. Again, some tips:

- Send your news item to every one on the list; there is no exclusivity.
- Guidance for writing news items for technical magazines is similar to that for popular newspapers, except that you should write your opening paragraph in a way that will appeal to technical readers.
- You may be more liberal, too, with the technical language, assuming that readers have some technical education. Define words and phrases that may be new to some readers but avoid condescension. We recently read, in a popular science magazine: 'The apparatus uses small particles, called electrons'.
- Read the news sections of technical magazines: study their style as well as their content.
- Unless you know the editor well or have a prior arrangement, you should submit the news items as hard copy, with a footnote that the text and graphics can be supplied in electronic form.
- Pictures and charts may be more technical than in a newspaper. Present them in reasonably high graphic files (jpeg, gif, tif) separate from the document file; the editor may want to edit them to suit his publication style.
- Do not expect to see proofs and do not expect to be sent press cuttings. It is up to you to discover whether or not your news has been used. There are press cuttings bureaux that you can pay to search the technical press for subjects.

15.5.1.2 Posters for conferences

Most conferences have poster sessions and the posters remain available for much of the conference. Poster presentations are useful for communicating with all conference delegates, particularly where a conference is streamed. They don't have the power of good presentations but can provide useful seeds for discussion.

The main issues are:

- A poster is available for a substantial part of the conference so that delegates may view them in their own time.
- They are available to all delegates in a streamed conference rather than those attending your stream.
- They are largely used by young researchers but sometimes by commercial companies keen to catch all eyes.
- More technical details can be included as people have time to understand them.
- They are less of a burden for the presenter, who has to talk to only one or two people at a time.
- They do not carry as much authority as a presentation.
- You cannot show your clear enthusiasm for the work to a large audience.

A poster can be good publicity for a presentation. You don't have the advantage of a captive audience so you have to attract passing traffic. People soon pass over a dull and boring poster session in favour of one that has more of the 'wow' factor. Only the cognoscenti will hover over a technical paper and try to engage you and they are often looking for holes and errors in your argument.

While there is more room for technical details than in a talk – a poster session is a little like a feature article – pay attention to presentation. So rather than talk about the reader or the audience, consider your targets as viewers.

An A4 poster will not attract attention. Make your poster at least A3, and preferably A2 or even A1 so that people can view it from a comfortable reading distance. Avoid many pieces of paper; if you have a single display board there should be no more than eight pages to view. It may be hard to condense your beloved project into such a small space but remember the key words of clarity and brevity. Use pictures to show the results and avoid too many words.

15.5.2 Medium-length items

We include here:

- feature articles for technical magazines;
- articles for company house journals;

- platform presentations for conferences;
- company training;
- training courses for wider industrial audiences.

Education and training are beyond the scope of this chapter, so we confine ourselves to the first three types of output.

15.5.2.1 Feature articles for technical magazines

Feature articles form the bulk of a technical magazine. Well illustrated case studies are usually preferred over general methodological or scientific expositions, but ask the editor what he would like. He will advise you about length and style as well as discussing what content will appeal to his readers and advertisers. He will also expect your article to be exclusive to him.

Read some articles in several issues of the magazines: study their style as well as their content before you start to write. You will find that, as with news stories, the first two or three lines are written in a way that will capture and hold the reader's interest. There is no preamble.

You may find it easy to write a preamble; faced with a blank sheet of paper, it may help you to start writing. Cross it out when you have finished the article. Your second paragraph, now the first, will make a much punchier introduction. Conclude the article with a short summarising paragraph that ties the whole story together, returning to the attention-grabbing point in the first paragraph.

And, please, do not write in that pseudo-objective style that is so unnecessarily common in academic papers. This is an opportunity for you to bring some humanity to your work.

15.5.2.2 Articles for company house journals

Most large companies have house journals. 'We are a big and happy family', they say. 'We want you all to know each other, to know what you are doing and why.' They use the journal to inform the staff about company success in sales and employment, about pensions, about safety, about arrivals and departures, promotions, births, marriages and deaths, about social and sporting activities, about home decorating, gardening, cooking and holidays. They use a light journalistic style; all articles are short and easy to understand.

So can you write an article to fit into such a scheme? Editors will welcome a good technical article provided it relates to the company's activities and it suits their house style, whether you are writing as a member of staff, a consultant, or a collaborating academic. Study the house journal and ask the editor what he will accept.

15.5.2.3 Platform presentations for conferences

Our advice for conference presentations to a technical audience follows very closely the advice for public consumption, although the level of the discussion will necessarily be more technical. The main point is not to try to say too much but to leave your audience wanting to know more; and include some guidance on where to get the information.

15.6 Specialist readers and audiences

Specialist people are our peers. They are highly competent in their own fields and are able to understand the nuances of statistical thought, provided these are relevant and clearly stated. They are naturally critical and can be a very difficult because that may come with a strong belief that they already know the answers. It is our task to shepherd them into the world of advanced statistical modelling and procedures.

15.6.1 Short items

Many companies have a short format for informing senior technical management that is the company internal memorandum. This should be no longer than a single side of paper, including the title, author, affiliation. The guidance given for a technical audience will probably suffice. Senior managers will not have time to read too much but they will, or should, be aware of all the technical issues.

If senior managers are not sufficiently technically qualified then approach this as a news item for a technical audience.

15.6.2 Medium-length items

We include here:

- internal technical reports for companies;
- academic papers;
- advanced training courses.

We consider only the first two types of output.

15.6.2.1 Internal technical reports for companies

A technical report published within a company should carry a warning: 'This report has not been peer reviewed'. Otherwise, it might be read as a paper in an academic journal.

Another distinction is that it will deal with a specific problem that has arisen within the company. Rarely will a company publish an internal report about a general theoretical or methodological issue. You may sometimes believe this is needed, but you should be aware of the cynicism of colleagues who may charge you with being remote from the company's practical needs.

A third distinction is that many companies will label internal reports as confidential. Perhaps they have a colour code for the level of confidentiality, such as white for open, green for restricted, yellow for highly confidential, and red for top secret. This means that, as you write, you must consider the readers you want to reach, those you need to inform, and moderate your style and content accordingly.

15.6.2.2 Academic papers

Apart from using formal English and subject to any instructions from the journal's editors, an academic paper should have the following form:

Title page	The title, author(s) and affiliation(s), email address(es), noting the principal or correspondence author. The journal will generally add submission and revision dates.
Abstract	Treat the abstract as a technical news item. Some search systems include the abstract, and on-line journals may publish the abstract as a taster to encourage purchase of the full item. It therefore needs to be concise. We suggest a maximum of 100 words.
Keywords	Keywords are needed for classification and for reference. Somebody sometime will be interested in your subject but will not know of your work. You want to help him to discover it through your library's on-line inquiry service. What words would his mind conjure if he wanted to find the best literature on the subject? Those must be your keywords.
Contents page	May not be needed.
Introduction	The motivation for the study with its aims in overview.
Main sections	There will generally be several main sections, such as Method and Results. It is generally better to place figures here rather than at the end, but this depends on the journal style.
Conclusions	Discussion of the findings, relevance to the problem and further work. Unlike internal technical reports, academic papers do not generally have numbered conclusions.
References	Organised alphabetically by author and referred to in the main text by Smith and Jones (1998b) or Williams *et al* (2000). The most usual standard is the Harvard system. The standard for a paper is: authors, date, title (in quotes), journal (in italics), volume (in bold), issue, pages. The standard for a book is: authors, date, title (in italics), publisher (including city).
Appendices	Describe the more technical stages of the work that are better not put in the main sections, such as an innovative mathematical derivation.

The need remains to write clearly and without unnecessary complication.

15.7 Web pages

There are many books about publication on the web, but there is little advice, if any, about scientific ideas, methods and research. Here are a few points for you to consider:

- If you just copy up an article, people see no more than a little of each page at any time. Construct your document so that the user can jump about using links, even within the same page, that include return facilities. Do not force readers to scroll down or, even worse, scroll across to read something.

- Limit the width of the page and add a small blank column to the right-hand side so that wrap-round does not exceed the screen. A suggested maximum for the whole screen is 800 pixels. You can place everything within a table with a single cell to enforce this.

- You need to ensure that all pages delivered are W3C (World Wide Web Consortium) compliant so that they can be viewed with all browsers.

- Do not require text processing plug-ins that your reader may not have. For example, while it may be easy for you to copy a Word document, not everyone will have Word on their machine or the right version of Word and it is not available on many Unix platforms (only on Apple Macs). In particular, avoid using tables in Word. It is better to convert everything to HTML format; Word makes a considerable mess when saving to HTML and will almost always need a lot of manual intervention.

- Spreadsheets are difficult to view, even if you have the plug-in. It is better to rearrange the spreadsheet contents so that they are viewed again in HTML. A small spreadsheet can be viewed as a table; a larger item may need some bespoke preparation to ensure simple navigation.

- If you want to be able to calculate things then you have to use forms and a server-side programming language such as PHP or a client-side programming language such as Javascript to do the calculations.

- Plug-ins such as Macromedia Flash or Real One Player that enable you to add voice and moving images can be useful, so you might ignore readers who do not have those plug-ins. For example a PowerPoint presentation that includes movies and a commentary may be a powerful training tool.

- Ensure that graphic images are not stored to excessive depth, and never put text or tables into graphics unless there is a really good reason. Use graphics for pictures and put text into HTML.

- Remember that search engines do not see any text within a graphic image or within a frame.

- Ensure that the page is printable and will not overrun the right-hand side. This is another reason to limit the width of a document.

- Remember that each web page delivered to your browser is an independent event. There are session variables and cookies that can communicate but the next page viewed will not contain any information from the present page, unless you create some sort of active page.

- If you have control of the website, consider using a contents management system so that the pages are delivered dynamically from a database. This can be complex but it will save a lot of time in the long run if you need regular maintenance of the site.

- If your web page is meant to be part of a permanent site, consider using a graphic designer and possibly a web developer who can port your document.

15.8 Good writing

The technical language used in most parts of the world – and most particularly in ENBIS – is English, so our guidance relates to English only.

Language rules are often broken. You will see many cases where a sentence is introduced with a conjunction. And even without a verb. This is always for effect so that when scanning the sentence the issues are separated. It is not formally correct. It applies principally when communicating with the public.

For technical articles, you may be expected to be a little more formal; for academic papers even more so. The level of formality required is a decision only you can take, but do not lose sight of the purpose of communication, which is to inform and educate.

There is a distinction between *jargon* and *technical language*. Jargon is pretentious syntax, vocabulary or meaning. Technical language is needed for specialist subjects. Use technical language but define any terms with which your readers may not be familiar. Do not use jargon.

Statistics is a pervasive subject: like sand in the desert, it gets everywhere. While this means that you need a broad set of skills, it also means that you should get a wide variety of interesting work and contacts. And every problem brings a unique set of issues, often in a specialist vocabulary. These have to be understood and translated into statistical ideas, concepts and models. In the end, however complex the statistical issues, you need to translate back into the terms of the application area and to write clearly so that your client will understand you. You should not provide answers in statistical language except where strictly necessary.

The style of your writing depends on the role you play. You may be any of the following:

- Originator: a statistician who has thought of a problem first and then gone on to find a solution. Other than being an academic, such a person could be an entrepreneur hoping to sell the derived process to customers.

- Collaborator: a statistician who works with other statisticians on a problem brought to them as a group. A statistical department running clinical trials would have such people and, while they may be responsible for a section of a report, they are unlikely to write the whole document.

- Consultant: a statistician who works directly with the applications area professionals and is free to pursue the best solution possible.

At the same time, you must act as an educator. Consider yourself an ambassador for statistics, explaining in as simple language as possible the design of your model, the interpretation of your results and how they relate to the problem. Every successful use of statistics will help to promote its use and although you may not personally benefit, it is all part of improving the relationship between business or industry and the quality of product or service. Every successful act of education will increase the use of statistics, so there is no reason to hide behind formulae or technical terms.

15.8.1 Spelling

Spelling is a constant problem in English: there seem to be many inconsistencies. These derive from the heritage of the language, leading, for example, to many more vowel sounds than there are in Latin languages such as Spanish and Italian.

The best-known difference is between British English and American English, which has its origins in the decision by Noah Webster in 1789 to define the then new American nation by means of language. Many of his attempts fell by the wayside, but there remain some distinctions such as *color* and *colour, favor* and *favour, center* and *centre*, the use of '*z*' rather than '*s*'. Be consistent and stick to one or the other. Phrases like *Generalized Linear Modelling* should be avoided (the spelling, not the technique; *generalized* is spelt in the American way and *modelling* in the British way). Use a spell-checker but ensure that your use is consistent; they are not perfect and do not trap the vocabulary changes such as *sidewalk* and *pavement, hood* and *bonnet, elevator* and *lift*.

Ensure also that the word means what you want it to mean and is not something that sounds the same. There are many such homonyms in English. Examples include *ascent* (the climb) and *assent* (to agree); *coarse* (rough) and *course* (path of travel), *principal* (a person or adjective meaning main) and *principle* (a theory) and so on. It is easy to make the mistake and, unless the wrong word is also grammatically incorrect, it will not be found by spell-checkers or grammar-checkers.

Every word processor now has a spell-checker, yet many published papers and theses are littered with misspellings and misprints. One reason is that the writers are too lazy or ignorant to use the spell-checker.

But another reason is that many words, when misspelled, are other correctly spelled words. The spell checker will not identify these. If there are any in your submission, you are telling the examiners that you have not read carefully what you have written. Here is a recent example: 'Improved understanding of these matters should acid cogent presentation ...'. The typist had misread 'aid' in poor handwriting as 'acid'.

The most common (anecdotally) of scientific misprints is the change from 'causal relationships' to 'casual relationships'. Other common ones, which the spell checker will not find, are:

- *fro* for *from* or *for*;
- *lead* for *led* (past tense of to lead);
- *gibe* for *give*;
- *correspondents* for *correspondence* or *corresponding* ('changes in watershed conditions can result in correspondents changes in stream flow').

Our favourite malapropism is: 'Sir Francis Drake circumcised the world with a one hundred foot clipper'.

15.8.2 Style

At the end of this chapter is a list of some good books about scientific and technical writing. Here are a few points to ponder.

15.8.2.1 The first person pronoun, I

There is continuing debate on whether to use the first person, singular, active I. One professional institute banned its use in all reports some years ago. The daughter of one of us, as an undergraduate, assured him that she would fail if she used it in her dissertation. The parent was shocked but didn't interfere.

It is not simply a matter of taste. It is a question of honesty and of the credit that you deserve and need to justify the award of a higher degree. *You* are responsible for ensuring that others recognise what *you* have done, what ideas *you* have had, what theories *you* have created, what experiments *you* have run, what analyses and interpretations *you* have made, and what conclusions *you* have reached. You will not succeed in any of this if you coyly state:

Poor style	*Write instead*
It was considered	I proposed or I thought
It was believed	I believed
It was concluded	I concluded
There is no doubt that	I am convinced
It is evident that	I think
It seems to the present writer	I think
The author decided	I decided

But be careful: too much use may seem conceited and arrogant.

15.8.2.2 Abbreviations, acronyms and points

Do *not* use points except as full stops at the ends of sentences or as decimal points:

- Write BSc, PhD, *not* B.Sc., PhD
- Write A Brown or Albert Brown *not* A. Brown.
- Write UK, UNO, WHO, ICI, *not* U.K., U.N.O., W.H.O., I.C.I.

Do *not* contract:

- department into dept.;
- institute into inst.;
- government into gov't.;
- professor into prof.

Do *not* use:

- such as (write 'such as' or 'for example');
- that is (write 'that is');
- *et al*(except in references);
- etc (put a full stop instead, otherwise the reader will wonder 'what are the etceteras?' and if you can't be bothered to tell him he may not want to read further).

There is a distinction between *abbreviations* and *acronyms*. An acronym is a set of letters that can be enunciated as a word to represent a full name. The plurals of acronyms may be written with the initials followed by a lower case 's' without an apostrophe (ROMs). Avoid possessives with acronyms (do not write NORWEB's) by recasting the sentence to eliminate the possessive.

Prefer single quotes to double quotes, but single and double are used when there is nesting of quotes.

15.8.2.3 Units

Use SI units with standard abbreviations without points. The base units are: m, kg, s, A, K, cd, and mol. Note that, by international agreement, including France and the USA, correct spellings are *gram* and *metre*.

In text, write out 0 to 10 (as zero, one, ..., ten) but for greater integers use figures (21). If an integer starts a sentence, write it in words.

Do not use a dash to denote an interval. Write:

- between 20 and 25 °C *not* 20–25 °C;
- from 10 to 15 September *not* 10–15 September.

15.8.2.4 Capitals

Like *The Times*, resist a tendency to a Germanic capitalisation of nouns by avoiding capitals wherever possible. Too many of them break the flow of the eye across a sentence. They also make pompous what need not be. The general rule is that proper names, titles and institutions require capitals, but descriptive appellations do not. Thus government needs no capital letter, nor does committee nor department. The same goes for jobs that are obviously descriptive, such as prime minister, foreign secretary, or even president unless it is used as a personal title ('President Washington' but 'the president'). There are a few exceptions such as Black Rod, The Queen, God. This is despite the instructions of many grammar-checkers.

Laws are lower case (second law of thermodynamics) unless they are named after somebody (Murphy's law).

15.8.2.5 Things to avoid

- *Ornate words and phrases* such as convey (take), pay tribute to (praise), seating accommodation (seats), utilise (use), Fred underwent an operation (Fred had an operation), we carried out an experiment (we did an experiment).
- *Needless prepositions* tacked on verbs: check up, try out, face up to.
- *Vague words* such as considerable, substantial, quite, very, somewhat, relatively, situation (crisis situation), condition (weather conditions), system.
- *Clichés* (last but not least, as a matter of fact).
- *Passive voice*, for example 'It was decided to' and 'As is shown in figure one...' (Figure one shows).
- *Obfuscation.* We found the following on the World Wide Web. It is attributed to Mark P Friedlander:

  Learn to obfuscate
  Children, children, if you please
  Learn to write in legalese,
  Learn to write in muddled diction,
  Use choice words of contradiction.
  Sentences must breed confusion,
  Redundancy and base obtusion,
  With a special concentration
  On those words of obfuscation.
  When you write, as well you should,
  You must not be understood.
  Sentences concise and clear

Will destroy a law career.
And so, my children, if you please,
Learn to write in legalese.
So that, my dears, you each can be
A fine attorney, just like me.

15.8.3 Plain english

The *Plain English Campaign* (www.plainenglish.co.uk) fights for crystal-clear language and against jargon, gobbledygook and other confusing language. Plain English is something that the intended audience can read, understand and act upon the first time they read it. Plain English takes into account design and layout as well as language.

The campaign also offers good advice on the design of websites: see www.plainenglish.co.uk/webdesign.

References

Altman DG, Gore SM, Gardner MJ, Pocock S J (1983) Statistical guidelines for contributors to medical journals. *British Medical Journal*, **286**, 1489–1493.

Ehrenberg ASC (1982) *A Primer in Data Reduction*. John Wiley & Sons, Ltd, Chichester.

Finney DJ (1995) Statistical science and effective scientific communication. *Journal of Applied Statistics*, **22**(2), 193–308.

Further reading

Barrass R (1978) *Scientists Must Write*. Chapman & Hall, London.

Blamires H (2000) *The Penguin Guide to Plain English.* Penguin, London.

Cooper BM (1975) *Writing Technical Reports*. Penguin, London.

Cresswell J (2000) *The Penguin Dictionary of Clichés.* Penguin, London.

Kirkman J (1992) *Good Style: Writing for Science.* E & FN Spon, London.

Lindsay D (1997) *A Guide to Scientific Writing.* Longman, Melbourne.

O'Connor M, Woodford FP (1978) *Writing Scientific Papers in English*. Pitman Medical, London.

Partridge E (1962) *A Dictionary of Clichés.* Routledge & Keegan Paul, London

Tufte ER (1997) *Visual Explanations: Images and Quantities, Evidence and Narrative*. Graphics Press, Cheshire, CT.

Pocket Style Book. The Economist, London.

Guide to English Style and Usage. The Times, London.

Index

When a word or phrase occurs several times in a chapter, the page number of only the first occurrence in that chapter is noted.

STATISTICS IN PRACTICE

Human and Biological Sciences

Berger – Selection Bias and Covariate Imbalances in Randomised Clinical Trials
Brown and Prescott – Applied Mixed Models in Medicine, Second Edition
Chevret (Ed) – Statistical Methods for Dose-Finding Experiments
Ellenberg, Fleming and DeMets – Data Monitoring Committees in Clinical Trials: A Practical Perspective
Hauschke, Steinijans & Pigeot – Bioequivalence Studies in Drug Development: Methods and Applications
Lawson, Browne and Vidal Rodeiro – Disease Mapping with WinBUGS and MLwiN
Lui – Statistical Estimation of Epidemiological Risk
Marubini and Valsecchi – Analysing Survival Data from Clinical Trials and Observation Studies
Molenberghs and Kenward – Missing Data in Clinical Studies
O'Hagan, Buck, Daneshkhah, Eiser, Garthwaite, Jenkinson, Oakley & Rakow – Uncertain Judgements: Eliciting Expert's Probabilities
Parmigiani – Modelling in Medical Decision Making: A Bayesian Approach
Pintilie – Competing Risks: A Practical Perspective
Senn – Cross-over Trials in Clinical Research, Second Edition
Senn – Statistical Issues in Drug Development, Second Edition
Spiegelhalter, Abrams and Myles – Bayesian Approaches to Clinical Trials and Health-Care Evaluation
Whitehead – Design and Analysis of Sequential Clinical Trials, Revised Second Edition
Whitehead – Meta-Analysis of Controlled Clinical Trials
Willan and Briggs – Statistical Analysis of Cost Effectiveness Data

Earth and Environmental Sciences

Buck, Cavanagh and Litton – Bayesian Approach to Interpreting Archaeological Data
Glasbey and Horgan – Image Analysis in the Biological Sciences
Helsel Nondetects and Data Analysis: Statistics for Censored Environmental Data
McBride – Using Statistical Methods for Water Quality Management
Webster and Oliver Geostatistics for Environmental Scientists, Second Edition
Wymer – Statistical Framework of Recreational Water Quality Criteria and Monitoring

Industry, Commerce and Finance

Aitken – Statistics and the Evaluation of Evidence for Forensic Scientists, Second Edition

Balding – Weight-of-evidence for Forensic DNA Profiles
Brandimarte – Numerical Methods in Finance and Economics: A MATLAB-Based Introduction, Second Edition
Brandimarte and Zotteri – Introduction to Distribution Logistics
Chan – Simulation Techniques in Financial Risk Management
Coleman, Greenfield, Stewardson and Montgomery (Eds) – Statistical Practice in Business and Industry
Frisen – Financial Surveillance
Lehtonen and Pahkinen – Practical Methods for Design and Analysis of Complex Surveys, Second Edition
Ohser and Mücklich – Statistical Analysis of Microstructures in Materials Science
Taroni, Aitken, Garbolino and Biedermann – Bayesian Networks and Probabilistic Inference in Forensic Science